This book describes the properties and device applications of hydrogenated amorphous silicon. It covers the growth, the atomic and electronic structure, the properties of dopants and defects, the optical and electronic properties which result from the disordered structure, and finally the applications of this technologically very important material. There is also a notable chapter on contacts, interfaces, and multilayers. The main emphasis of the book is on the new physical phenomena which result from the disorder of the atomic structure.

The development of a new material with useful technological applications is relatively rare. The special attribute of a-Si:H which makes it useful is the ability to deposit the material inexpensively over large areas, while retaining good semiconducting properties. The material is gaining increasing use in photovoltaic solar cells and in large area arrays of electronic devices. Only the future years will reveal how extensive are the applications of this promising material.

The book will be of major importance to those who are researching or studying the properties and applications of a-Si:H. It will have a wider interest for anyone working in solid-state physics, semiconductor devices or electronic engineering in general.

Hydrogenated amorphous silicon

Cambridge Solid State Science Series

EDITORS:
Professor R. W. Cahn, FRS
Department of Materials Science and Metallurgy, University of Cambridge
Professor E. A. Davis
Department of Physics, University of Leicester
Professor I. M. Ward, FRS
Department of Physics, University of Leeds

Titles in print in this series

B. W. Cherry
Polymer surfaces
D. Hull
An introduction to composite materials
S. W. S. McKeever
Thermoluminescence of solids
D. P. Woodruff and T. A. Delchar
Modern techniques of surface science
C. N. R. Rao and J. Gopalakrishnan
New directions in solid state chemistry
P. L. Rossiter
The electrical resistivity of metals and alloys
D. I. Bower and W. F. Maddams
The vibrational spectroscopy of polymers
S. Suresh
Fatigue of materials
J. Zarzycki
Glasses and the vitreous state
R. A. Street
Hydrogenated amorphous silicon

R. A. STREET
Xerox Corporation, Palo Alto Research Center

Hydrogenated amorphous silicon

CAMBRIDGE UNIVERSITY PRESS
Cambridge
New York Port Chester Melbourne Sydney

CAMBRIDGE UNIVERSITY PRESS
Cambridge, New York, Melbourne, Madrid, Cape Town, Singapore, São Paulo

Cambridge University Press
The Edinburgh Building, Cambridge CB2 2RU, UK

Published in the United States of America by Cambridge University Press, New York

www.cambridge.org
Information on this title: www.cambridge.org/9780521371568

© Cambridge University Press 1991

This publication is in copyright. Subject to statutory exception
and to the provisions of relevant collective licensing agreements,
no reproduction of any part may take place without
the written permission of Cambridge University Press.

First published 1991
This digitally printed first paperback version 2005

A catalogue record for this publication is available from the British Library

Library of Congress Cataloguing in Publication data
Street, R. A.
Hydrogenated amorphous silicon/R. A. Street.
(Cambridge solid state science series)
Includes bibliographical references and index.
ISBN 0 521 37156 2 (hardcover)
1. Silicon. 2. Amorphous semiconductors. 3. Surface chemistry.
I. Title. II. Series.
QC611.8.S5S76 1991
621.381′52 – dc20 90-2387 CIP

ISBN-13 978-0-521-37156-8 hardback
ISBN-10 0-521-37156-2 hardback

ISBN-13 978-0-521-01934-7 paperback
ISBN-10 0-521-01934-6 paperback

Contents

To Angela, Johanna and Timothy

Preface

This book describes the material properties and physical phenomena of hydrogenated amorphous silicon (a-Si:H). It covers the growth of material, the atomic structure, the electronic and optical properties, as well as devices and device applications. Since it focusses on the specific properties of one amorphous material, there is a considerable emphasis on describing and interpreting the experimental information. Familiarity with semiconductor physics is assumed, and the reader is also referred to the excellent books by Mott and Davis, Elliott, and Zallen for further information about the general properties of amorphous semiconductors and glasses.†

Research into amorphous semiconductors is directed towards an understanding of how the structural disorder leads to their unique properties. A-Si:H has sometimes been regarded as a derivative of crystalline silicon, in which the disorder simply degrades the electronic properties. Much of the interest in a-Si:H comes from the realization that this is not an accurate view. The disordered atomic structure and the presence of hydrogen combine to give new phenomena which are strikingly different from those in the crystalline semiconductors. The structural disorder results in the localized band tail states characteristic of amorphous materials, which are reflected in the optical, transport and recombination properties, while the hydrogen gives unique defect, doping and metastability effects.

The development of a new material with useful technological applications is relatively rare. Hydrogenated amorphous silicon is gaining increasing use in photovoltaic solar cells and in large area arrays of electronic devices. In order for any new material to be technologically useful it must have either electronic properties or material attributes which cannot be provided by other materials. The special attribute of a-Si:H is the ability to deposit the material inexpensively over large areas. Thus, whereas crystalline silicon is the material of choice for integrated circuits, it is unsuitable for making a

† Mott, N. F. and Davis, E. A. (1979) *Electronic Processes in Non-crystalline Materials*, (Oxford University Press, Oxford).
Elliott, S. R. (1983) *Physics of Amorphous Materials*, (Longman, New York).
Zallen, R. (1983). *The Physics of Amorphous Solids*, (Wiley, New York).

large display. The future years will reveal how extensive are the applications of this promising material.

I would like to express my appreciation to the Xerox Corporation for their support in this project, and to the many colleagues at the Xerox Palo Alto Research Center, with whom it has been my pleasure to work on amorphous silicon. Particular thanks go to D. K. Biegelsen, M. Hack, A. C. Street and S. Wagner for their comments on the text.

Palo Alto R. A. S.
1990

1 Introduction

1.1 Early research

Hydrogenated amorphous silicon (a-Si:H) was a late arrival to the research on amorphous semiconductors, which began to flourish during the 1950s and 1960s; studies of insulating oxide glasses, of course, go back much further. Interest in the amorphous semiconductors developed around the chalcogenides, which are materials containing the elements sulfur, selenium and tellurium; examples are As_2Se_3, GeS_2 etc. The chalcogenides are glasses which may be formed by cooling from the melt, with structure similar to the oxides but with smaller energy band gaps. Research in these amorphous semiconductors addressed the question of how the disorder of the non-crystalline structure influences the electronic properties. The study of chalcogenides was further promoted by the introduction of xerographic copying machines. Xerography was invented in 1938 and the first successful copier was made in 1956, using selenium as the photoconductive material.

A-Si:H was first made in the late 1960s. Before that time there was research on amorphous silicon without hydrogen, which was prepared by sputtering or by thermal evaporation. The unhydrogenated material has a very high defect density which prevents doping, photoconductivity and the other desirable characteristics of a useful semiconductor. Electronic measurements were mostly limited to the investigation of conduction through the defect states.

Chittick and coworkers in the UK were the first to make a-Si:H, using glow discharge as the deposition technique (Chittick, Alexander and Sterling 1969). Silane gas (SiH_4) is excited by an electrical plasma which causes the gas molecules to dissociate the deposit on heated substrates. The technique is essentially the same as is used currently, although the design of the deposition systems has evolved. The first reactor was inductive: the plasma is excited by a coil outside the quartz chamber. Most reactors now are capacitative, consisting of two parallel electrodes within a stainless steel chamber, but the deposition mechanism is not substantially different.

These early experiments demonstrated the deposition of silicon films, the lack of conduction in defect states (implying a low defect density) and increased conduction due to impurities (Chittick *et al.* 1969,

Chittick and Sterling 1985). The infra-red (IR) vibrations of silicon–hydrogen bonds were observed, although they were not recognized as such, and also some metastable phenomena which are now being widely studied. However, the significance of this early work was evidently missed by the sponsors of the research and the project was terminated.

The significance of the results was not lost on Spear at the University of Dundee, who saw the promise of this new method of making amorphous silicon and arranged to take Chittick's reactor to Dundee. In the following years this group documented in considerable detail the superior properties of the amorphous silicon made by this deposition technique (Spear 1974). The research showed that the material had good electrical transport properties with a fairly high carrier mobility (LeComber and Spear 1970) and also strong photoconductivity resulting from a very low defect density (Spear, Loveland and Al-Sharbaty 1974). A major turning point in the development of a-Si:H was the report in 1975 of substitutional n-type or p-type doping by the addition of phosphine or diborane to the deposition gas (Spear and LeComber 1975). The significance of all these observations was widely acknowledged and the subsequent years saw a period of rapidly increasing interest in this form of amorphous silicon.

The essential role of the hydrogen in a-Si:H was first recognized by Paul's group at Harvard, who had studied sputtered amorphous silicon and germanium since the late 1960s (Lewis *et al.* 1974). They understood that the high defect density of amorphous silicon and amorphous germanium prevented these materials from being useful for electronic devices and tried to find ways of eliminating the defects, eventually succeeding by introducing hydrogen into the sputtering system. The hydrogen caused a similar improvement in the material properties as was found for glow discharge a-Si:H, with a high photoconductivity, low defect density and doping. The Harvard group demonstrated that the hydrogen concentration in the films was about 10 atomic %, by observing its characteristic IR vibration, which has a frequency close to 2000 cm^{-1} for Si–H bonds and 1800 cm^{-1} for Ge–H bonds. These were actually the same absorption lines seen earlier by Chittick, but not identified as hydrogen vibrations. Shortly after the Harvard experiments, it was confirmed that the glow discharge material also contained hydrogen (Fritzsche 1977). This is now recognized as an essential component of the films which is responsible for suppressing defects. Research on both glow discharge and sputtered a-Si:H continued, but the glow discharge technique has become increasingly dominant because it seems to give slightly better material.

A-Si:H device research was started by Carlson and Wronski (1976) at RCA Laboratories with the development of photovoltaic devices. They demonstrated the feasibility of a-Si:H solar cells and initially obtained conversion efficiencies of 2–3%. Subsequent research by RCA and by many other groups, increased the cell efficiency by roughly 1 percentage point each year to 12–14% in 1989. Although RCA began the development of solar cells, Sanyo in Japan was the first company to market devices. RCA was primarily interested in large scale electric power production, a topic which was popular in the oil crises of the late 1970s. Economically viable solar cells for power production need to have high conversion efficiency (although the actual value was debated for several years). Sanyo, however, recognized that even quite low efficiency cells could power hand-held calculators. In fact the efficiency of a-Si:H solar cells is larger under fluorescent light compared to sunlight, so that the calculators work well in an office environment. Sanyo began producing cells in 1979, and Japanese companies have dominated this market for a-Si:H ever since. Solar cells for large scale power production are still not in significant use, although the intermediate scale power markets are developing.

Research on large area electronic arrays of a-Si:H devices started a few years later after the first field effect transistors were reported (Snell *et al.* 1981). These devices take advantage of the capability to deposit and process a-Si:H over large areas. Applications include liquid crystal displays, optical scanners and radiation imagers. Present devices contain up to 10^6 individual elements and are presently used in hand-held televisions and FAX machines.

1.2 Basic concepts of amorphous semiconductors

The disorder of the atomic structure is the main feature which distinguishes amorphous from crystalline materials. It is of particular significance in semiconductors, because the periodicity of the atomic structure is central to the theory of crystalline semiconductors. Bloch's theorem is a direct consequence of the periodicity and describes the electrons and holes by wavefunctions which are extended in space with quantum states defined by the momentum. The theory of lattice vibrations has a similar basis in the lattice symmetry. The absence of an ordered atomic structure in amorphous semiconductors necessitates a different theoretical approach. The description of these materials is developed instead from the chemical bonding between the atoms, with emphasis on the short range bonding interactions rather than the long range order.

The structural disorder influences the electronic properties in several different ways which are summarized in Fig. 1.1. The similarity of the covalent silicon bonds in crystalline and amorphous silicon leads to a similar overall electronic structure – amorphous and crystalline phases of the same material tend to have comparable band gaps. The disorder represented by deviations in the bond lengths and bond angles broadens the electron distribution of states and causes electron and hole localization as well as strong scattering of the carriers. Structural defects such as broken bonds have corresponding electronic states which lie in the band gap. There are also new phenomena which follow from the emphasis on the local chemical bonds rather than the long range translational symmetry. The possibility of alternative bonding configurations of each atom leads to a strong interaction between the electronic and structural states and causes the phenomenon of metastability.

The following brief survey of the properties of amorphous semiconductors is intended to provide an introduction to the subsequent more detailed discussion.

1.2.1 *Atomic structure*

Amorphous semiconductors are not completely disordered. The covalent bonds between the silicon atoms are much the same as in crystalline silicon, with the same number of neighbors and the same average bond lengths and bond angles. The disorder is represented by the atom pair distribution function, which is the probability of finding an atom at distance R from another atom. Schematic pair distribution functions for crystalline, amorphous (or liquid) and gaseous phases are illustrated in Fig. 1.2. The relative positions of atoms in a dilute gas are random (except at very close spacings), whereas a perfect crystal is completed ordered to large pair distances. The amorphous material has the same short range order as the crystal but lacks the long range order. The first few nearest neighbor distances are separately distinguished, but the correlation between atom pairs loses structure after a few interatomic spacings. The material properties of amorphous semi-

Fig. 1.1. The correspondence between features of the atomic structure and the resulting electronic properties.

STRUCTURE		ELECTRONIC PROPERTIES
Bonding disorder	⟶	Band tails, localization, scattering
Structural defects	⟶	Electronic states in the band gap
Alternative bonding configurations	⟶	Electronically induced metastable states

conductors are similar to their crystalline counterparts because they share the same local order. In contrast, there is often little connection between the properties of gaseous and condensed phases.

The short range order and long range disorder lead to the model of the continuous random network, introduced by Zachariasen (1932) to describe glasses such as silica. The periodic crystalline structure is replaced by a random network in which each atom has a specific number of bonds to its immediate neighbors (the coordination). Fig. 1.3. is a two-dimensional illustration of such a network, containing atoms of different coordination (4, 3 and 1). The random network has the property of easily incorporating atoms of different coordination, even in small concentration. This is in marked contrast to the crystalline lattice in which impurities are generally constrained to have the coordination of the host because of the long range ordering of the lattice. This difference is most distinctly reflected in the doping and defect properties of a-Si:H, discussed in Chapter 4 and 5.

A real crystal contains defects such as vacancies, interstitials and dislocations. The continuous random network may also contain defects, but the definition of a defect has to be modified. Any atom

Fig. 1.2. Schematic diagram of the atom pair distribution functions for a crystalline and amorphous solid and a gas, scaled to the average separation of nearest neighbor atoms, R_{av}, showing the different degree of structural order.

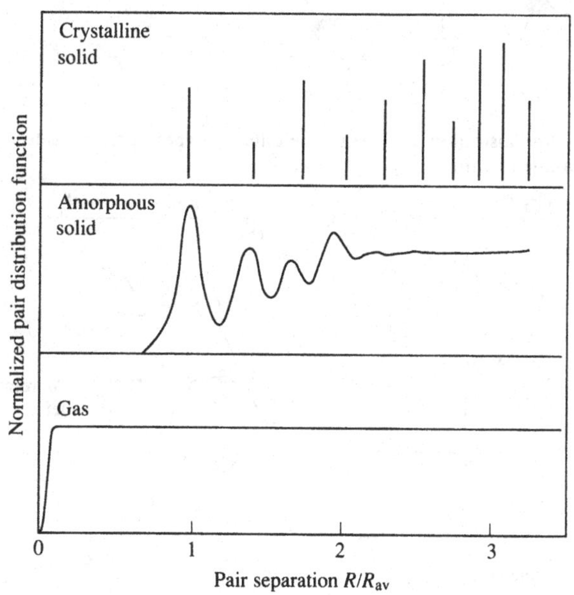

which is out of place in a crystal is a defect – the simplest such defects are vacancies and interstitials. The only specific structural feature of a random network is the coordination of an atom to its neighbor. Thus the elementary defect of an amorphous semiconductor is the coordination defect, when an atom has too many or too few bonds. The ability of the disordered network to adapt to any atomic coordination

Fig. 1.3. An example of a continuous random network containing atoms of different bonding coordination, as indicated.

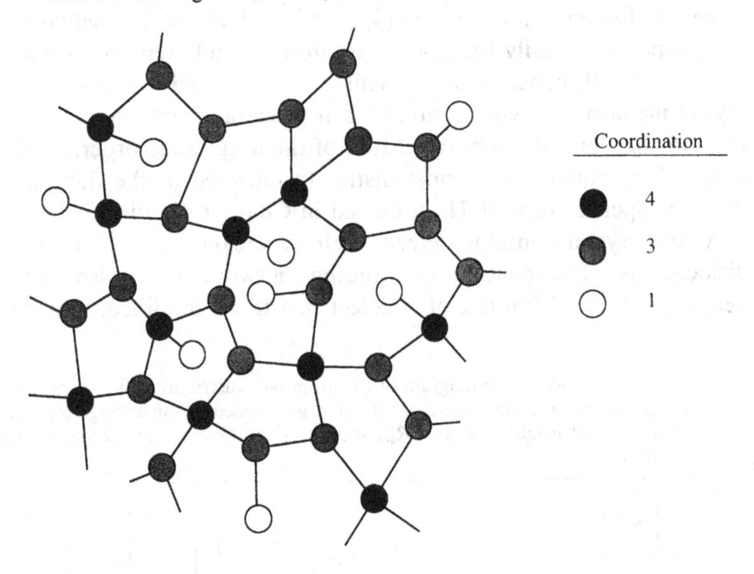

Fig. 1.4. An illustration contrasting the different types of simple defects in (*a*) crystalline and (*b*) amorphous networks.

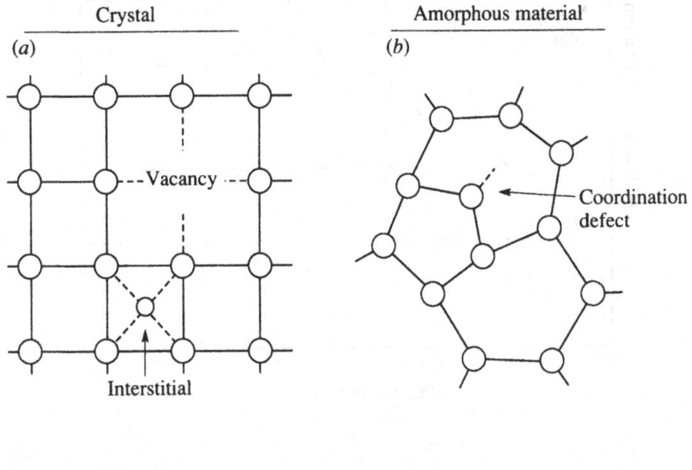

Crystal
(*a*)
Vacancy
Interstitial

Amorphous material
(*b*)
Coordination defect

allows an isolated coordination defect, which is not possible in a crystal. The contrasting defects are illustrated in Fig. 1.4, and are discussed further in Chapter 4.

The intrinsic disorder of the continuous network is less easily classified in terms of defects. The network has many different configurations, but provided the atomic coordination is the same, all these structures are equivalent and represent the natural variability of the material. Since there is no correct position of an atom, one cannot say whether a specific structure is a defect or not. Instead the long range disorder is intrinsic to the amorphous material and is described by a randomly varying disorder potential, whose effect on the electronic structure is summarized in Section 1.2.5.

1.2.2 *Chemical bonding, the 8 − N rule and defect reactions*

The continuous random network model places emphasis on the local chemical bonding of the atoms. A-Si:H and most other amorphous semiconductors are covalently bonded, with well-defined bonding geometries and coordination. A molecular orbital model for silicon and selenium is illustrated in Fig. 1.5 and this type of diagram is used elsewhere in this book. The electrons of an isolated silicon atom occupy two 3s and two 3p states, in addition to the deeper core states which are not involved in the bonding. When the atoms combine to form a solid, the electron interaction splits the valence states into bonding and anti-bonding levels, as illustrated in Fig. 1.5. Chemical bonding occurs because the bonding state has a lower energy than the

Fig. 1.5. Illustration of the bonding configuration of (*a*) silicon and (*b*) selenium atoms constructed from hybridized molecular orbitals. The position of the Fermi energy, E_F, is indicated.

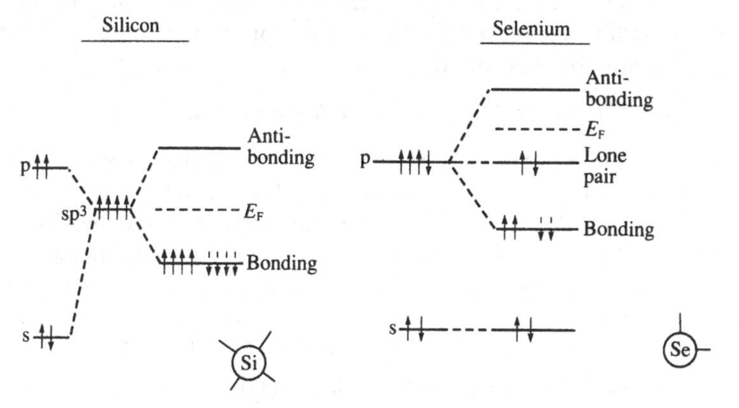

isolated atomic levels and the material has the lowest total energy when the maximum number of electrons occupy bonding states. Their number is constrained by the Pauli exclusion principle which prevents more than two electrons occupying one state. To optimize the number of bonding states, the atomic wavefunctions combine to form hybrid molecular orbitals, described by,

$$\Phi_{hyb} = a\Phi(3s) + b\Phi(3p) \qquad (1.1)$$

where a and b are constants. The four silicon valence electrons combine to give four sp^3 orbitals. Each orbital comprises $\frac{1}{4}$ of an s state and $\frac{3}{4}$ of one of the three equivalent p states. These four orbitals form bonds to adjacent atoms and since silicon has four valence electrons, all the bonds are occupied by two electrons, one from each atom forming the bond. The hybridization minimizes the total energy by arranging as many electrons as possible in bonding orbitals.

It is not always possible to arrange all the valence electrons in bonding orbitals, because four is the maximum number of orbitals which can be made from s and p states. In selenium, which contains six valence electrons, sp^3 hybridization is no longer the lowest energy configuration because this constrains some electrons to occupy the high energy anti-bonding states. Instead there is no hybridization and the s state and one of the p states are filled with electron pairs, forming non-bonding states known as lone pairs. The remaining two singly occupied p states form covalent bonds, splitting into bonding and anti-bonding orbitals, which are then fully occupied by the remaining electrons (Fig. 1.5). The atomic coordination is 2, and the top of the valence band is formed from the lone pair p state electrons, with the bonding p states deeper in the valence band.

Similar diagrams can be constructed for other elements. For example, group V elements bond in p^3 configurations, with s electrons forming the non-bonding pairs. It follows that the optimum number, Z, of covalent bonds for elements is,

$$Z = 8 - N \text{ (for } N \geqslant 4 \text{ and } Z = N \text{ for } N < 4) \qquad (1.2)$$

where N is the number of valence electrons. This prediction of the atom coordination is known as the '$8-N$' rule (Mott 1969). The essential point is that chemical bonds are formed such that the maximum number of electrons are paired in bonding orbitals, the remaining electrons are paired in non-bonding states and the anti-bonding states are empty. The continuous random network allows atoms to take their preferred coordination. Thus the network in Fig. 1.3 might be composed of silicon, phosphorus and hydrogen.

The different roles of local chemistry in the amorphous and crystalline networks are highlighted by considering the bonding of an impurity atom. Since every atomic site in a crystal is defined by the lattice, the impurity either substitutes for the host, adapting itself to the chemistry of the host or occupies a position which is not a lattice site, forming a defect. A substitutional impurity such as phosphorus is four-fold coordinated and acts as a donor because one of its electrons is not involved in bonding and is released into the conduction band. An amorphous material has no rigidly defined array of lattice sites, so that an impurity can adapt the local environment to optimize its own bonding configuration, while also remaining a part of the host atomic network. The $8-N$ rule suggests that phosphorus in amorphous silicon is three-fold coordinated and therefore inactive as an electronic dopant. Indeed it seems to follow from the $8-N$ rule that substitutional doping must be impossible in an amorphous semiconductor. Actually, the chemical bonding does not forbid, but does severely constrain, the doping in a-Si:H, as is explained in Chapter 5.

The $8-N$ rule also suggests the importance of electronically induced structural reactions. The excitation of an electron from one state to another changes the occupancy of bonding and anti-bonding states. The $8-N$ rule predicts that such an electronic excitation destabilizes the atomic bond and induces a change in coordination. Such reactions are usually prevented in crystalline semiconductors by the long range order of the lattice and the extended electron wavefunction. They are promoted in amorphous materials by the adaptibility of the continuous random network and by the localization of electronic carriers. Electronically induced structural changes are an important and fascinating feature of all amorphous semiconductors (see Chapter 6).

1.2.3 *Electronic structure*

One of the fundamental properties of a semiconductor or insulator is the presence of a band gap separating the occupied valence band from the empty conduction band states. According to the free electron theory, the band gap is a consequence of the periodicity of the crystalline lattice. In the past, there was considerable debate over the reason that amorphous semiconductors had a band gap at all, let alone one that is similar to that in the corresponding crystal. Subsequent work explained that the band gap is equivalently described by the splitting of the bonding (or lone pair) and anti-bonding states of the covalent bond (Fig. 1.5). The bands are most strongly influenced by the short range order, which is the same in amorphous and crystalline silicon and the absence of periodicity is a small perturbation.

These results were most clearly stated by Weaire and Thorpe (Weaire 1971, Thorpe and Weaire 1971), who described the bonding by a tight binding Hamiltonian of the form

$$H = V_1 \sum_{j \neq j'} |\Phi_{ij}\rangle\langle\Phi_{ij'}| + V_2 \sum_{i \neq i'} |\Phi_{ij}\rangle\langle\Phi_{i'j}|. \quad (1.3)$$

The wavefunctions, Φ, are the sp^3 hybrid orbitals of the tetrahedral silicon bonding. The first term in Eq. (1.3) is a sum over interactions for which the two wavefunctions Φ_{ij} belong to the same atom and the second term sums pairs of orbitals that belong to the same bond. This Hamiltonian describes the short range bonding information, but contains no information about the long range order, and so applies equally to amorphous and crystalline silicon. Weaire and Thorpe showed that there are ranges of the interaction strength V_1/V_2 for which the conduction and valence bands are separated by a band gap, irrespective of the long range structure.

The three principal features of the structure of amorphous semiconductors are the short range order of the ideal network, the long range disorder and the coordination defects. The preservation of the short range order results in a similar overall electronic structure of an amorphous material compared to the equivalent crystal. Thus, silicon dioxide is an insulator in both its crystalline and amorphous forms and silicon is a semiconductor. The abrupt band edges of a crystal are replaced by a broadened tail of states extending into the forbidden gap, which originates from the deviations of the bond length and angle arising from the long range structural disorder. The band tails are most important despite their relatively small concentration, because electronic transport occurs at the band edge. Electronic states deep within

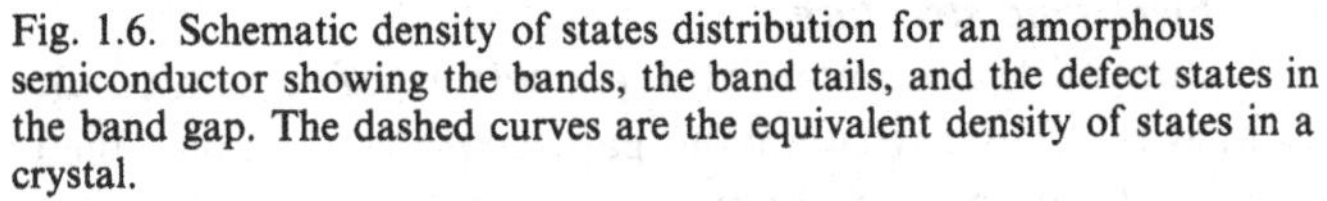

Fig. 1.6. Schematic density of states distribution for an amorphous semiconductor showing the bands, the band tails, and the defect states in the band gap. The dashed curves are the equivalent density of states in a crystal.

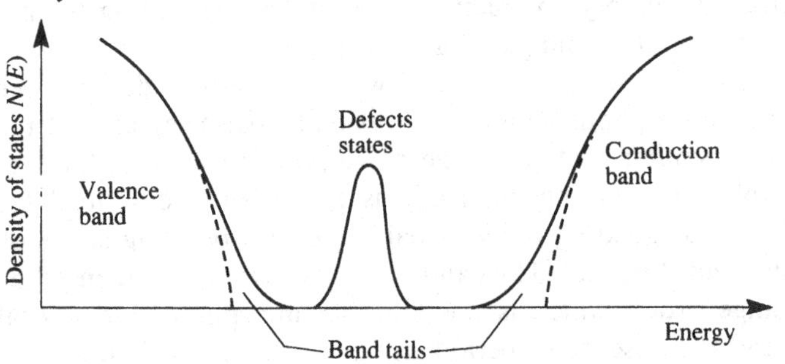

the band gap arise from departures from the ideal network, such as coordination defects. These defects determine many electronic properties by controlling trapping and recombination. The electronic structure of an amorphous semiconductor (see Fig. 1.6) comprises the bands, the band tails and the defect states in the gap and the correspondence between the structure and the electronic properties is summarized in Fig. 1.1.

1.2.4 *Electronic properties*

The wavefunctions of the electronic states are the solutions to Schrödinger's equation,

$$-\frac{h^2}{2m}\nabla^2\Phi + V(\mathbf{r})\Phi = E\Phi \tag{1.4}$$

where E is the electron energy and $V(\mathbf{r})$ is the potential energy arising from the atomic structure. The periodic potential of the ordered crystal leads to the familiar Bloch solutions for the wavefunction,

$$\Phi(\mathbf{r}) = \exp(i\mathbf{k}\cdot\mathbf{r})\,U_k(\mathbf{r}) \tag{1.5}$$

where $U_k(\mathbf{r})$ is the periodicity of the lattice. There is a constant phase relation between the wavefunction at different lattice sites. The wavefunction has a well-defined momentum, k, and extends throughout the crystal. The energy bands are described by energy–momentum

Fig. 1.7. Illustration of the wavefunctions of extended and localized states of an amorphous material, compared to the extended states of a crystal.

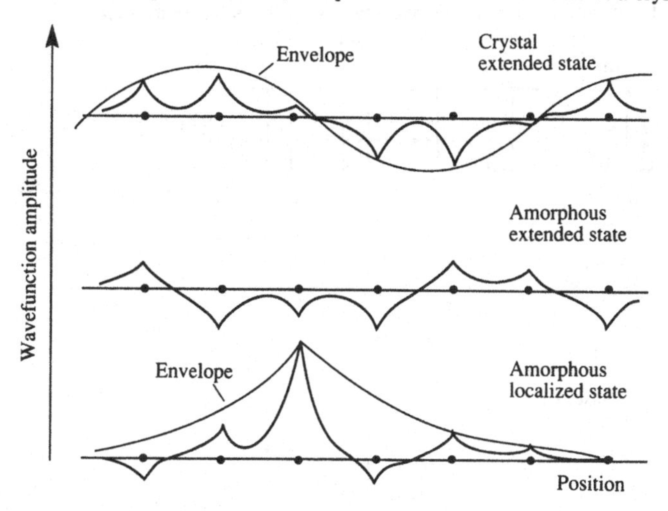

(E–$\mathbf{k}$) dispersion relations, which, in turn, determine the effective mass, electronic excitations, etc.

These solutions to Schrödinger's equation do not apply to an amorphous semiconductor because the potential $V(\mathbf{r})$ is not periodic. A weak disorder potential results in only a small perturbation of the wavefunction and has the effect of scattering the electron from one Bloch state to another. The disordering effect of an amorphous semiconductor is strong enough to cause such frequent scattering that the wavefunction loses phase coherence over a distance of one or two atomic spacings. Fig. 1.7 illustrates the wavefunction of extended electron states in crystalline and amorphous semiconductors and shows the rapid change of phase induced by the disorder.

The strong scattering causes a large uncertainty in the electron momentum, through the uncertainty principle,

$$\Delta k = \hbar/\Delta x \approx \hbar/a_0 \approx k \tag{1.6}$$

where Δx is the scattering length and a_0 is the interatomic spacing. The uncertainty in k is similar to the magnitude of k, so that momentum is not a good quantum number and is not conserved in electronic transitions.

The loss of k-conservation is one of the most important results of disorder and changes much of the basic description of the electronic states. There is a greater emphasis on the spatial location of the carrier than on its momentum. Some consequences of the loss of k-conservation are:

Fig. 1.8. The Anderson model of the potential wells for (a) a crystalline lattice and (b) an amorphous network. V_0 is the disorder potential.

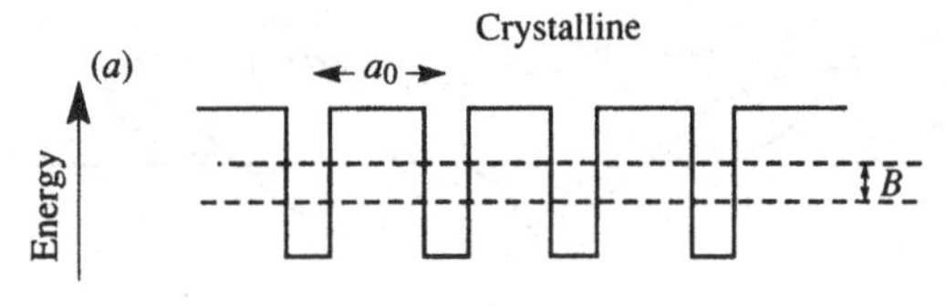

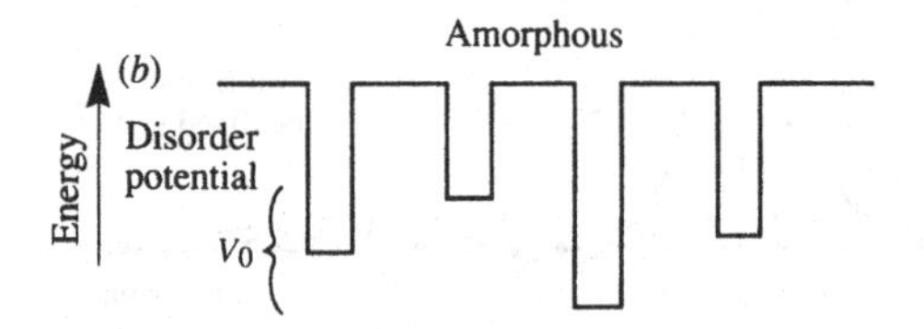

(1) The energy bands are no longer described by the E–$\mathbf{k}$ dispersion relations, but instead by a density-of-states distribution $N(E)$, illustrated in Fig. 1.6. Also the electron and hole effective masses must be redefined as they are usually expressed as the curvature of $E(\mathbf{k})$.

(2) The conservation of momentum selection rules does not apply to optical transitions in amorphous semiconductors. Consequently, the distinction is lost between a direct and an indirect band gap, the latter being those transitions which are forbidden by momentum conservation. Instead transitions occur between states which overlap in real space. This distinction is most obvious in silicon which has an indirect band gap in its crystalline phase but not in the amorphous phase.

(3) The disorder reduces the carrier mobility because of frequent scattering and causes the much more profound effect of localizing the wavefunction.

1.2.5 *Localization, the mobility edge and conduction*

An increasing disorder potential causes first strong electron scattering and eventually electron localization, in which the wavefunction is confined to a small volume of material rather than being extended. The form of the localized wavefunction is illustrated in Fig. 1.7. Anderson's theory of localization uses the model illustrated in Fig. 1.8 (Anderson 1958). The crystal is described by an array of identical atomic potential wells and the corresponding band of electronic states is broadened to a band width B by the interaction between atoms. The disordered state is represented by the same array of sites to which a random potential with average amplitude V_0 is added. Anderson showed that when V_0/B exceeds a critical value, there is zero probability for an electron at any particular site to diffuse away. All of the electron states of the material are localized and there is no electrical conduction at zero temperature.

The critical value of V_0/B for complete localization is about three. Since the band widths are of order 5 eV, a very large disorder potential is needed to localize all the electronic states. It was apparent from early studies of amorphous semiconductors that the Anderson criterion for localization is not met. Amorphous semiconductors have a smaller disorder potential because the short range order restricts the distortions of the bonds. However, even when the disorder of an amorphous semiconductor is insufficient to meet the Anderson criterion, some of the states are localized and these lie at the band edges. The center of the band comprises extended states at which there is strong scattering and states at the extreme edges of the bands are localized. The extended and localized states are separated by a mobility edge at energy E_C, which derives its name because at zero temperature, only electrons above E_C are mobile and contribute to the conduction. This is the essence of the standard model of amorphous semiconductors proposed by Mott (e.g. Mott and Davis (1979) Chapter 1).

The electronic structure is illustrated in Fig. 1.9. The energy of the mobility edge within the band depends on the degree of disorder and is typically 0.1–0.5 eV from the band edge in all amorphous semiconductors. The properties of states near the mobility edge are actually more complicated than in this simple model of an abrupt mobility edge and are described in more detail in Chapter 7. Nevertheless, the model of Fig. 1.9 provides a good description of amorphous semiconductors.

Fig. 1.9. The density of states distribution near the band edge of an amorphous semiconductor, showing the localized and extended states separated by the mobility edge.

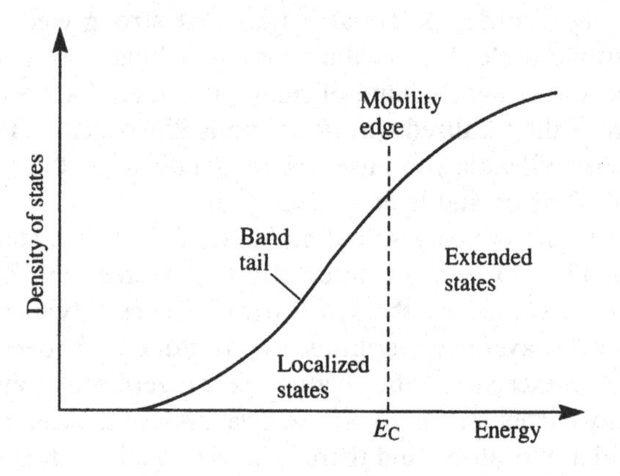

Fig. 1.10. Model showing the tunneling between two localized states separated by distance R and energy E_{12}; R_0 is the localization length.

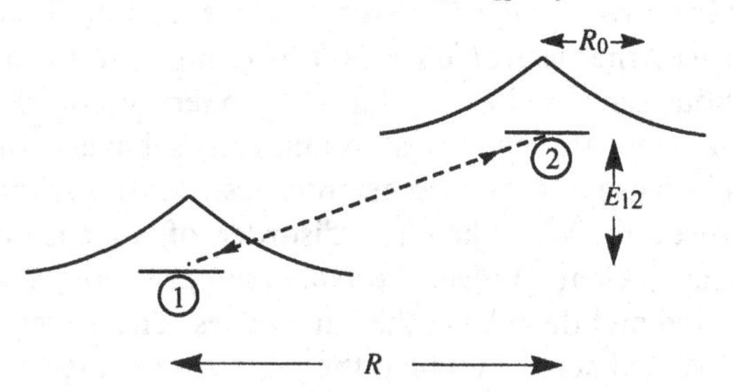

In addition to creating the distinction between extended and localized states, the disorder also influences the mobility of the electrons and holes above the mobility edges. The carrier mobility is reduced by scattering, which increases with the degree of disorder. Under conditions of weak scattering, the mobility is,

$$\mu = e\tau/m = eL/mv_c \tag{1.7}$$

where τ is the scattering time, L the mean free path and v_c is the electron thermal velocity. The room temperature mobility of crystalline silicon is about 1000 cm^2 V^{-1} s^{-1} which corresponds to a scattering length of about 1000 Å. The interatomic spacing is a lower limit to the scattering length in a disordered solid and reduces the carrier mobility to about 2–5 cm^2 V^{-1} s^{-1} according to Eq. (1.7). Although this is approximately the actual magnitude, the simple transport equations do not apply to strong disorder and the calculation of the mobility and conductivity near the mobility edge is complicated (see Chapter 7).

Although there is no macroscopic conduction in localized states at zero temperature, tunneling transitions occur between localized states and give rise to conduction at elevated temperatures. The spatial extent of the wavefunction allows tunneling between neighboring localized states as illustrated in Fig. 1.10 for two states of separation R and energy difference E_{12}. The probabilities of tunneling between states 1 and 2, and vice versa are respectively,

$$P_{12} = \omega_0 \exp(-2R/R_0)\exp(-E_{12}/kT) \tag{1.8}$$

$$P_{21} = \omega_0 \exp(-2R/R_0) \tag{1.9}$$

Fig. 1.11. Illustration of the three main conduction mechanisms expected in an amorphous semiconductor.

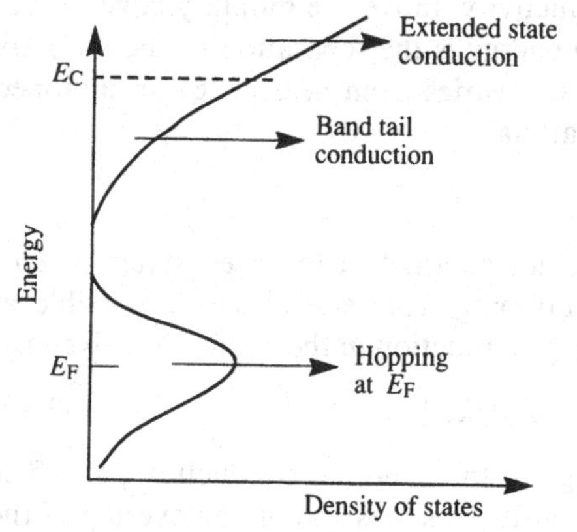

where R_0 is the localization length. The energy-dependent term is included whenever the transition requires the absorption of thermal energy. The prefactor $\omega_0 \sim 10^{13}\ \mathrm{s}^{-1}$ is approximately a phonon frequency. Transitions between localized states effectively cease when $R > 10R_0$ because of the R-dependent exponential term. The localization length varies with the binding energy of the state and is roughly,

$$R_0(E) = \left[\frac{\hbar^2}{m(E-E_C)}\right]^{\frac{1}{2}} \tag{1.10}$$

where E_C is the mobility edge, which is the energy of the unbound electron. Deep states are more strongly localized than shallow ones; the localization length is about 3 Å at a binding energy of 0.5 eV, but is more extended nearer the mobility edge. Tunneling therefore occurs between states which are less than about 50–100 Å apart and so hopping conduction is significant whenever the density of states exceeds about $10^{18}\ \mathrm{cm}^{-3}$.

So far no amorphous semiconductors have been made with a Fermi energy in the extended states beyond the mobility edge. The Fermi energy of doped a-Si:H moves into the band tails, but is never closer than about 0.1 eV from the mobility edge. There is no metallic conduction, but instead there are several other possible conduction mechanisms, which are illustrated in Fig. 1.11.

(1) *Extended state conduction*

Conduction is by thermal activation of carriers from E_F to above the mobility edge and follows the relation,

$$\sigma_{ext} = \sigma_{oe} \exp[-(E_C - E_F)/kT] \tag{1.11}$$

where σ_{oe} is the average conductivity above the mobility edge (about $100\ \Omega^{-1}\ \mathrm{cm}^{-1}$). The activation energy is the separation of the mobility edge from the Fermi energy, and varies from nearly 1 eV in undoped a-Si:H to 0.1 eV in n-type material.

(2) *Band tail conduction*

Although carriers cannot conduct in localized states at zero temperature, conduction by hopping from site to site is possible at elevated temperatures. Hopping conduction in the band tail is given by,

$$\sigma_{tail} = \sigma_{ot} \exp[-(E_{CT} - E_F)/kT] \tag{1.12}$$

where E_{CT} is the average energy of the band tail conduction path. The prefactor σ_{ot} depends on the density of states and on the overlap of the wavefunction and is smaller than σ_{oe}. On the other hand, E_{CT} is closer to E_F than is E_C, so that the exponential term offsets the smaller prefactor, particularly at low temperature.

(3) *Hopping conduction at the Fermi energy*

Conduction at the Fermi energy occurs when the density of states is large enough for significant tunneling of electrons. The conductivity is small but weakly temperature-dependent and consequently this mechanism tends to dominate at the lowest temperature. The states at E_F usually originate from defects and so the conductivity varies greatly with the defect density. For example, the addition of hydrogen to amorphous silicon reduces the defect density and almost completely suppresses the Fermi energy hopping conduction.

2 Growth and structure of amorphous silicon

Amorphous materials do not have the regular atomic structure characteristic of a crystal. Instead, the specific bonding arrangement within a particular volume of material represents one of many alternative configurations. Hydrogenated amorphous silicon has the added variability of a hydrogen content which can reach 50 at%. Most features of the a-Si:H network structure are defined at the time of growth and therefore depend on the details of the deposition process. Thus it is anticipated that the electronic properties vary with the growth conditions and that a detailed understanding of the growth mechanisms is essential for the optimization of the electronic properties. Indeed, a-Si:H does exhibit a great range of specific properties. However, the optimization of the growth process produces films which are remarkably independent of the detailed growth process. The best films are all similar, while low quality films are defective in many different ways. It is now recognized that the electronic structure is influenced by defect reactions taking place within the material after growth, which are largely independent of the growth process. These are described in Chapter 6 and explain why the properties of low defect density material are not so sensitive to the deposition method.

2.1 Growth of a-Si:H

The usual method of depositing a-Si:H is by plasma decomposition of silane gas, SiH_4, with other gases added for doping and alloying. Silane decomposes in the absence of the plasma above about 450 °C and high temperature pyrolitic decomposition is used to make polycrystalline or epitaxial silicon. Amorphous films can be grown in this way if the temperature is less than about 550 °C, but these films are mostly of low quality because the temperature is too high to retain the hydrogen. The deposition of hydrogenated films at lower temperatures requires a source of energy to dissociate the silane molecule and this is the role of the plasma. The first plasma deposition system for amorphous silicon was developed by Chittick *et al.* (1969), and was a radio frequency (rf) inductive system in which an induction coil outside the quartz deposition chamber created the plasma. Most subsequent reactors have been designed in a diode configuration in which the plasma is confined between two parallel electrodes. This type of reactor is illustrated in Fig. 2.1. Deposition usually takes place at a gas pressure of 0.1–1 Torr, which is the optimum pressure to sustain the plasma. The reactor consists of a gas inlet arrangement, the deposition chamber which holds the substrate, a pumping system, and the source of power for the discharge. The deposition process is referred to as plasma enhanced chemical vapour deposition (PECVD).

Despite being a relatively simple reactor, there are many variables in the deposition process which must be controlled to give good material. The gas pressure determines the mean free path for collisions of the gas molecules and influences whether the reactions are at the growing surface or in the gas. The gas flow rate determines the residence time of the gas species in the reactor. The rf power controls the rate of dissociation of the gas and therefore also the film growth rate, and the temperature of the substrate controls the chemical reactions on the growing surface.

There are numerous modifications which can be made to the basic reactor design. The initial dissociation of the silane gas can be by ultraviolet (UV) light illumination (photo-CVD), which either excites the silane directly, or by energy transfer from mercury vapour introduced into the chamber. Photo-CVD reactors eliminate the electric discharge and prevent the bombardment of the growing film by ions from the plasma, which may be a source of defects. A different way of reducing bombardment is to separate the plasma from the growing

Fig. 2.1. Schematic diagram of a typical rf diode plasma reactor for depositing a-Si:H and alloys.

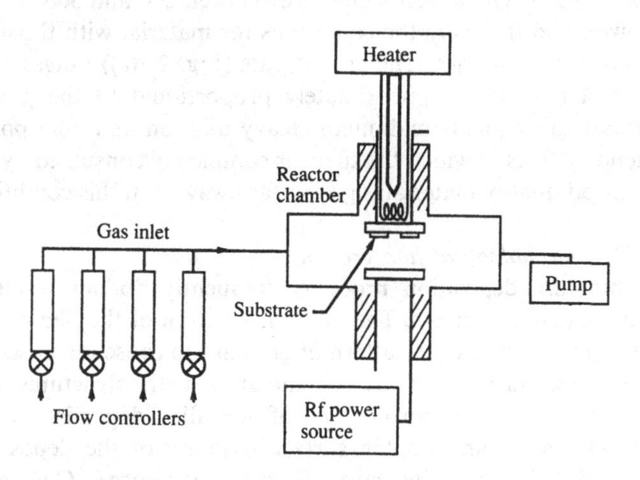

surface, either using a metal grid in the reactor – giving a triode structure – or by separating the plasma completely from the deposition and only introducing the gas radicals into the chamber. Alternatively the silane can be thermally dissociated in a hot part of the reactor, then directed to the cooler substrate.

Reactive sputtering is a method of depositing a-Si:H films, in which a silicon target is sputtered, usually with argon ions, but also in the presence of hydrogen. The sputtered silicon reacts with the atomic hydrogen in the plasma, forming SiH_x radicals from which deposition takes place in much the same way as in the glow discharge process. Indeed, the sputtered films have essentially the same properties as the plasma deposited material, although they may suffer more from ion bombardment damage because of the higher energy of the ions reaching the surface.

All these growth techniques give a-Si:H material with the same general properties. This book emphasizes the glow discharge material as this is the most common deposition technique, but the following description of the structure, growth mechanisms and atomic bonding applies broadly to all other deposition techniques.

Fig. 2.2 summarizes some typical properties of PECVD a-Si:H films (Knights and Lucovsky 1980). The hydrogen content ranges from 8 to 40 at% and decreases as the substrate temperature is raised. The hydrogen content also depends on the rf power in the plasma and on the composition of the gas. Fig. 2.2(*b*) shows the variation when silane is diluted to 5% concentration in argon. The defect density also depends on the substrate temperature and power and can vary by more than a factor 1000. The lowest values are between 200 and 300 °C and at low power, and these are the conditions for material with the most useful electronic properties. The growth rate (Fig. 2.2(*c*)) ranges from about 1–10 Å s^{-1} and is approximately proportional to the power, provided that the silane is undiluted. Heavy dilution and high power give depletion effects in which the silane is completely consumed by the reaction. Good quality material is grown far away from this condition.

2.1.1 *The morphology of film growth*

Thin film deposition processes frequently do not result in uniform homogeneous layers. The initial nucleation of the film on the substrate and the subsequent pattern of growth can cause macroscopic inhomogeneities, such as voids, columnar growth structures and surface roughness. The morphology of the film depends on the deposition chemistry and on the surface mobility of the depositing molecules and is a sensitive indicator of the growth process. Control of

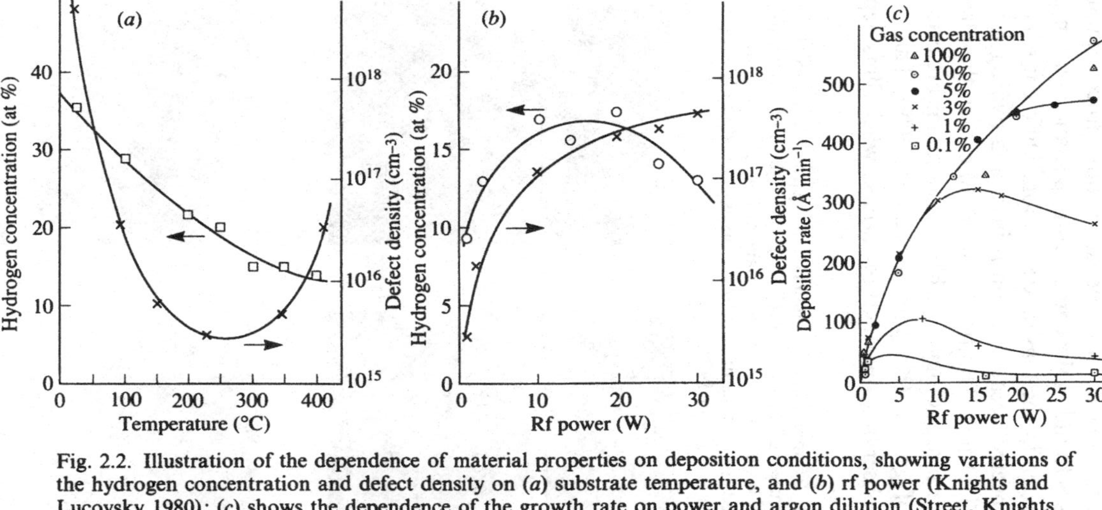

Fig. 2.2. Illustration of the dependence of material properties on deposition conditions, showing variations of the hydrogen concentration and defect density on (*a*) substrate temperature, and (*b*) rf power (Knights and Lucovsky 1980); (*c*) shows the dependence of the growth rate on power and argon dilution (Street, Knights and Biegelsen 1978).

the morphology is important because there is good reason to expect that inhomogeneities will degrade the electronic properties of the material.

A useful illustration of how the deposition process influences the growth of films is found in the comparison of silane dilution with either argon or hydrogen in the plasma deposition. Different effects of these diluents are expected because argon is inert and hydrogen is an active part of the a-Si:H film. Dilution with argon causes the films to grow with a columnar morphology, in which the columns are oriented along the direction of growth (Knights and Lujan 1979). This structure is easily seen by observing the fracture surface, when films deposited on

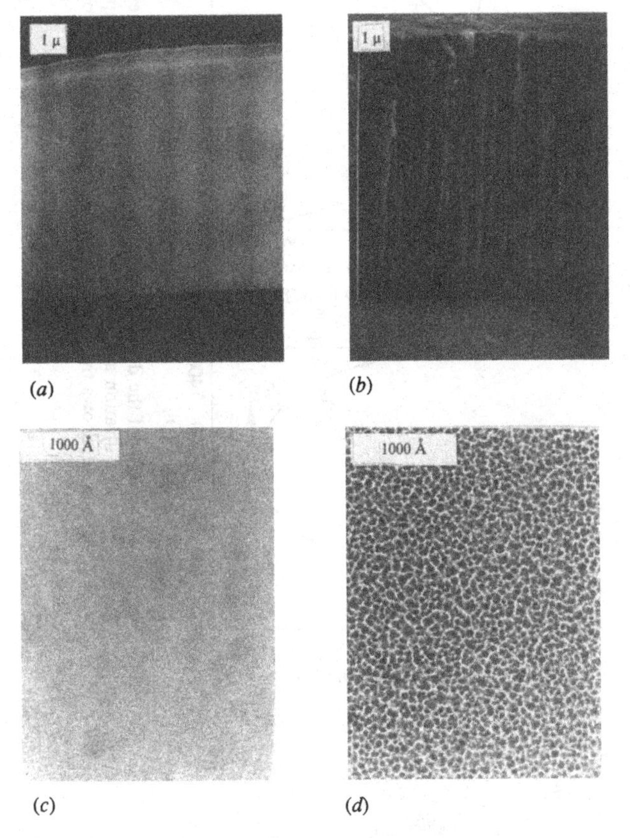

Fig. 2.3. Examples of the columnar microstructure of a-Si:H films seen in the fracture surface ((*a*) and (*b*)) and in the plane of growth ((*c*) and (*d*)). Film (*a*) is deposited from pure silane at low rf power and film (*b*) is deposited from silane diluted to 5% in argon and at high rf power (Knights and Lujan 1979).

glass are broken and viewed from the side. Examples of material grown with and without argon dilution are shown in Fig. 2.3. The fracture surface of material made from pure silane has no visible structure but has the smooth conchoidal surface typical of glasses. The columnar structure in the films made with argon is, however, clearly visible in the micrograph. There is no single characteristic size of the columns, although the minimum dimension is about 100 Å. When viewed at lower magnification, features up to about 1 micron are visible. Messier and Ross (1982) identified four different length scales for the columnar structure, increasing with the thickness of the film.

The columnar structure occurs in films deposited with the other noble gases and is more pronounced with krypton but less with helium and neon (Knights *et al.* 1981). The effect also increases with the rf power in the plasma, as is indicated in Fig. 2.4, which maps the dependence of the columnar structure on power and argon dilution and shows the division between the columnar and the non-columnar growth regimes. The transition between the two regimes is gradual – the columns become more indistinct as the dividing line is crossed and eventually fade away. Films deposited without argon dilution and at low power

Fig. 2.4. Diagram showing the deposition conditions for columnar and non-columnar a-Si:H films deposited from silane/argon mixtures at different rf power.

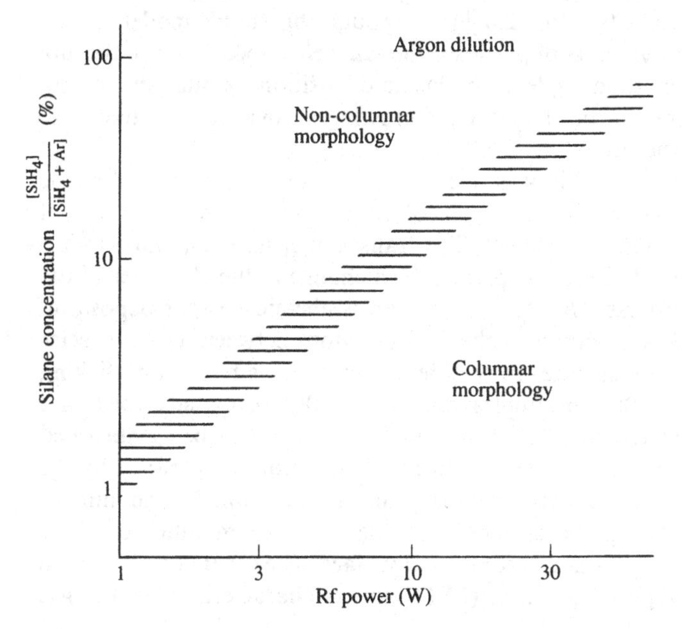

are furthest from the columnar region and also have the best electronic properties.

Hydrogen dilution of silane has a completely different effect on the growth. A high concentration of hydrogen causes the deposited films to become crystalline rather than amorphous. The crystallite size is small, often less than 100 Å, so that the material is termed microcrystalline silicon. Fig. 2.5 shows the power–dilution diagram for hydrogen, indicating the transition to microcrystallinity. Dilution to about 5% silane is needed to cause crystallization and the effect is enhanced by high rf power. The transition is abrupt, and a mixed phase material is only observed in a narrow region near the dividing line. It has been suggested that a real thermodynamic transition is involved, but the experimental evidence for this conjecture is incomplete (Veprek *et al.* 1988).

The columnar structure in material prepared under argon dilution is a well-known phenomenon in the vapor deposition of thin films and is caused by shadowing. Atomic scale shadowing is modeled by assuming that the atoms are hard spheres, with zero surface mobility when they hit the surface and is illustrated in Fig. 2.6. The columns occur because an atom on the surface shadows an area extending to a distance of twice its radius, so that an incoming atom attaches to the first one rather than forming a continuous surface layer. The growth of chain-like structures results, with frequent branches in the direction of the flux of atoms (see Fig. 2.6(*b*)). Although this simple model applies to a unidirectional flux of particles, the same effect occurs when the flux is from a large solid angle as in plasma deposition. Similar shadowing by undulations in the shape of the growing surface causes larger size columnar structure.

A low surface mobility is a necessary condition for the columnar growth, otherwise the diffusion of atoms across the surface will mitigate the shadowing effect. The transition from columnar to non-columnar morphology is explained by a change in the character of the deposition process. One type of growth is chemical vapor deposition (CVD) which is governed by chemical reactions between the gas species and the growing surface. In the ideal case, the gas has a low sticking coefficient, so that molecules are continually being adsorbed and released from the surface before finally reacting and being absorbed into the growing film. The growth rate is therefore determined by the rate of the surface chemical reaction, rather than by the flux of molecules striking the surface. The high surface mobility of CVD growth results in a smooth conformal surface. The other type of growth is physical vapor deposition (PVD) and is characterized by the gas having a high sticking coefficient when each molecule remains where it first strikes the surface. In this case the growth rate is determined by the flux of molecules striking the surface. The PVD mechanism gives columnar morphology.

Plasma deposition tends to be intermediate between the CVD and PVD extremes. When highly reactive gas species are formed in the plasma and the deposition rate is high, then a PVD mechanism is favored, whereas less reactive species favor CVD. The gas dilution

Fig. 2.5. Diagram showing the typical deposition conditions for microcrystalline silicon and a-Si:H films deposited from silane/hydrogen mixtures at different rf power.

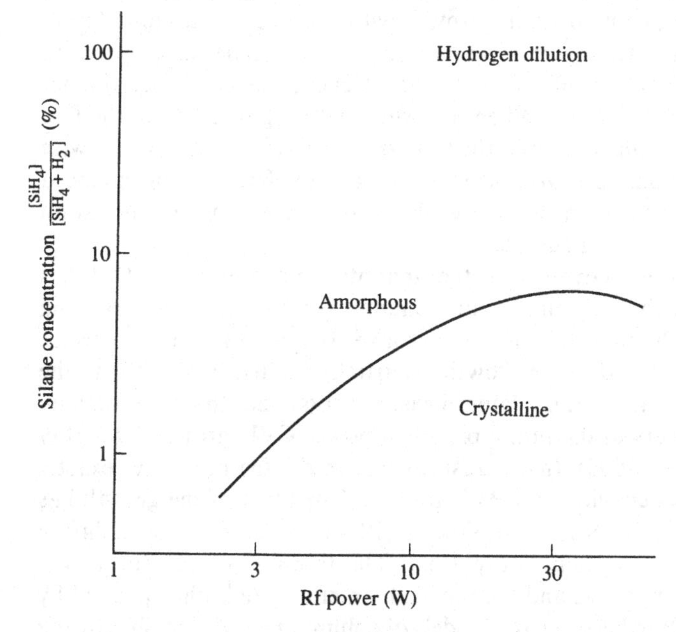

Fig. 2.6. Illustration of (*a*) the shadowing effect of atoms on the growth surface, and (*b*) the resulting chain-like growth morphology when the sticking coefficient is high.

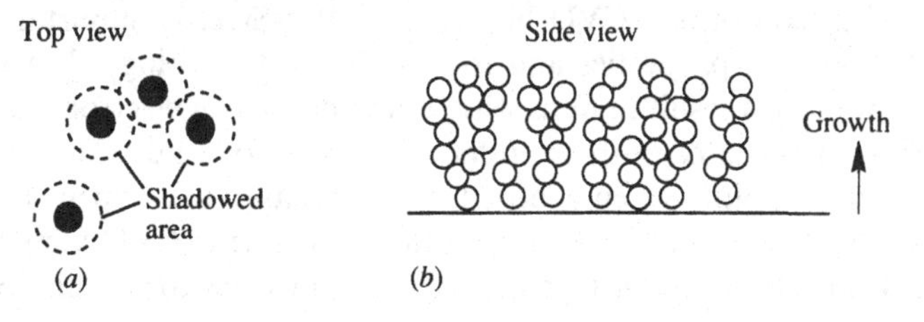

results in Fig. 2.3 show that the addition of argon into the plasma and a high rf power cause the deposition to be more PVD-like, giving the columnar growth.

One way to observe and quantify the type of growth is to deposit material into a deep trench cut into the substrate. In the PVD process the total deposit in the trench is proportional to the area of the top opening because it is flux-limited, and when the trench is deep and narrow, the film will be thin. In contrast, the deposit in a CVD process is proportional to the substrate surface area, so that the film in the trench is the same thickness as on the top surface. These two situations are illustrated in Fig. 2.7.

Fig. 2.7 also shows cross-section views of a-Si:H films deposited into such trenches, comparing the growth with and without argon dilution (Tsai, Knights, Chang and Wacker 1986a). The material grown with undiluted silane is uniform within the trench, but is slightly thinner than on the top surface. When deposited with argon dilution, the film in the trench is much thinner than on top, and is also non-uniform with a pin-cushion shape at the bottom of the groove. There is a pronounced columnar structure on the side walls, with the columns oriented at an angle to the growing surface.

The conformal coverage of the undiluted silane growth establishes that the growth mechanism is predominantly CVD with a low sticking coefficient. The thinner film in the groove compared to the top surface in Fig. 2.7 shows that the growth is partially limited by the flux to the surface. This result and later measurements find that the sticking coefficient is about 0.4 in a typical low power CVD growth (Doughty and Gallagher, 1990). In contrast, the argon-diluted films have exactly the expected behavior of a PVD process. The form of the growth has been modeled by Ross and Vossen (1984) for the case of a diffuse deposition source such as a plasma. The thinning of the film at the bottom of the groove and the pin-cushion effect are both explained by shadowing. Furthermore the models of columnar growth by shadowing predict that for oblique incidence the columns will be oriented away from the surface normal towards the flux of particles and this effect is also apparent in Fig. 2.7.

The flat, conformal, CVD-like growth of a-Si:H is important both for the material properties and for the device technology. Invariably, PVD conditions result in films with a high density of electronic defects, which are associated with the internal surfaces of voids. The columns also oxidize rapidly when exposed to air, because their open structure allows the diffusion of oxygen from the atmosphere. CVD films have a much lower defect density and no oxidation except for a thin layer on the surface. The ability to grow conformal films is necessary in the technology of electronic devices to ensure that complex structures maintain film continuity and uniformity. Argon dilution is therefore generally detrimental, as is a high growth rate. The best a-Si:H films are deposited from pure silane and with a low rf power.

The structure and morphology of the a-Si:H films obviously depend greatly on the form of the growing surface. *In situ* ellipsometry is a powerful experimental tool which gives information about the surface layer during the deposition, by measuring the optical constants of the film. The effective dielectric function of the material is obtained from

Fig. 2.7. Predicted (upper) and observed (lower) growth morphology of deposits made over a trench cut in a substrate, showing the difference between (*a*) CVD and (*b*) PVD growth processes (Tsai *et al.* 1986a).

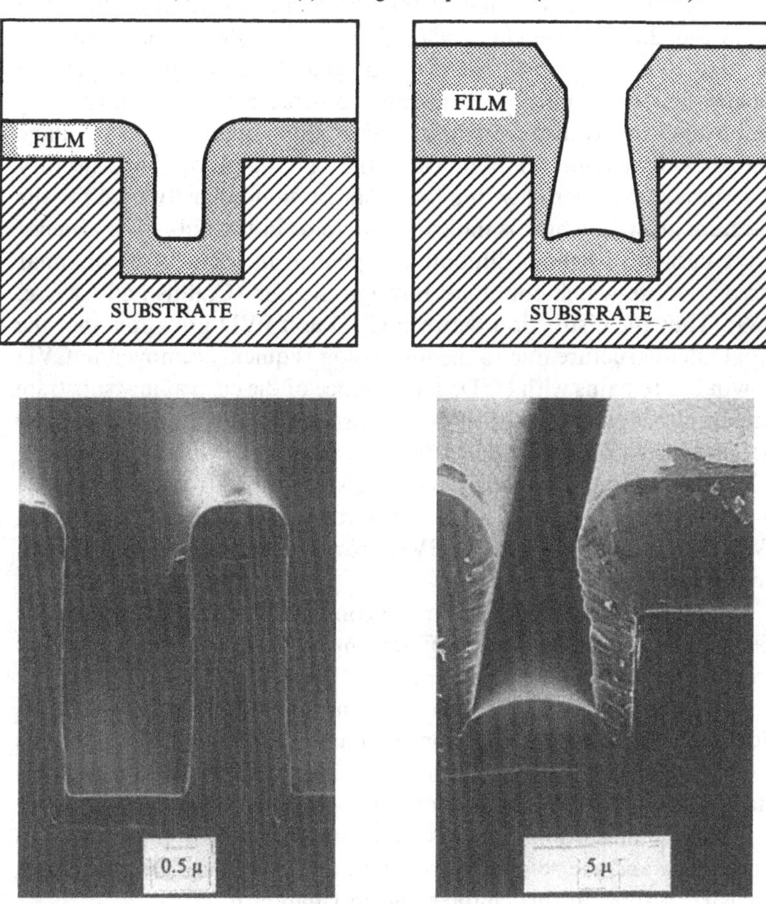

the intensity and polarization of light reflected from the surface. The evolution of the real and imaginary parts of the dielectric function can be followed as the film growth occurs; an example of such a trajectory is shown in Fig. 2.8 (Collins and Cavese, 1989). At the beginning of the measurement there is no deposit and the optical constants are those of the substrate. When a thickness of about 1000 Å is reached there is no further influence of the substrate and any change in the optical constants reflects a change in the growth process. At intermediate thicknesses, there are contributions from both the film and substrate.

The dashed line in Fig. 2.8 is the predicted trajectory of the dielectric constants if the a-Si:H is assumed to grow as a uniform film. This model evidently does not explain the data, particularly the cusp which occurs when the film thickness is about 50 Å. The cusp structure results from non-uniform nucleation of the film and is modeled by the solid line in Fig. 2.8. The best fit is when 70% of the substrate is covered in the first 10 Å of deposit, but full coverage is not complete until a thickness of 50 Å. When the substrate temperature is reduced, the cusp disappears, and is also absent in a-Si:H films deposited from silane diluted with helium. In both cases the end point of the trajectory corresponds to material with a lower silicon bond density than for the optimum growth, which is indicative of voids in the bulk. The absence of the cusp with helium dilution is explained in terms of a columnar microstructure which continues through the film (Collins and Cavese, 1989). Thus the distinction between CVD and PVD growth is that the initial microstructure due to the nucleation is quickly removed in CVD growth but remains with PVD. The absence of the cusp at low substrate temperature is explained by the lower surface mobility of the depositing species which suppresses the island formation in the initial nucleation.

The surface roughness of the films is also obtained from ellipsometry. Typical roughness values of 10 Å are found for films with optimum CVD growth conditions, while PVD growth conditions lead to a much greater roughness.

The growth of microcrystalline silicon by hydrogen dilution of the plasma gases is a completely different mechanism from the columnar structure resulting from argon dilution. PVD deposition and columnar growth arise from a very non-equilibrium growth situation with high sticking coefficients and low surface mobility. Growth of crystalline silicon represents a higher degree of ordering than the amorphous phase, and presumably results from a deposition process that is closer to thermal equilibrium. The degree to which the a-Si:H films are in equilibrium and the role of hydrogen in the growth are central themes discussed further in this chapter and in Chapter 6.

2.1.2 *Growth mechanisms*

Most a-Si:H material is made by PECVD and the following discussion concentrates on this technique. Although the principle of the deposition is quite simple, the physical and chemical processes which take place are exceedingly complex, and it is difficult to identify the dominant reaction paths from the many possibilities. A discharge is sustained by the acceleration of electrons by the electric field. These electrons collide with the molecules of the gas causing ionization (amongst other processes) and releasing more electrons. The ionization energy of typical molecules is in the range 10–20 eV. The average electron energy is given by $eE\lambda_e$, where E is the electric field and λ_e is the mean free path of the electrons for collisions with molecules. A high gas pressure which results in a small λ_e, therefore requires a high field to cause ionization. However, a low pressure reduces the number of collisions which occur before the electron reaches the electrode, and a high field is again needed to increase the ionization rate and sustain the plasma. These effects are the basis of the Paschen curve of sustaining voltage, V_s, versus the product, pd, of gas pressure and electrode spacing, which has a minimum value of V_s near 300 V occurring at about $pd = 1$ cm Torr. Accordingly plasma deposition is usually performed at a pressure of 0.1–1 Torr to minimize the voltage necessary to sustain the plasma.

Fig. 2.8. Trajectories of the dielectric function obtained by *in situ* ellipsometry during growth of a-Si:H. The predicted trajectories are shown for uniform growth (dashed) and island growth (solid) (Collins and Cavese 1989).

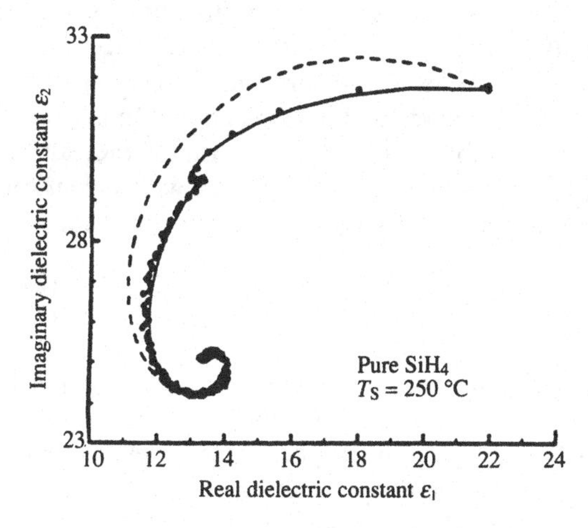

There is negligible acceleration of the ions in the plasma by the electric field because of their large mass, so that the energy of the plasma is acquired by the electrons. This difference, together with the electrical asymmetry of the reactor, causes a dc potential to develop within the plasma, despite being powered by an alternating rf voltage. The negative dc bias of the powered electrode can reach 200–400 V, and the presence of this bias can greatly affect the material properties of the a-Si:H films.

The collisions of the energetic electrons with the gas molecules cause many processes in addition to the ionization. Gas molecules may be excited into a higher electronic state, from which recombination to the ground state results in the emission of photons and is the origin of the plasma glow, although many of the transitions are in the UV region of the spectrum. The gas molecules are also excited into higher vibrational or rotational states – in a typical plasma the vibrational temperature of silane has been measured to be 850 K. However, neither of these excitations directly causes deposition of the film. More important is the dissociation of the gas either as neutral radicals or ions. Examples of silane dissociation reactions which require low energies are (Knights, 1979),

$$SiH_4 \rightarrow SiH_2 + H_2 \qquad 2.2\ \text{eV} \tag{2.1}$$

$$SiH_4 \rightarrow SiH_3 + H \qquad 4.0\ \text{eV} \tag{2.2}$$

$$SiH_4 \rightarrow Si + H_2 \qquad 4.2\ \text{eV} \tag{2.3}$$

There are many other possible reactions involving increasingly higher energies.

At normal deposition pressures, the mean free path of the gas molecules is 10^{-3}–10^{-2} cm and is much smaller than the dimensions of the reactor, so that many intermolecular collisions take place in the process of diffusion to the substrate. An understanding of the growth is made particularly difficult by these secondary reactions. In a typical low power plasma, the fraction of molecular species that is radicals or ions is only about 10^{-3}, so that most of the collisions are with silane. An important process is the formation of larger molecules, for example

$$SiH_4 + SiH_2 \rightarrow Si_2H_6 \tag{2.4}$$

$$Si_2H_6 + SiH_2 \rightarrow Si_3H_8 \tag{2.5}$$

etc. These molecular species have been detected in the reactor exhaust. High rf power in the reactor promotes the formation of macroscopic particles in the gas which interfere with the growth and these particles are believed to be nucleated by the above reactions.

The secondary reactions greatly modify the mix of radicals within the plasma (Gallagher 1988). Their concentration is described by the diffusion–reaction equation,

$$G(x) = -D\frac{d^2n}{dx^2} + knN \tag{2.6}$$

where $G(x)$ is the rate of creation of the radical at position x, n is the concentration, D the diffusion coefficient and k the reaction rate with silane molecules present with concentration N. The concentration of radicals decays from a source at $x = 0$ as,

$$n(x) = n_0 \exp(-x/R) \qquad R = (D/kN)^{1/2} \tag{2.7}$$

Within a uniform plasma, the radical concentration is $G\tau$, where the lifetime, τ, is $(kN)^{-1}$. Those radicals with a high reaction rate have a low concentration and a short diffusion length and so are less likely to reach the growing surface. The least reactive species survive the collisions longest and have the highest concentrations, irrespective of the initial formation rates. One such radical is SiH_3, which does not react with SiH_4, unlike radicals with fewer hydrogens, because Si_2H_n structures are possible only with $n \leqslant 6$. Thus the plasma contains a combination of long-lived primary radicals, and the secondary products of the more reactive gas species.

Much of the direct experimental information about the radicals and ions in the silane plasma comes from mass spectrometry. Fig. 2.9 shows the concentrations as a function of argon dilution for a typical low power plasma. Gallagher and Scott (1987) find that SiH_3 accounts for at least 80% of the gas radicals in a pure silane plasma. Argon dilution increases the concentration of other radicals and these eventually dominate the plasma.

The growth of a-Si:H films involves two further processes beyond the primary excitation of ions and radicals, and their secondary reactions. First is the adsorption of the molecular fragments onto the growing surface. There is still considerable doubt about exactly which species cause deposition (Veprek 1989), but the common view is that under low power deposition with undiluted silane gas, it is SiH_3. The second process is the release of atoms or molecules from the surface. The resulting a-Si:H material has a hydrogen content of about 10 at%, but the radicals arriving at the surface bring far more hydrogen with them. Clearly hydrogen must be released from the surface during growth.

Fig. 2.10 illustrates some of the processes which may occur at the growing surface of an a-Si:H film (Gallagher 1986). A surface which is completely terminated by Si—H bonds will not take up SiH_3 radicals for the same reason that it is unreactive with silane – no Si–Si bond can be formed. However, bonding of SiH_3 can occur at any unterminated silicon bond. The removal of hydrogen from the surface is therefore a

Fig. 2.9. The density of various radicals relative to the silane concentration for low power discharges in pure silane and silane/argon mixtures (Gallagher and Scott 1987).

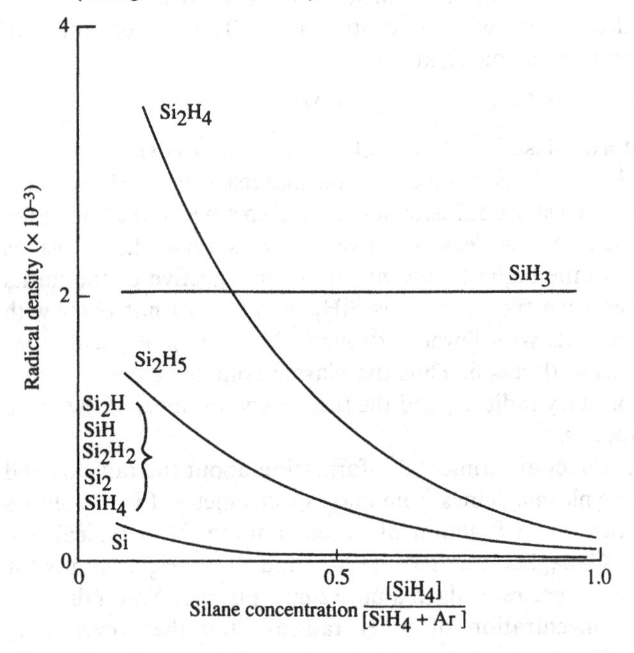

Fig. 2.10. Illustration of some possible processes taking place at the a-Si:H surface during growth.

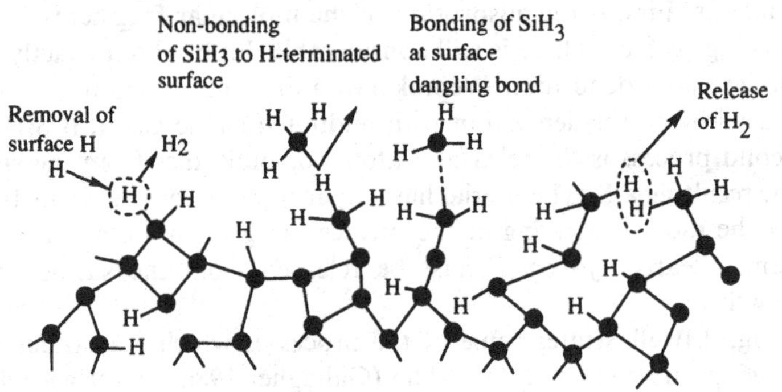

necessary step in the deposition of films from SiH_3. The hydrogen can be released spontaneously by thermal excitation either as atomic or molecular hydrogen, or it can be stripped from the surface by gas radicals or ions. For example

$$\equiv Si—H + SiH_3 \rightarrow \equiv Si— + SiH_4 \tag{2.8}$$

or

$$\equiv Si—H + H \rightarrow \equiv Si— + H_2 \tag{2.9}$$

where $\equiv$Si— represents silicon at the surface which is bonded into the network. The second of these reactions is observed directly and is weakly temperature-dependent (Muramatsu and Yabumoto 1986).

An illustrative model of growth is one in which SiH_3 radicals both remove hydrogen from the surface and bond to surface dangling bonds. The rate equation for the concentration N_{SD}, of surface dangling bonds is,

$$\frac{dN_{SD}}{dt} = \alpha(N_0 - N_{SD})F_{SiH_3} - \beta N_{SD} F_{SiH_3} \tag{2.10}$$

where N_0 is the total number of surface atoms, F_{SiH_3} is the flux of SiH_3 radicals hitting the surface, and α and β are the probability that SiH_3 reacts with the surface to remove hydrogen or to form a bond. In steady state,

$$N_{SD} = \alpha N_0/(\alpha + \beta) \tag{2.11}$$

and the growth rate, r, is,

$$r = \beta F_{SiH_3} = \text{const. } W_{rf} \tag{2.12}$$

where the flux of radicals is set proportional to the ionization rate and hence to the rf power in the plasma, W_{rf}.

This simple model accounts for the basic observation that the deposition rate is proportional to the rf power and almost independent of temperature. It also predicts that the hydrogen coverage of the surface is independent of power, which is consistent with the weak dependence of hydrogen concentration on power. The slow decrease of the hydrogen content with increasing temperature (see Fig. 2.2) is probably because some part of the hydrogen release is thermally activated.

The removal of hydrogen from the surface is more complicated than in the above model, because many dangling bonds which result from the release of hydrogen reconstruct into Si—Si bonds which form the amorphous network. The final hydrogen content is set by a delicate balance between the removal of hydrogen into the gas, the attachment of radicals from the gas, and the reconstruction of dangling bonds. As growth proceeds and a particular silicon atom moves deeper below the surface, its ability to reconstruct or take up gas molecules decreases. The trick in making good material is to ensure that all these processes occur without resulting in undesirable structures, such as unterminated dangling bonds in the bulk.

Of all the gas radicals near the surface, atomic hydrogen can penetrate farthest into the material. The diffusion of hydrogen and its removal and adsorption at the surface, described in Section 2.3.3, show that, at the deposition temperature, interstitial hydrogen can move quite rapidly into the bulk where it readily attaches to silicon dangling bonds. Hydrogen therefore has the fortunate property of being able to remove any subsurface defects left by the deposition process.

The diffusion of silicon in the bulk is very much slower than that of hydrogen, largely because silicon has four covalent bonds but hydrogen only one. There is a corresponding difference in the time and length scales over which the hydrogen and silicon structures are established. The majority of silicon bonds are fixed at the surface by the initial deposition step and by reconstruction when hydrogen is removed. However, the hydrogen bonding is more easily changed even within the bulk material after deposition. This distinction is important in understanding the role of defect equilibrium in the electronic properties and is developed further in Chapter 6.

Returning to the plasma reactions, different conditions occur when the silane is diluted with argon and the rf power is high. There are secondary processes in which energy is transferred from the excited argon to the silane, causing dissociation. The higher power and lower partial pressure of silane promote the involvement of more highly reactive species in the deposition as is seen in Fig. 2.9. SiH_3 is no longer the dominant radical and the plasma contains many others, such as Si, SiH, SiH_2. Each of these is able to react directly with the hydrogen-terminated surface, inserting into the Si—H bonds. The result is a high sticking coefficient and low surface mobility. These are exactly the characteristics of a PVD deposition process, so that the observation of a columnar growth morphology is no surprise.

2.2 The silicon bonding structure

The atomic structure of the a-Si:H films has features with a range of length scales. The shortest length is that of the atomic bonds and the structure is defined by the orientation of the bonds and the coordination of each atom to its nearest neighbors. The intermediate range order involves the topology of the network at the level of second, third or fourth nearest neighbors and is described by the size and distribution of rings of silicon atoms. The next largest length scale is the void structure, which has typical dimensions of 5–10 Å in CVD material, and finally there is the growth morphology, which can have feature sizes of upwards from 100 Å. These different structural features are almost independent. For example, a determination of the coordination of silicon and hydrogen atoms gives little information about the existence or other properties of voids. The hydrogen bonding structure is described in Section 2.3 and gives complementary information about the various structural features.

2.2.1 *Silicon–silicon atomic bonding*

Most of the information about the local order of silicon atoms comes from the radial distribution function (RDF) obtained from X-ray or neutron scattering. The RDF is the average atomic density at a distance r from any atom. The diffraction pattern of amorphous silicon has the diffuse rings characteristic of all amorphous materials, with radial intensity $I(k)$ for the scattering wave vector $k = (2\pi \sin\theta)/\lambda$, where θ is the scattering angle and λ is the wavelength. Analysis of the data starts with the reduced scattering function

$$F(k) = k[I(k)/f^2(k) - 1] \tag{2.13}$$

where $f(k)$ is the known atomic form factor. The Fourier transform of $F(k)$ gives the reduced RDF.

$$G(r) = 4\pi r[\rho(r) - \rho_0] = \frac{2}{\pi}\int F(k)\sin(kr)\,\mathrm{d}k \tag{2.14}$$

where ρ_0 is the average atomic density. The RDF, $J(r)$, is obtained from Eq. (2.14) by,

$$J(r) = 4\pi r^2 \rho(r) \tag{2.15}$$

The reduced RDF of a-Si:H shown in Fig. 2.11 (Schulke 1981) has sharp structure at small interatomic distances, progressively less well-defined peaks at larger distances, and is featureless beyond about 10 Å. This reflects the common property of all covalent amorphous semiconductors, that there is a high degree of short range order at the first and second neighbor distances, but then the spatial correlations decrease rapidly.

The Si—Si bond length, a, given by the first peak in the RDF is equal to that of crystalline silicon, 2.35 Å, and the intensity of the first peak confirms the expected four-fold coordination of the atoms. The width of the peak should give the extent of bond length distortions, but thermal vibrations and limitations of the data also contribute to the measured width. The bond length distribution is evidently small, but has not been measured accurately from the RDF.

The second peak in the RDF arises from second neighbor atoms at a distance $2a \sin(\theta/2)$, where θ is the bond angle (see Fig. 2.11). The distance of 3.5 Å is the same as in crystalline silicon, giving an average bond angle of 109°, and establishing the tetrahedral bonding of a-Si:H. The width of the second peak is significantly broader than that of the crystalline RDF, indicating that there is bond angle disorder of about $\pm 10°$.

The third neighbor peak in the RDF is even broader. The distance depends on the dihedral angle, φ, (see Fig. 2.11) and an almost random distribution of angles is deduced from the position and width of the peak. The third neighbor peak and the more distant shells overlap, so that less detailed information can be deduced.

2.2.2 *Intermediate range order, network voids and stress*

The RDF reveals that the short range bonding of a-Si:H is tetrahedral and that there is a long range randomness of the network, but does not give much information about the interesting intermediate region which relates to the structural topology. The diffraction data contain only directionally averaged data and cannot be inverted to deduce the topology. The usual approach to this problem is to construct structural models of continuous random networks (CRN),

Fig. 2.11. Example of the RDF of a-Si:H obtained from X-ray scattering. The atomic spacings which correspond to the RDF peaks are indicated (Schulke 1981).

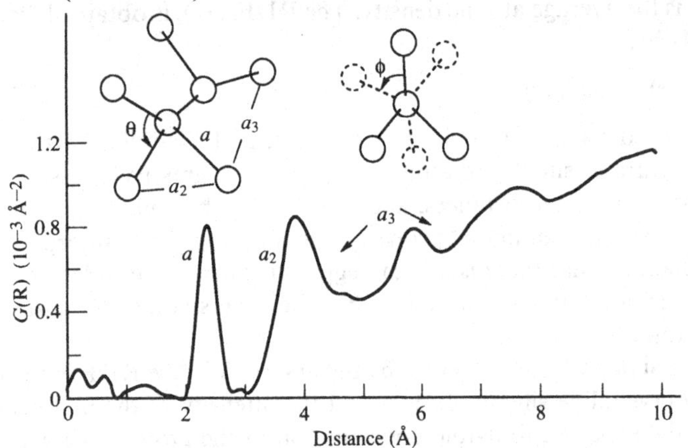

and to compare their calculated diffraction with the X-ray scattering data. Many models of amorphous silicon, with and without hydrogen, have been constructed by hand or by computer, e.g. Polk (1971), Connell and Temkin (1974) and Wooten, Winer and Weaire (1985). The models generally confirm that the X-ray diffraction results are fully consistent with the CRN structures, but it is not possible to make a unique determination of the actual intermediate range order.

An important aspect of the structural topology is the ring statistics. All the atoms in crystalline silicon lie in six-membered rings, but there is no reason to suppose that the same is true of amorphous silicon. Most of the structural models contain different fractions of five-, six-, or seven-membered rings, but at least one has been constructed with no odd-membered rings. Although the calculated diffraction patterns are slightly different, no specific model fits the data much better than another. The hydrogen bonding must affect the ring statistics, because it changes the average coordination of the network, but the X-ray data have been unable to resolve the details. The lack of detailed information about intermediate range order is unfortunate because the bonding disorder is the origin of the band tails of localized electronic states which, for example, control the electron and hole mobility.

The presence of the hydrogen imposes a distinction between the atomic and the network coordination. All the silicon atoms are four-fold coordinated, apart from the small concentration of coordination defects. However, the network coordination is reduced by the presence of hydrogen which forms a single bond to the silicon and so does not help to link the network together. The average network coordination is,

$$Z_n = 4 - f/(1-f) \qquad (2.16)$$

where f is the atomic concentration of hydrogen (e.g. the composition $Si_{1-f}H_f$).

The major source of the disorder energy is the bond strain within the random network. Phillips (1979) proposed a model to explain the relation between network coordination and disorder. A four-fold continuous random network is overcoordinated, in the sense that there are too many bonding constraints compared to the number of degrees of freedom. The constraints are attributed to the bond stretching and bending forces, so that for a network of coordination Z_n, their number, $N_C(Z_n)$ is,

$$N_C(Z_n) = Z_n/2 + Z_n(Z_n - 1)/2 = Z_n^2/2 \qquad (2.17)$$

The factors of one half arise because each bond constrains two atoms. The first term is for bond stretching and the second term is from the bond bending forces, which involve three atoms. When N_C is set equal to the three degrees of freedom for a three-dimensional solid, an 'ideal' coordination of 2.45 results. A-Si:H, with a network coordination of close to 4, is overcoordinated.

Ball and stick models contain only bond length and angle constraints and substantiate this result. It is easy to see that such a model of a two-fold coordinated material will collapse because there is nothing holding the chains apart. (In reality, of course, additional constraints are provided by the weaker van der Waals forces between chains.) On the other hand the common experience of those who have constructed amorphous silicon models is that a large bond strain tends to accumulate in the process of satisfying each silicon atom's four bonds.

More detailed calculations of the elastic properties of model networks have confirmed Phillips' model. The coordination dependence of the elastic modulus is shown in Fig. 2.12 (He and Thorpe 1985). Both the modulus C_{11} and the number of zero frequency vibrational modes, f, drop to zero at the critical coordination of 2.4, as predicted by Eq. (2.17). The properties are explained in terms of percolation of rigidity. The coordination of 2.4 represents the lowest network coordination for which locally rigid structures are fully connected, so that the entire network is rigid, but only just so. The elastic modulus is therefore non-zero and continues to increase as the network becomes more connected. The four-fold amorphous silicon network is far from the critical coordination and is very rigid.

The topological constraints of the overcoordinated network promote processes which reduce the network coordination. Breaking bonds is the only way that this can be achieved in pure amorphous silicon, but the hydrogen in a-Si:H provides an alternative. The hydrogen reduces the network coordination while retaining the local four-fold bonding of the silicon. The hydrogen therefore not only terminates broken bonds, but it also allows a more 'ideal' structure. However, a hydrogen content of more than 50 at% is needed to achieve a network coordination of 2.45, whereas the best films have only about 10 at%. Thus a-Si:H remains a rigid structure and is mechanically hard. Also, nothing in the above reasoning requires that any reduction of the network coordination occurs uniformly throughout the material. Both pure and hydrogenated amorphous silicon are structurally inhomogeneous and contain voids.

The inhomogeneity of the structure is completely missed in the X-ray RDF measurements, but can be observed by small angle X-ray or neutron diffraction, and also by nuclear magnetic resonance (NMR) measurements of the hydrogen distribution, described in Section 2.3.2.

A scattering angle of order λ/R is needed to observe a feature of size R, where λ is the X-ray or neutron wavelength, and hence small angle scattering is needed to observe large features. Fig. 2.13 shows X-ray data for a-Si:H and a-Si:C:H alloys in the form of a Guinier plot (Mahan, Nelson, Crandall and Williamson 1989),

$$I(k) = I_0 \exp(-k^2 R_g^2/3) \tag{2.18}$$

where R_g is the radius of gyration, or effective size, of the inhomogeneity. The increasing density of voids in the a-Si:C:H alloys is reflected in stronger small angle scattering. A-Si:H grown under PVD conditions also exhibits strong small angle scattering from the columnar structure (Leadbetter *et al.* 1980). Material grown under CVD conditions has weaker scattering, but still indicates a void structure which is isotropic with a characteristic dimension of about 5 Å.

The internal stress in the a-Si:H films is a further manifestation that the silicon network is rigid and overcoordinated. Typical films grown under CVD conditions have a compressive stress of about 2 kbar, which is observed as a visible curvature of the substrate. The different thermal contractions of the film and substrate during cooling contribute only a fraction of the stress, most of which is an intrinsic property of the film. Such a stress is common in deposited films and is a

Fig. 2.12. Calculated dependence of the elastic modulus, C_{11}, of a random network on coordination number Z_n. The inset shows the average number of zero frequency modes per atom (He and Thorpe 1985).

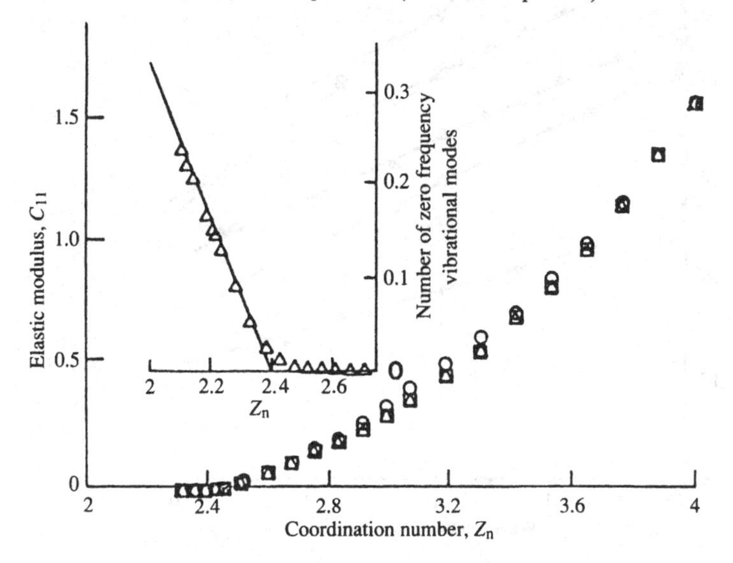

consequence of a network which does not relax significantly after growth. The stress is highly dependent on the growth conditions and is largest in high quality CVD material. A-Si:H films exhibiting a columnar growth morphology have a low stress because of stress relaxation at the open intercolumnar regions.

2.2.3 *Network vibrations*

To the extent that the local bonding of a-Si:H is similar to that of crystalline silicon, the phonon spectrum is also little different. Some broadening of the phonon density of states is expected because of the disorder in bond strength. The hydrogen bonding introduces additional

Fig. 2.13. Small angle neutron scattering of a-Si:H and a-$Si_{1-x}C_x$:H alloys showing the presence of voids with a density which increases with carbon concentration (Mahan *et al.* 1989).

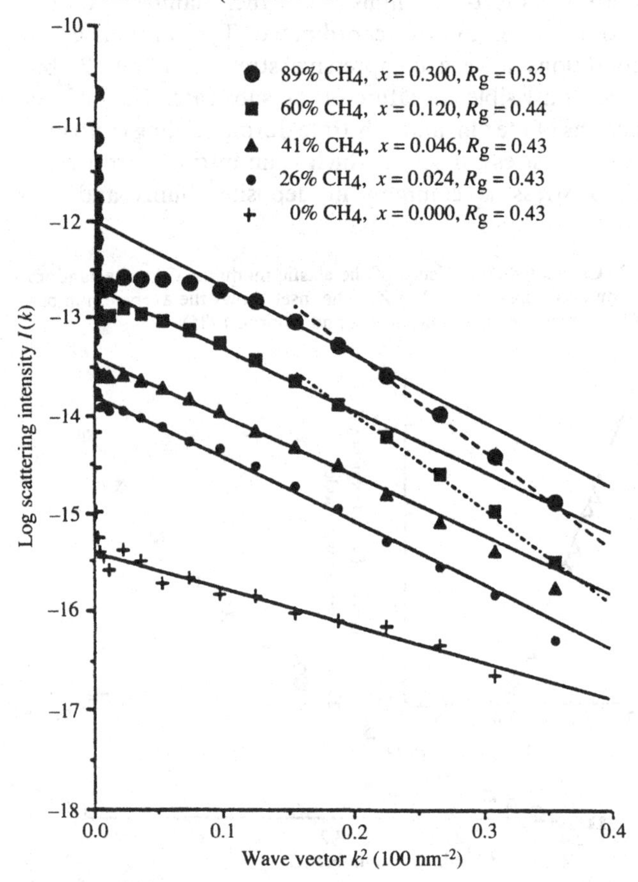

phonon modes which are discussed in Section 2.3.1. There are also excess low frequency modes which are associated with disordered materials and are common in glasses.

The experimental techniques most commonly used to measure the phonon distributions are IR absorption, Raman scattering and neutron scattering. The IR and Raman spectra of crystalline silicon reflect the selection rules for optical transitions and are very different from the phonon density of states. The momentum selection rules are relaxed in the amorphous material so that all the phonons contribute to the spectrum.

In absorption, the photon couples to the dipole moment induced by the phonon vibration and the absorption spectrum, $\alpha(\omega)$, is given approximately by,

$$n\alpha(\omega) = \pi^2(4\pi Ne^2/M)f(\omega)g(\omega) \qquad (2.19)$$

where $f(\omega)$ is the oscillator strength, $g(\omega)$ is the phonon density of states, N is the density of oscillators, M their mass, and n the refractive index. The symmetry of the modes is such that the transitions are forbidden in crystalline silicon and there is no first order absorption. However, the modes become active, although weak, in amorphous silicon because of the disorder.

In Raman scattering, the excitation light couples to changes in the polarizability and first order transitions are allowed in crystalline silicon. (See Lannin (1984) for a discussion of Raman scattering in amorphous silicon.) The scattering intensity, as a function of the phonon frequency, ω, is approximately

$$I(\omega) = C(\omega)[1+n(\omega)]g(\omega)/\omega \qquad (2.20)$$

where $C(\omega)$ is the coupling constant, $g(\omega)$ is the phonon density of states and $n(\omega)$ is the phonon occupancy factor. The coupling constant depends on the polarization, and contains a factor ω_s^4, where ω_s is the scattered light frequency. C is also roughly proportional to ω^2 for low frequencies, so that the Raman intensity is related to the phonon density of states by

$$g(\omega) = \text{const.}\ I(\omega)/\omega[1+n(\omega)] \qquad (2.21)$$

It is generally the case that IR absorption and Raman scattering give complementary information about the phonons.

Fig. 2.14 compares the Raman and neutron spectra of the amorphous silicon network modes, which occur at wavenumbers up to 500 cm^{-1}, with the calculated phonon density of states of crystalline silicon. It should not be surprising that the amorphous spectrum is a broadened version of the crystalline density of states, because the vibrational modes are determined largely by bond length and bond angle force constants and these reflect the local order which is the same in the two phases. The cusp-like structure in the spectrum results from the crystalline periodicity and is smoothed out in the amorphous spectra.

The relaxation of the momentum selection rules is most easily seen in the Raman spectrum in the vicinity of the transverse optical modes near 500 cm^{-1}. The crystalline peak at 520 cm^{-1} is replaced in the amorphous film by a band at 480 cm^{-1}. The shift is not due to a weakening of the force constants, but to the loss of k-conservation, so that the peak is an average over the transverse optic (TO) band, instead of being its upper $k = 0$ point. The disorder relaxes the selection rules and mixes in more of the TO band, broadening it and moving it to lower frequency. This effect is used to measure the degree of disorder in a-Si:H. For example, the peak is at higher energy and has a narrower line width in a-Si:H compared with unhydrogenated silicon, showing that the disorder is less. A quantitative measure of disorder is obtained in this way (Lannin, 1987). There is a correlation between the disorder

Fig. 2.14. The phonon density of states measured by Raman (Tsai and Nemanich 1980) and neutron scattering (Kamitakahara *et al.* 1984), compared to the calculated phonon spectrum of crystalline silicon (Weber 1977). The transverse (T) and longitudinal (L) modes of the acoustic (A) and optical (O) phonons are indicated.

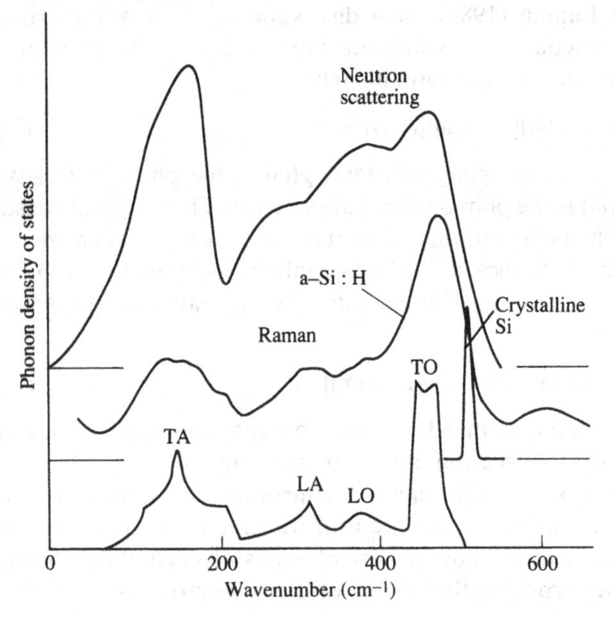

broadening of the TO mode and the broadening of the optical absorption edge, showing the relation between electronic and structural disorder.

Many amorphous materials exhibit at very low frequencies the rather mysterious excitations known at tunneling modes (Phillips 1985). These are modes whose structural origin is unclear even though their presence is well documented. They cause an excess specific heat below about 10 K and are also observed in acoustic experiments and in the spin relaxation of paramagnetic states. The tunneling modes are represented by a two-level system in which the atomic configurations can exist in one of two states differing by a small energy and separated by a larger energy barrier. Fig. 2.15 illustrates such an energy–configuration plot of undefined structural origin. A transition from one well to the other is presumed to correspond to a rearrangement of a group of atoms into a different configuration having roughly the same energy. The density of states of the low frequency modes is generally found to be nearly constant in energy. Since the network phonons increase in density as E^2, the two-level system modes dominate at low energy.

Fig. 2.15. Temperature dependence of the heat capacity fitted to the form of Eq. (2.22). The shaded region represents the excess heat capacity from the tunneling modes (Graebner *et al.* 1984).

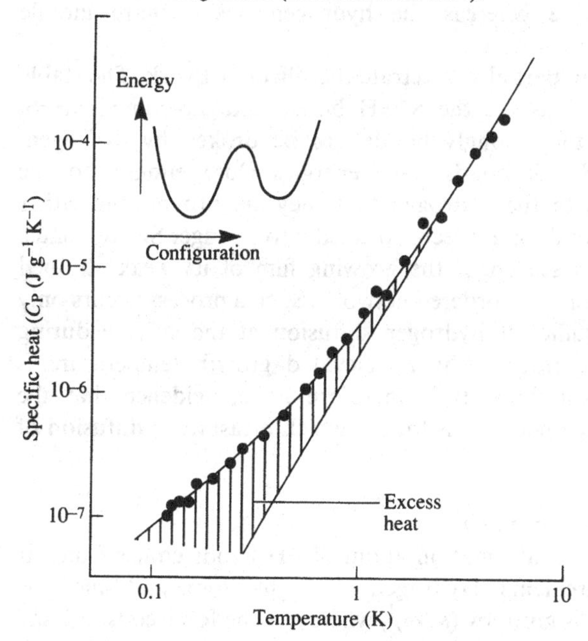

Fig. 2.15 shows the temperature dependence of the heat capacity of a-Si:H (Graebner *et al.* 1984). The data are fitted to an expression of the form

$$C_p = c_1 T^{\alpha} + c_3 T^3 \qquad (2.22)$$

with $\alpha = 1.34$. The T^3 term is the usual specific heat of the phonons. The excess specific heat given in the first term is caused by excitations of the two-level system, and both the magnitude and temperature dependence of the excess heat are about the same as in silicon dioxide.

2.3 The hydrogen bonding structure

The hydrogen and the silicon in a-Si:H have different bonding properties. The silicon structure is described in terms of a rigid overcoordinated network containing a high strain energy. There is negligible diffusion of the silicon and no apparent relaxation of the strain after growth. In contrast, the hydrogen is more weakly bound and can diffuse within the material and across the surface. The hydrogen bonding is very sensitive to the deposition conditions and its diffusion is closely associated with the many metastable changes of the electronic properties described in Chapter 6. As growth proceeds and the silicon is bonded into the network, the different characters of the silicon and hydrogen begin to develop. The silicon forms a rigid non-equilibrium structure whereas the hydrogen has a more mobile equilibrium structure.

The hydrogen can partially penetrate the silicon network. The stable bonding configurations are the Si—H bonds and unstrained Si—Si bonds which are too strongly bonded to be broken by hydrogen. Highly strained Si—Si bonds have energies close enough to the chemical potential of the hydrogen that they are broken and either remain as Si—H bonds or are reconstructed into stronger Si—Si bonds. Thus the hydrogen scavenges the growing film of its weak strained bonds, resulting in a more ordered network. Such a process occurs only when there is a sufficient hydrogen diffusion at the surface during growth and is the reason why an elevated growth temperature is needed for the best films. It is therefore no coincidence that the optimum growth temperature is that at which measurable diffusion of hydrogen occurs.

2.3.1 *Silicon–hydrogen bonds*

Much of the information about Si—H bonds comes from IR absorption measurements. Hydrogen is a light atom, and since the phonon frequency is given by $(k/m)^{\frac{1}{2}}$, where k is the force constant and

m is the reduced mass, the frequencies of the hydrogen modes are above the silicon network modes, making them easy to observe. The vibration is almost entirely confined to the hydrogen atom, so that the analysis of the modes is relatively simple. Fig. 2.16 shows a set of IR spectra for samples prepared under different growth conditions (Lucovsky, Nemanich and Knights 1979). Phonon modes occur in three energy bands; a broad peak at 630 cm^{-1} which is always present; a group of sharp lines at 800–900 cm^{-1} whose shape and intensity depend on deposition conditions; and modes in the range 2000–2200 cm^{-1} with a similar dependence on deposition. All can be identified conclusively as hydrogen vibrations from the isotope shift which is observed when hydrogen is replaced by deuterium. The shift in frequency of close to $\sqrt{2}$ also confirms that the vibration is confined to the hydrogen atom.

The normal modes of various Si—H bonding configurations are shown in Fig. 2.17. A single hydrogen atom bonded as ≡Si—H has three degrees of freedom. The corresponding modes are a bond stretching vibration in which the hydrogen moves along the direction of the Si—H bond and two perpendicular, degenerate, bond bending modes. The first mode has a frequency near 2000 cm^{-1} and the second at 630 cm^{-1}, the difference being because the bond bending force is weaker than the bond stretching.

Fig. 2.16. Examples of the IR transmission spectra for a-Si:H samples deposited at different growth conditions. The deposition power is indicated; A and C refer to deposition on the anode and cathode (Lucovsky *et al.* 1979).

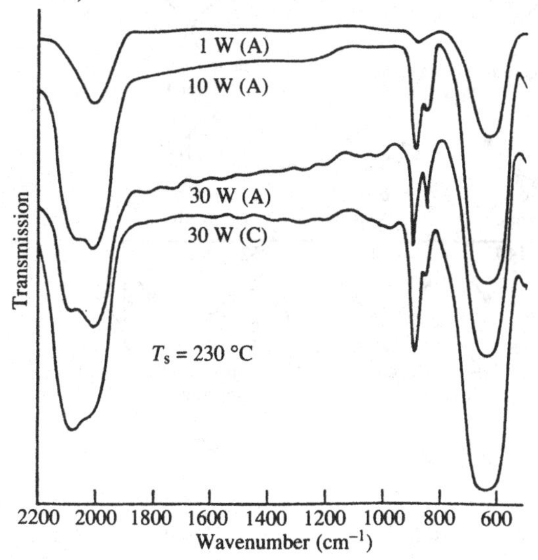

The modes at 800–900 cm^{-1} originate from =Si=H_2 or —Si≡H_3 bonding configurations. The extra degrees of freedom introduced by the additional hydrogen allow many more vibrational modes of both the bond stretching and bending types and these are listed in Fig. 2.17. The figure also gives the calculated frequencies of all the modes of the various Si—H configurations. Of the bending modes, those labeled wag and twist have a significantly higher frequency than the rocking modes and account for the modes at 800–900 cm^{-1}. The reason for the higher energy is that the combined displacement of the two bonds increases the force constant of each. It is worth noting that the variation in the peaks near 2000 cm^{-1} is not due to differences in the type of mode – both the symmetric and the asymmetric stretch modes are degenerate. Rather the difference is due to a chemically-induced change in the force constants due to the different bonding which causes a small change in the bond charge. Similar shifts occur when one of the silicon back

Fig. 2.17. The set of Si—H vibrational modes for SiH, SiH_2 and SiH_3 groups, with calculated frequencies as indicated. The frequencies in brackets are estimates (Lucovsky *et al.* 1979).

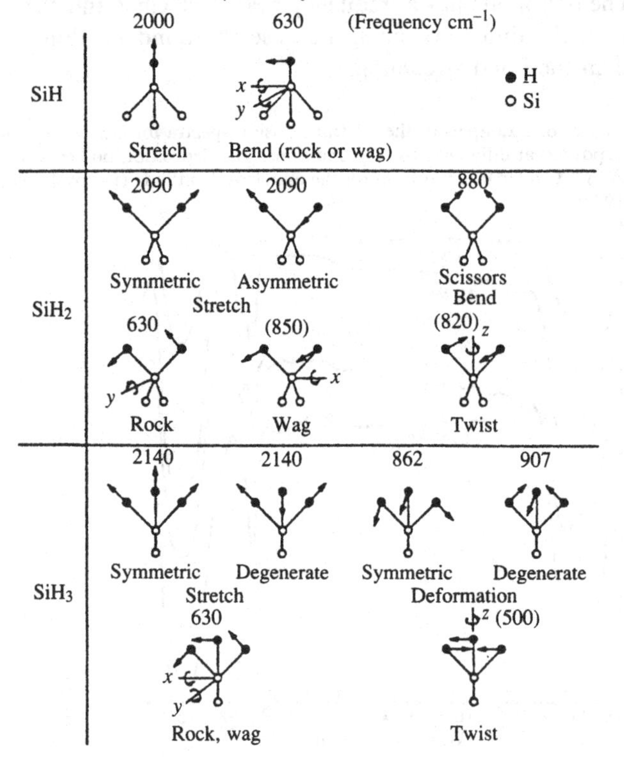

bonds is replaced by another atom. Thus in alloys with oxygen and nitrogen etc., the Si—H stretch modes shift by 100–200 cm^{-1}.

It is well established that Si—H configurations result in modes at 630 cm^{-1} and 2000 cm^{-1} and that the 800–900 cm^{-1} and 2100 cm^{-1} modes arise from structures involving Si=H_2 and Si≡H_3. There are some reports of a mode at 2100 cm^{-1} which is not accompanied by the 800–900 cm^{-1} lines, and it has been suggested that Si—H can also give the mode at 2100 cm^{-1} when in some different environment, possibly the surface of a large void.

The IR absorption at frequency ω is proportional to the concentration of hydrogen, these being related by

$$N_{\mathrm{H}} = f^2 A \int \left[\frac{\alpha(\omega)}{\omega}\right] d\omega \qquad (2.23)$$

f is the local field factor, which is present because the Si—H bond experiences a local field which is different from the externally applied electro-magnetic field and A is a constant. The local field correction is sensitive to the environment and has a different magnitude for the various Si—H configurations. Studies in which the IR absorption is calibrated with the total hydrogen content measured by hydrogen evolution find that the integrated intensity of the 630 cm^{-1} line is the best measure of the concentration and that the other lines are less reliable (Shanks *et al.* 1980).

The IR and Raman experiments are used to explore how the deposition conditions influence the hydrogen bonding. The weak absorption at 800–900 cm^{-1}, and predominance of the 2000 cm^{-1} stretching modes in the low rf power data in Fig. 2.16 show that the CVD growth conditions of low power and undiluted silane result in the predominance of Si—H bonding. On the other hand the PVD conditions which give a columnar morphology are associated with large concentrations of SiH_2 and/or SiH_3 structures. Since the columnar structure is undesirable, the minimization of the SiH_2 modes has become an important figure of merit in the empirical optimization of the material. However, although the SiH_2 vibrations are always present in columnar material, their observation is not proof of this type of microstructure. Rather the modes seem to be associated with material with a high hydrogen content and with an extensive void structure, but not necessarily in the form of columns. Likewise the absence of SiH_2 bonding structures does not imply that the material is void-free and homogeneous, as is apparent in the next section.

2.3.2 *The hydrogen local order*

While IR experiments focus on the local Si—H bonding, NMR of protons gives more information about the local environment in which the hydrogen atoms reside. NMR arises from transitions between the different spin states of the nucleus, which are split by an applied magnetic field. An isolated proton has a precisely defined resonance frequency, but the interactions between atoms in a solid modify the resonance by a variety of mechanisms. The NMR experiment is fairly complex and no attempt is made here at a detailed treatment.

Examples of the frequency response of the resonance spectra for various a-Si:H films are shown in Fig. 2.18 (Reimer, Vaughan and Knights, 1980). The line is broadened by the short lifetime of proton spin orientation. There are two distinct components to the spectrum, corresponding to fast and slowly relaxing states, with line widths of 22–27 kHz, and 4–5 kHz (compared to the resonance frequency of 56 MHz). The magnetization decay is due to dipolar interactions of the observed proton with neighboring hydrogen atoms. An immediate deduction of the two component spectrum is that there are two distinct

Fig. 2.18. Proton NMR spectra of a-Si:H samples grown with different deposition conditions, showing the clustered (broad line) and dilute (narrow line) hydrogen sites. The hydrogen concentration ranges from 32 at% (*a*) to 8 at% (*d*) (Reimer *et al.* 1980).

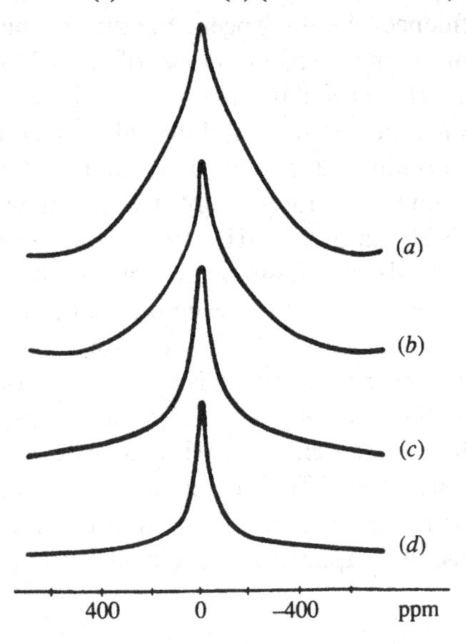

local environments for the hydrogen in a-Si:H where the dipolar couplings differ. The dipole interaction is proportional to r^{-6}, where r is the distance between hydrogen atoms, so that the broader line corresponds to more clustered hydrogen. The dipolar broadening of SiH_2 or SiH_3 configurations is calculated to be at least 14 kHz, so that the narrow line can only be due to Si—H structures which are well separated from each other. On the other hand the broad line must arise from small clusters of hydrogen atoms which are either bonded to the same silicon atom or to neighbors within 2–3 Å. A void with a hydrogenated surface is the obvious interpretation of the clustered hydrogen. Furthermore there are NMR experiments which find that the number of hydrogen atoms which interact strongly is limited to about 5–7, suggesting a divacancy type of void structure (Baum, Gleason, Pines, Garroway and Reimer, 1986).

The interesting changes in the concentration of hydrogen in the broad and narrow lines at different deposition conditions are shown in Fig. 2.18. In all cases the narrow line contains 2–3 at% of hydrogen and the broad line contains the reminder. This is true for hydrogen concentrations from less than 10 at% up to 50 at%. One can also compare the NMR and the IR data on the same material. For a-Si:H

Fig. 2.19. Temperature dependence of the NMR spin-lattice relaxation time, T_1, showing the minimum characteristic of hydrogen molecules (Carlos and Taylor 1982).

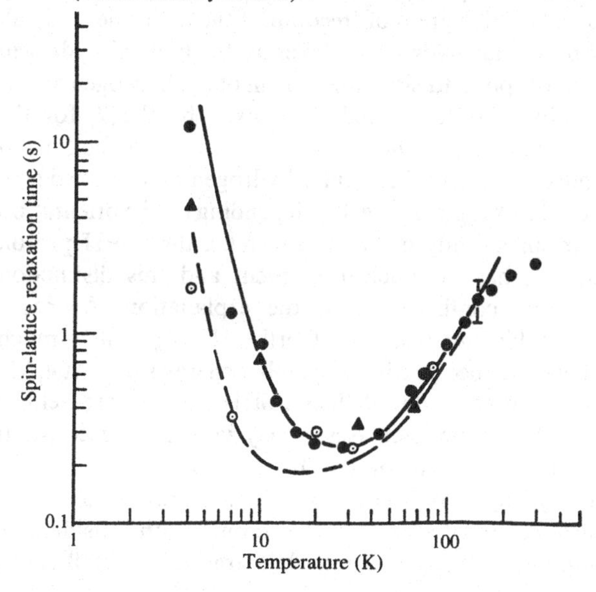

made under optimal conditions, the IR data show a predominance of Si—H bonds whereas the NMR find that more than half the hydrogen is in the broad line.

Evidently a-Si:H has a well-defined two-phase inhomogeneous hydrogen bonding structure. The data are interpreted in terms of hydrogenated voids embedded in an amorphous matrix in which hydrogen is randomly dispersed with concentration 2–3 at%. The void surface comprises either Si—H or Si—H_2 bonding configurations and the density of voids varies with the deposition conditions. The macroscopic columnar morphology is one extreme manifestation of the two phase material, but in material without this type of structure there still remain small voids which are more isotropic in size and distribution. Small angle X-ray scattering also finds isotropic voids in non-columnar material.

NMR measurements of the spin relaxation time have identified a small concentration of molecular hydrogen contained within the a-Si:H material. The nuclear spin excitation relaxes to the ground state by transferring its energy to the lattice with a time constant denoted by T_1. The temperature dependence of T_1 is shown in Fig. 2.19 and has a minimum value at a temperature of about 30 K (Carlos and Taylor 1980). The relaxation is caused by a small density of hydrogen molecules, trapped in the amorphous silicon network (Conradi and Norberg 1981). The hydrogen molecules cause rapid spin-lattice relaxation because the energy of the magnetization quantum is easily taken up by the rotational degrees of freedom of the molecule. A small concentration of molecular hydrogen relaxes all the bonded hydrogen through the process of spin diffusion in which all of the hydrogen atoms are in communication. The measured T_1 is given by the T_1 for the hydrogen molecules reduced by their relative concentration compared to that of the bonded hydrogen. Molecular hydrogen exists in two spin states, spin 1 ortho-H_2 and spin 0 para-H_2, depending on the orientation of the two nuclear spins. Only ortho-H_2 can relax the Si—H proton spin because para-H_2 has no nuclear moment and this distinction has provided a direct confirmation of the explanation. At room temperature, the equilibrium fraction of ortho-H_2 is $\frac{3}{4}$, but is much lower at low temperature, because it is the higher energy state. At 4.2 K the conversion time exceeds a day. When a-Si:H samples are held at 4.2 K, the T_1 relaxation time decreases slowly with a time constant exceeding a day (Carlos and Taylor 1982).

The molecular hydrogen is present at a concentration of about 0.1 at%, and is also seen in IR experiments, although with considerable difficulty. The molecular hydrogen is immobile, trapped in small voids, and, in fact, a surprising amount is known about it (Boyce and Stutzmann 1985). It is estimated that there are about 10 hydrogen molecules in each void. Annealing to high temperature which tends to remove bonded hydrogen from the a-Si:H network, increases the amount of molecular hydrogen in the voids. At temperatures below about 20 K, the hydrogen freezes, going through the same order–disorder transition that is seen in normal molecular hydrogen, but shifted in temperature, apparently because the pressure of the hydrogen gas in the voids is about 2 kbar.

2.3.3 *Hydrogen diffusion, evolution and rehydrogenation*

The ability of hydrogen to move into, out of, and within a-Si:H has both beneficial and undesirable properties. The low defect density is a beneficial result of the bonding of hydrogen to weak or

Fig. 2.20. The temperature dependence of the hydrogen diffusion coefficient at different doping levels as indicated (Street, Tsai, Kakalios and Jackson 1987b), including data from Carlson and Magee (1978).

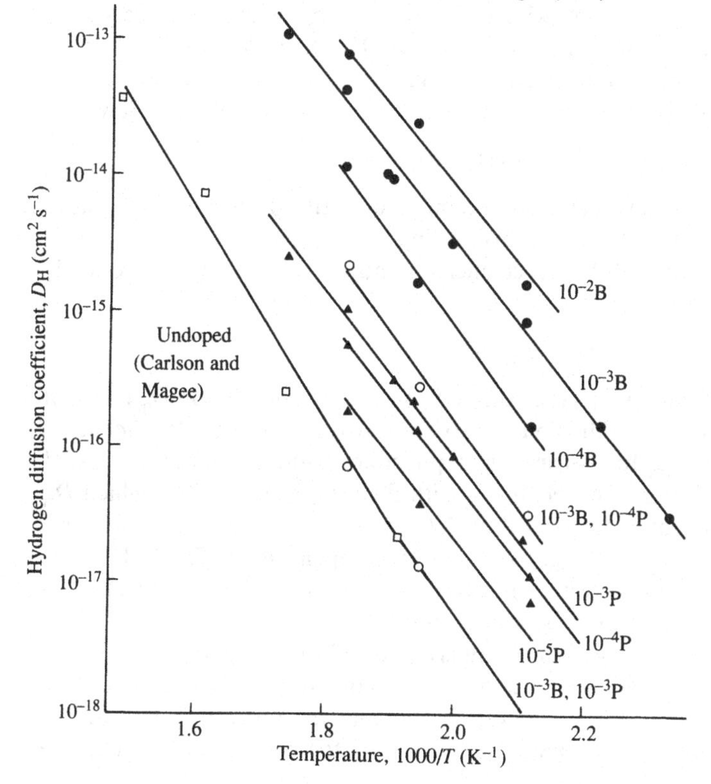

broken silicon bonds. Hydrogen is, however, responsible for the instability of a-Si:H at elevated temperatures. Hydrogen is completely removed from a-Si:H above about 400 °C, degrading the properties in a way that cannot be reversed unless hydrogen is deliberately reintroduced. At intermediate temperatures (100–300 °C), the mobility of hydrogen causes reversible metastable phenomena which are described in Chapter 6.

Diffusion occurs when there is a gradient of the chemical potential. In a uniform bulk material, the chemical potential depends only on the concentration, C, resulting in Fick's law of diffusion,

$$dC/dt = D d^2C/dx^2 \qquad (2.24)$$

The diffusion coefficient of hydrogen, D_H, is determined from the time evolution of a concentration profile. The common techniques are to use secondary ion mass spectrometry (SIMS) or nuclear reaction profiling. In the SIMS experiment, the atomic fragments sputtered from the surface by an ion beam are analyzed by a mass spectrometer. The usual procedure for measuring the hydrogen profile is to grow a film containing a layer in which part of the hydrogen is replaced by deuterium (Carlson and Magee 1978). The hydrogen and deuterium become intermixed during annealing, and the deuterium profile gives a measure of the diffusion. The appropriate solution to Fick's law is,

$$C_H(x) = \tfrac{1}{2}C_H(0)[1 - \mathrm{erf}(x/k)] \qquad (2.25)$$

where $\mathrm{erf}(x/k)$ is the error function and the diffusion depth, k, is $(4D_H t)^{1/2}$ where t is the diffusion time.

The hydrogen diffusion coefficients shown in Fig. 2.20 are thermally activated,

$$D_H = D_0 \exp(-E_D/kT) \qquad (2.26)$$

where the energy E_D is about 1.5 eV, and the prefactor D_0 is about 10^{-2} cm^2 s^{-1}. The diffusion is also quite strongly doping-dependent, as seen in the data for boron-doped, phosphorus-doped and compensated material. The greatest change is with the p-type material for which D_H is increased by a factor 10^3 at 250 °C. The increase is less in n-type material and no change is found in compensated a-Si:H. The low diffusion in compensated material suggests that the doping effect is electronic rather than structural in origin.

Thermally activated diffusion is explained by a trapping mechanism. Almost all the hydrogen is bonded to silicon atoms which do not diffuse significantly at these temperatures, so that hydrogen diffusion occurs by breaking a Si—H bond and reforming the bond at a new site. The obvious bonding sites are the silicon dangling bonds and the mobile hydrogen moves in an interstitial site. The interstitial is the bond center site in crystalline silicon and is most probably the same in a-Si:H. For this model the diffusion is given by,

$$D_{\mathrm{H}} = \tfrac{1}{6}a_{\mathrm{D}}{}^{2}\omega_{0}\exp(-E_{\mathrm{HD}}/kT) \tag{2.27}$$

where $\omega_0 \sim 10^{13}$ s^{-1} is the attempt-to-escape frequency for the excitation out of the trapping site, a_{D} is the distance between sites, and E_{HD} is the energy to release hydrogen from a Si—H bond. A diffusion prefactor of 10^{-3} cm^2 s^{-1} results from a hopping distance of about 3 Å.

Although Eq. (2.27), when applied to hydrogen hopping between unterminated silicon atoms, gives a correct general description of the diffusion, it is unsatisfactory as a detailed model. For example, low defect density material contains only about 10^{15} cm^{-3} unterminated silicon atoms, so that a_{D} should be about 300 Å, giving a much larger prefactor than is observed. Furthermore, doped material has a much larger defect density than undoped a-Si:H and these extra trapping sites should reduce the diffusion, rather than increase it.

The diffusion data indicate that there must be many more bonding sites than the few broken bond defects that are present. It is proposed that these traps are the weak Si—Si bonds which result from the disorder of the network (Street *et al.* 1987b). Fig. 2.21 shows schematically the potential profile through which the hydrogen moves, the distribution of traps representing Si—H bonds and weak Si—Si bonds. When hydrogen inserts itself into a weak bond, the result is a strong Si—H bond and a neighboring dangling bond which can also bond to hydrogen. This diffusion mechanism does not conserve the number of dangling bonds in the material, but in steady state there

Fig. 2.21. Schematic diagram of the possible hydrogen diffusion mechanism: (*a*) the potential wells corresponding to the trapping sites and the energy of the mobile hydrogen; (*b*) the motion of the hydrogen through the Si—Si bonds.

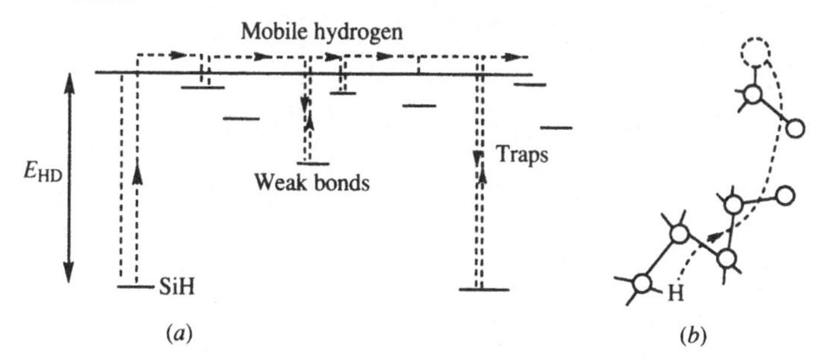

must be as many hopping steps in which the weak bonds are reformed as are broken. The breaking and reforming of Si—Si bonds means that hydrogen is capable of modifying the silicon bonding network. This occurs during deposition when the release of hydrogen from the surface promotes the reconstruction of the silicon bonds. The hydrogen diffusion also causes a temperature-dependent defect density and doping efficiency, as described in Section 6.2.3.

The 1.5 eV diffusion activation energy, needed to release hydrogen from the Si—H bond into an interstitial site, is smaller than the 2.5–3 eV energy of the Si—H bond, E_{SiH}. The diffusion energy is lower because interstitial hydrogen has a considerable binding energy to the silicon, E_{HI}, so that

$$E_{HD} \approx E_{SiH} - E_{HI} \tag{2.28}$$

The interstitial binding energy is calculated to be about 1–1.5 eV in crystalline silicon (Johnson, Herring and Chadi 1986) and is expected to be about the same in the amorphous material. Thus the measured diffusion energy is consistent with the expected bond strengths. The further lowering of the diffusion energy by doping is explained by an electronic effect. The dangling bond which is left when a hydrogen atom is excited out of a Si—H bond is an electrically active state in the band gap which can be occupied by zero, one or two electrons, depending on the doping. The electronic energy of the dangling bond

Fig. 2.22. The power law decrease in the time dependence of the hydrogen diffusion coefficient of p-type a-Si:H (Street *et al.* 1987b).

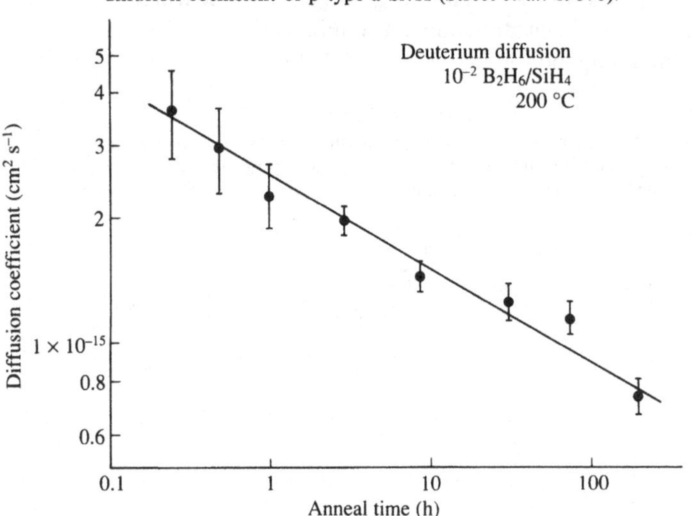

contributes to the total energy needed to break the bond and depends on the doping. This aspect of the diffusion is described in Chapters 5 and 6.

The hydrogen diffusion coefficient is not constant, but decreases with time (Street *et al.* 1987b). The data in Fig. 2.22 show a power law decrease in p-type a-Si:H of the form $t^{-\alpha}$, with $\alpha \approx 0.2$ at the measurement temperature of 200 °C. The time dependence is associated with a distribution of traps originating from the disorder. A similar effect is found in the trap-limited motion of electrons and holes and is analyzed in Section 3.2.1. The time dependence of D_H is reflected in the kinetics of structural relaxation discussed in Section 6.3.1.

The nuclear reaction technique for measuring the hydrogen profile makes use of a resonant reaction between ^{15}N and ^{1}H. Incident high energy nitrogen atoms lose energy by electron–hole excitation until they reach the resonance energy at a well-defined depth from the surface. Fig. 2.23 shows the profile of hydrogen obtained from a sample which has been annealed to 500 °C (Reinelt, Kalbitzer and Moller 1983). Hydrogen has evolved from the material and the shape of the profile is determined by the relative bulk diffusion and the surface evolution rates. A higher evolution rate results in a low hydrogen concentration at the surface and a higher diffusion rate gives a flat hydrogen profile. The actual profile shows that the rates are comparable

Fig. 2.23. Hydrogen profiles of a-Si:H obtained by the nuclear reaction technique, after an anneal at 500 °C. Curves 1 and 2 are calculated assuming different rates of hydrogen desorption from the surface (Reinelt *et al.* 1983).

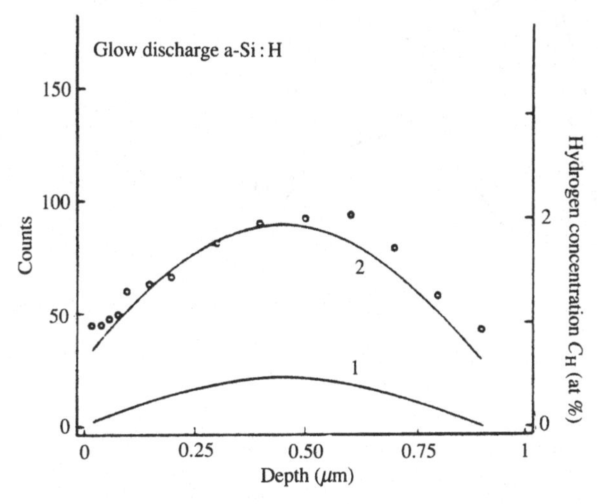

and an activation energy of 1.7 eV is found for the evolution, slightly larger than the bulk diffusion activation energy.

The more common way of measuring the hydrogen evolution is from the rate of pressure increase of gas released into a known volume, for samples heated at a constant rate. The sensitivity of the evolution to the deposition conditions is illustrated in Fig. 2.24 (Biegelsen, Street, Tsai and Knights 1979). Material with a high degree of columnar morphology or extensive internal inhomogeneity evolves a large amount of hydrogen between 300–400 °C with further evolution near 600 °C, while samples made under optimum CVD growth conditions have only the higher temperature peak. The temperature at which the evolution occurs also depends on the thickness of the sample because diffusion within the bulk plays a large part in determining the evolution temperature. Indeed Beyer and Wagner (1982) were able to determine the diffusion coefficient from the evolution data and were the first to show the doping dependence of D_H. The lower temperature peak in the

Fig. 2.24. Temperature dependence of the rate of hydrogen evolution in material deposited under different conditions: (*a*) low power CVD growth; (*b*) PVD columnar material; (*c*) deposited at room temperature and with a high hydrogen content (Biegelsen *et al.* 1979).

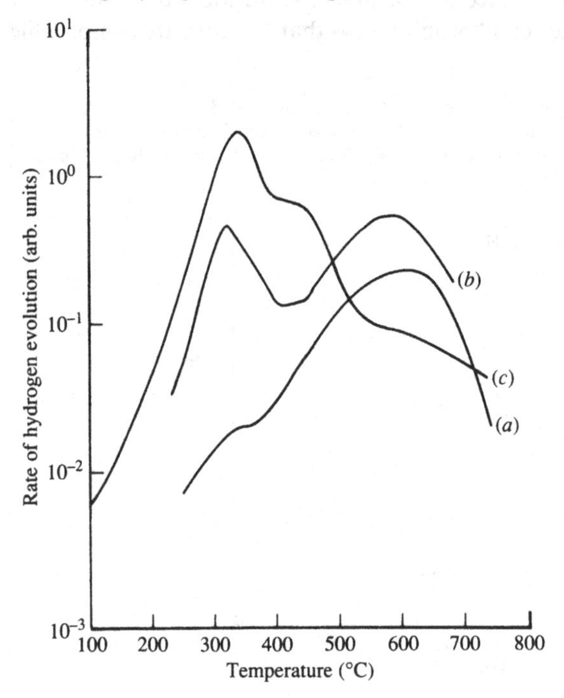

columnar material is explained by hydrogen released from internal surfaces, which can diffuse easily through the void network.

As hydrogen evolution proceeds, the density of electronic defects increases, again showing the connection between the hydrogen bonding and the defect states. Far more hydrogen is released than defects created, so there must be extensive reconstruction of the network. The electronic properties reveal an increase in the structural disorder during this process – presumably the formation of weak Si–Si bonds. The degree of local reconstruction changes the energy needed to release the hydrogen. It therefore seems likely that the low temperature evolution reflects structures where reconstruction is easy and which are absent in good quality material because such reconstruction has already taken place during the growth process.

The opposite process, the hydrogenation of films, is possible by the exposure of the surface to atomic hydrogen, although an elevated temperature is needed for the hydrogen to diffuse a significant distance into the film. Several atomic per cent of hydrogen can be introduced in this way and this causes a decrease in the defect density of initially unhydrogenated material. The amount of hydrogen greatly exceeds the initial defect density, indicating an interaction with the silicon network and the removal of weak bonds.

Hydrogenation occurs only with atomic hydrogen, not with

Fig. 2.25. Schematic diagram of the hydrogen energy levels in the plasma and in a-Si:H.

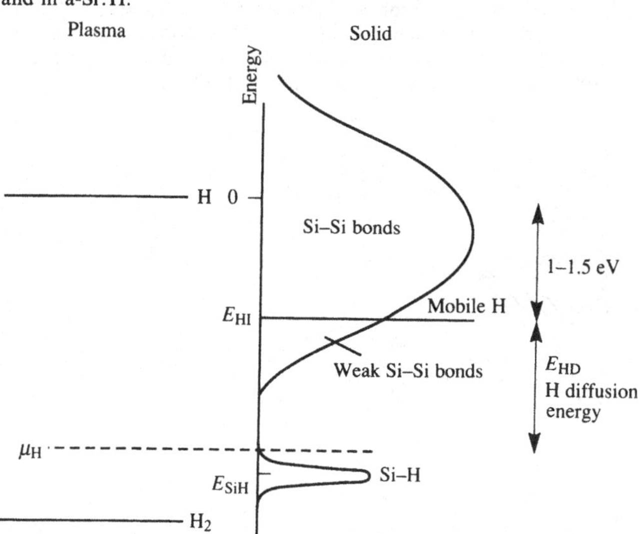

molecular hydrogen. Molecular hydrogen is unreactive with the a-Si:H surface because of its strong binding energy. Fig. 2.25 shows a schematic diagram of the different hydrogen energy levels in the plasma and in the a-Si:H film. The Si—H bond energy is about 3 eV below that of free hydrogen in the gas. The interstitial level in the bulk is about 1.5 eV below the free hydrogen level and the energy difference is the diffusion energy. Molecular hydrogen has a lower energy than Si—H and does not react with a broken silicon bond. In the absence of a plasma, the chemical potential of the hydrogen gas is well below the molecular hydrogen level and no atomic hydrogen is present. The plasma artificially raises the chemical potential so that it is above the Si—H bonding level but below that of the Si—Si bonds. Weak bonds in the vicinity of the chemical potential are broken and reconstruct to the lower energy Si—H bonds or Si—Si bonds, leaving no states near the chemical potential.

2.3.4 *The role of hydrogen in the growth of a-Si:H*

Fig. 2.26 summarizes the interactions of hydrogen with the surface of a-Si:H, illustrated by measurements on p-type material at 250 °C. Exposure of the surface to atomic hydrogen increases the surface concentration and causes hydrogen to diffuse into the film. The shape of the concentration profile is the same as for the normal hydrogen diffusion, indicating that the extra hydrogen is bonded to

Fig. 2.26. Hydrogen profiles of p-type a-Si:H comparing hydrogenation of a film exposed to atomic hydrogen at 250 °C, hydrogen–deuterium exchange when the sample is exposed to deuterium, and hydrogen evolution by annealing in vacuum (Street 1987a).

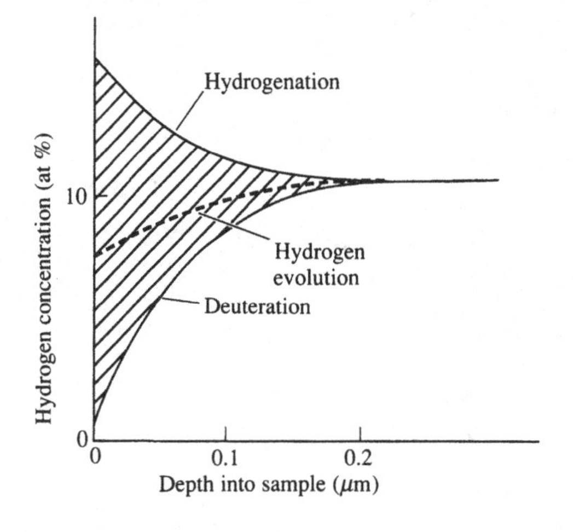

silicon atoms. The excess Si—H bonds can only come from the breaking of weak Si—Si bonds as there are not enough dangling bonds in the film to absorb the extra hydrogen. Annealing in vacuum depletes the surface of hydrogen, again with a concentration profile corresponding to bulk hydrogen diffusion. Exposure of the surface to atomic deuterium reduces the hydrogen concentration more rapidly than the vacuum anneal, by a process of deuterium exchange with the hydrogen. The hatched region in Fig. 2.26 indicates the hydrogen which is replaced by deuterium.

These processes illustrate that atomic hydrogen can move in and out of the surface during growth of films. The diffusion allows the hydrogen to react with the silicon network in the subsurface region, after the silicon has attached to the surface. The hydrogen terminates dangling bonds and removes weak bonds while excess hydrogen is evolved from the film. The free exchange of the hydrogen between the film and the plasma establishes (approximate) equilibrium between the plasma gas and the film. Thus the concentration of hydrogen in the film and the reactions with the network depend on the chemical potential of the hydrogen in the plasma. Weak Si—Si bonds which lie below the hydrogen chemical potential are broken while stronger bonds remain.

In a simple material, the concentration of hydrogen should increase as the chemical potential is raised. The concentration of hydrogen in the a-Si:H films actually decreases when hydrogen is added to the plasma (Johnson *et al.* 1989), and is an indication of the complex interaction between the hydrogen and the silicon bonding structure. For example, the reconstruction of the silicon network may move binding sites from below to above the chemical potential, in effect replacing weak bonds with strong ones.

This approach suggests a close connection between the hydrogen in the plasma and the disorder of the a-Si:H film. The optimum growth temperature of 200–300 °C is explained by the hydrogen interactions, at least in general terms. At much lower temperatures, the diffusion coefficient is too low to allow structural equilibration, while at much higher temperatures, the chemical potential is lower and hydrogen does not remain in the film and does not remove weak bonds. Both conditions result in a more disordered structure. It is also known that excess hydrogen in the plasma induces a transition to microcrystalline silicon (see Section 2.1.1). A possible explanation is that the minimum disorder of an amorphous network contains too many weak bonds to be consistent with the chemical potential, thus inducing crystallization. The most ordered structure should occur when the hydrogen chemical potential is the highest and this corresponds to the conditions for microcrystalline growth as indicated in Fig. 2.5. The role of hydrogen in the growth is discussed further in Section 6.2.8.

2.3.5 *Hydrogen in amorphous and crystalline silicon*

Over recent years much information has been gathered about hydrogen in crystalline silicon which pertains to the amorphous counterpart. There are many similarities between the effects of hydrogen in the two materials. Hydrogen passivates point defects, usually by the terminating silicon bonds, and also causes the passivation of dopants (Pankove, Carlson, Berkeyheiser and Wance 1983). The effect is most pronounced in p-type crystalline silicon; in the specific case of boron acceptors, hydrogen breaks the Si—B bond, leaving boron three-fold coordinated and forming an adjacent Si—H bond. Neither of these states is electrically active, so that the acceptor state is completely eliminated. A similar passivation of donors is also found, but the precise bonding mechanism is less clear (Johnson *et al.* 1986).

Another effect of hydrogen in crystalline silicon is to break Si—Si bonds. After exposure of the surface to atomic hydrogen, extended defects are found in the surface region, typically to a depth of about 1000 Å (Johnson, Ponce, Street and Nemanich 1987). These defects have no Burgers vector and are therefore not dislocations, but rather appear to be microcracks, in which the (111) planes of the crystal are pushed apart. A plausible explanation of the crack is that the silicon atoms are terminated by hydrogen and so are pushed apart. The presence of Si—H bonds is confirmed by Raman scattering. Hydrogen therefore can break Si—Si bonds and has a tendency to disorder the crystal.

The diffusion of hydrogen in crystalline silicon is thermally activated and occurs by motion between bond center sites, interrupted by trapping processes in which hydrogen is bonded to silicon or to dopants. The magnitude of D_H is larger than that found in a-Si:H but also depends on doping, again because the trapping and release energy is related to the occupancy of the electronic states in the gap and so to the doping.

The three effects of hydrogen in the crystal, namely defect passivation, dopant passivation and microcrack formation, have their counterparts in a-Si:H. The passivation of dangling bonds is the primary beneficial effect of the hydrogen in a-Si:H. However, hydrogen also causes a reconstruction of the network, breaking and removing weak Si—Si bonds, particularly during growth. Chapter 5 describes how the doping properties of a-Si:H are understood by the ability of the dopant atoms to be either three-fold or four-fold coordinated. It is unclear whether the different coordination is specifically a consequence of the hydrogen, or whether it is a more general property of the amorphous network, but hydrogen is certainly implicated in the thermal changes of the doping efficiency that occur.

3 The electronic density of states

The description of the electronic properties of a-Si:H starts with the energy distribution of electronic states. Depending on their energy and character, the different states determine the electrical transport, recombination and doping etc. Some effects of disorder in a-Si:H on the electronic states are the broadening of the density of states distribution compared to the crystal to form the band tails; the localization of the band tail states; the reduction of the scattering length to atomic distances; and the loss of momentum conservation in the electronic transitions. The last of these necessitates the replacement of the energy–momentum band structure of a crystalline semiconductor by an energy-dependent density of states distribution, $N(E)$. It is convenient to divide $N(E)$ into three different energy ranges; the main conduction and valence bands, the band tail region close to the band edge, and the defect states in the forbidden gap (see Fig. 1.6). This chapter describes the first two types of state and defects are dealt with in Chapter 4. The distribution of states, $N(E)$, is derived in this chapter and is used throughout the remainder of the book in the analysis of experimental results.

One conclusion from the structure studies in Chapter 2 is that the bonding disorder of a-Si:H is relatively small. The silicon atoms have the same tetrahedral local order as crystalline silicon, with a bond angle variation of about 10% and a much smaller bond length disorder. Fig. 3.1 shows the calculated dependence of the bond energy on bond length and bond angle (Biswas and Hamann 1987). The distortions of the network correspond to a disorder energy of order 0.1 eV, compared to a bond strength of 2.5 eV. Therefore, in common with most other covalent amorphous semiconductors, the overall shapes of the valence and conduction bands of a-Si:H are hardly different from a smoothed crystalline density of states. The most significant difference between the crystal and amorphous phases comes at the band edges where the disorder creates a tail of localized states extending into the gap. The width of the tail depends on the degree of disorder and on the bonding character of the states. The localized tail states are separated from the extended band states by the mobility edge, the properties of which are described in Chapter 7.

3.1 The conduction and valence bands

A molecular orbital representation of the conduction and valence bands of a-Si:H is illustrated in Fig. 3.2. The s and p states combine to form the sp^3 hybrid orbitals of the tetrahedral silicon bonding. These orbitals are split by the bonding interactions to form the valence and conduction bands. In terms of the Weaire–Thorpe Hamiltonian of Eq. (1.3), the splitting of the bands is determined mostly by the energy V_2 and the broadening of the bands by V_1. The valence band comprises $\frac{3}{4}$ p states and $\frac{1}{4}$ s states, with the p states lying at the higher energies, as in the isolated atom. The ordering is the same in the conduction band, with the s states lying at the band edge. Other higher

Fig. 3.1. Calculation of the dependence of Si—Si bond energy on (*a*) bond length and (*b*) bond angle. The average bond angle distortions in a-Si:H are shown (Biswas and Hamann 1987).

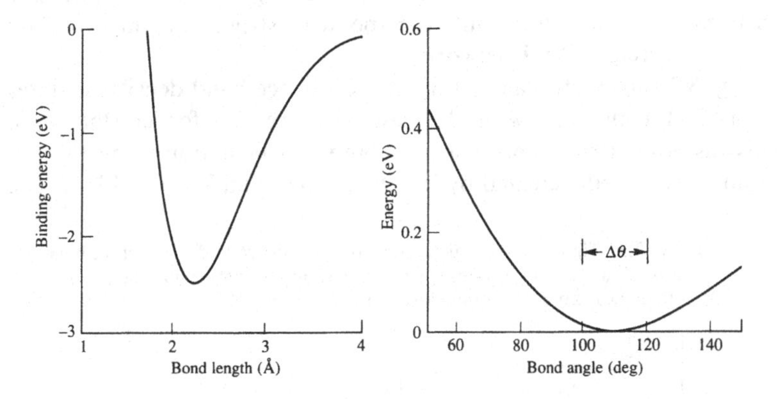

Fig. 3.2. Schematic molecular orbital model of the electronic structure of amorphous silicon and the corresponding density of states distribution.

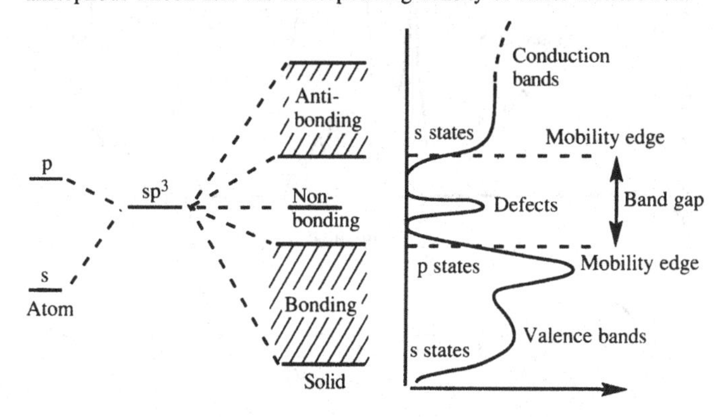

energy atomic levels mix in with the upper conduction band states. Any non-bonding silicon orbitals, such as dangling bonds, are not split by the bonding interaction and give states in the band gap.

The model in Fig. 3.2 is sufficient to predict the general features of $N(E)$, but much more detailed calculations are needed to obtain an accurate density of states distribution. Present theories are not yet as accurate as the corresponding results for the crystalline band structure. The lack of structural periodicity complicates the calculations, which are instead based on specific structural models containing a cluster of atoms. A small cluster gives a tractable numerical computation, but a large fraction of the atoms are at the edge of the cluster and so are not properly representative of the real structure. Large clusters reduce the problem of surface atoms, but rapidly become intractable to calculate. There are various ways to terminate a cluster which ease the problem. For example, a periodic array of clusters can be constructed or a cluster can be terminated with a Bethe lattice. Both approaches are chosen for their ease of calculation, but correspond to structures which deviate from the actual a-Si:H network.

Fig. 3.3 shows one calculation of the valence band density of states of a-Si:H compared with the equivalent results for crystalline Si (Biswas *et al.* 1990). This is a tight binding calculation using the 216 atom cluster model created by Wooten, Winer and Weaire (1985). The

Fig. 3.3. Tight binding calculations of the valence and conduction bands of crystalline and amorphous silicon. The result for amorphous silicon is obtained from a 216 atom cluster (Biswas *et al.* 1990).

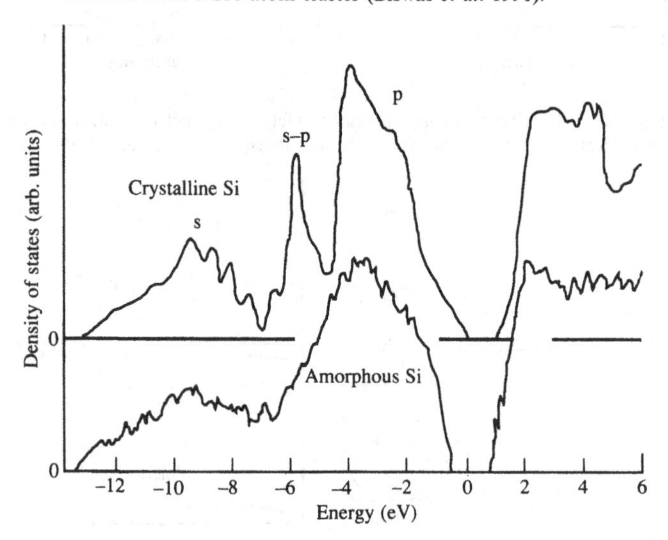

cluster does not contain hydrogen but correctly reproduces the RDF and the phonon density of states. The crystalline valence band in Fig. 3.3 contains three peaks; the uppermost peak is predominately p-like in character, the lowest is s-like, and the middle peak is of mixed s–p character. The upper and lower peaks remain in the calculated amorphous silicon valence band, but the s–p peak has merged into the other two bands. Apart from this difference, the density of states of the amorphous silicon network is little different from a broadened version of the crystal. The merging of the two lower peaks was originally attributed to the presence of odd-membered rings in the network topology. However, later calculations found the same effect with a broad distribution of dihedral angles, without considering the specific ring topology (Hayes, Allen, Beeby and Oh 1985).

The Si–H bonds add further features to the density of states distribution. The Si–H bonding electrons are more localized than the Si–Si states and lead to sharper features in the density of states. One set of calculations, shown in Fig. 3.4, finds a large hydrogen-induced peak

Fig. 3.4. Calculations of the electronic states derived from SiH_2 and SiH_3 bonds (Ching *et al.* 1980). The dashed lines are the experimental results shifted by 2 eV (see Fig. 3.7).

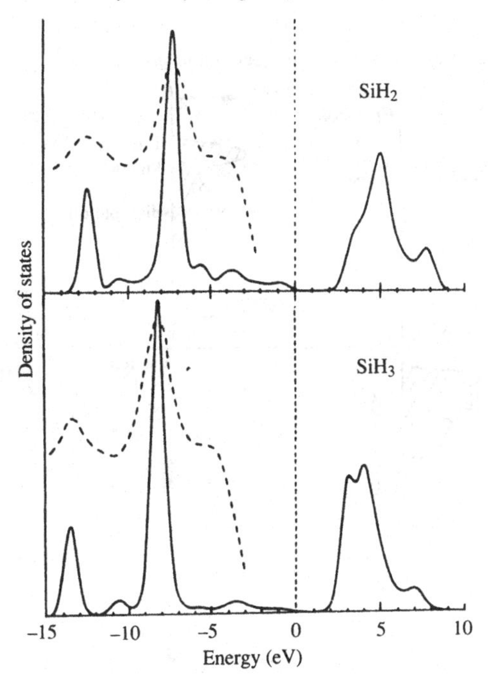

at about 5–10 eV from the top of the valence band and others at larger binding energies (Ching, Lam and Lin 1980). The different Si-H_x structures give slightly different valence band structures but it has not proved possible to use this as a reliable way of extracting Si–H bonding information.

3.1.1 *Measurements of the conduction and valence band density of states*

A direct experimental measure of the density of states in the bands is by optical spectroscopy, particularly photoemission (Ley 1984). In this experiment, illustrated in Fig. 3.5, a photon of energy $\hbar\omega$ excites an electron from the valence to the conduction band. Provided that the electron energy is high enough to overcome the work function energy, Ψ, and no inelastic scattering takes place, the electrons are ejected from the surface with kinetic energy,

$$E_{kin} = \hbar\omega - \Psi - E_{VB} \tag{3.1}$$

where E_{VB} is the binding energy of the valence band states with respect to the Fermi energy. Measurement of the kinetic energy distribution of the ejected electrons then gives the valence band density of states. Inelastic scattering of the electrons does occur, but is a smooth

Fig. 3.5. A schematic diagram of the photoemission process.

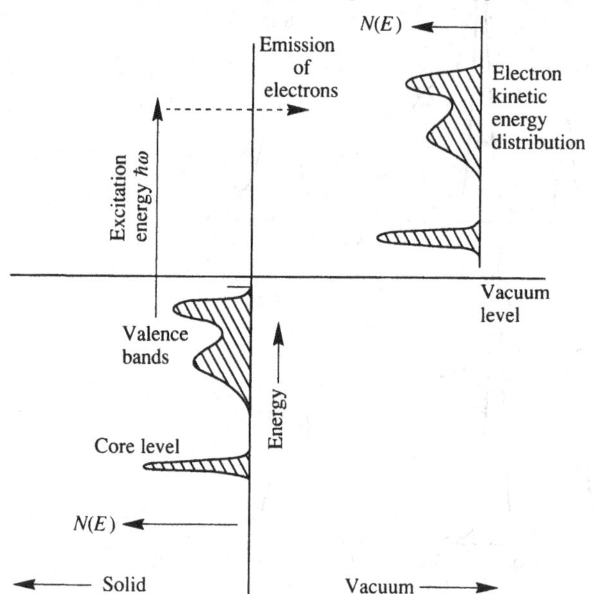

background to the spectrum which can be subtracted. The photoemission spectrum does not immediately give the true shape of the valence band spectrum, because the cross-section for emission depends on the specific atom and on the character of the wavefunction. The cross-section varies with the incident photon energy, which is a useful feature, as it provides a method of determining the character of the wavefunction. For X-ray photoemission (XPS) at the common energy of 1486 eV, the cross-section of the silicon 3s states is about 3.5 times larger than the 3p states and hydrogen has a very small cross-section. The XPS spectrum therefore enhances the deeper s states and suppresses hydrogen states which are only observed using UV photoemission (UPS).

The empty conduction band states are observed by the inverse of the

Fig. 3.6. The valence and conduction bands of crystalline silicon and a-Si:H measured by XPS and inverse XPS. The results can be compared to the calculations in Fig. 3.3 (Jackson *et al.* 1985).

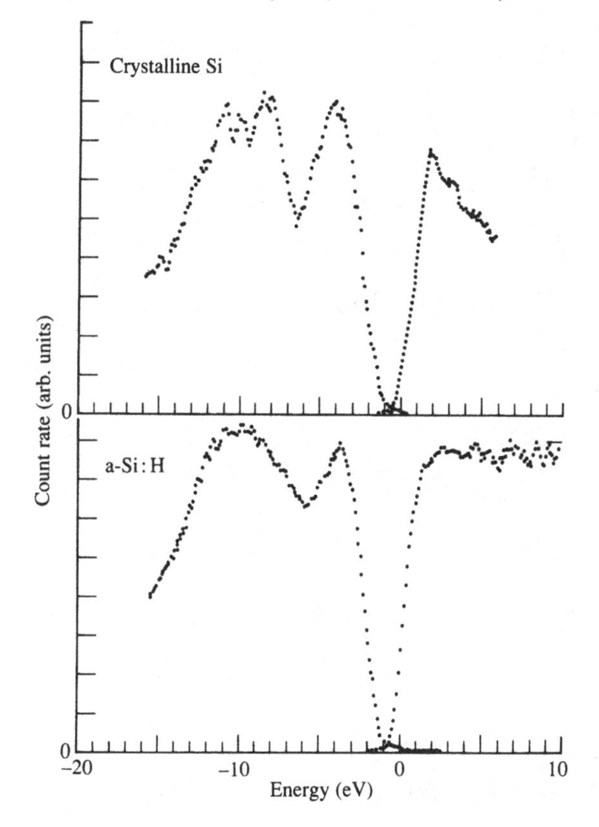

photoemission process, in which the sample is exposed to an electron beam of known energy E_I. The electrons make transitions down to empty states within the material, emitting photons of energy $E_I - E_{CB} + \Psi$, where E_{CB} is the energy of the conduction band states from E_F. In practice, inverse photoemission is usually measured by varying the electron energy and keeping the photon energy constant.

Photoemission experiments are sensitive only to states that are close to the surface because of the short escape depth of the electrons. The escape depth varies with energy and is only aobut 10 Å at 10–100 eV, increasing to 40–50 Å above 1 keV, so that the larger energies tend to be more appropriate for the study of bulk properties. Surface states usually extend no more than 5–7 Å into the bulk and so their contribution should be small in the XPS spectra. However, in view of the growth process of a-Si:H and the effects of hydrogen near the

Fig. 3.7. UPS valence band spectra of a-Si:H showing features at 6 eV and 11 eV due to Si–H bonds (von Roedern *et al.* 1977).

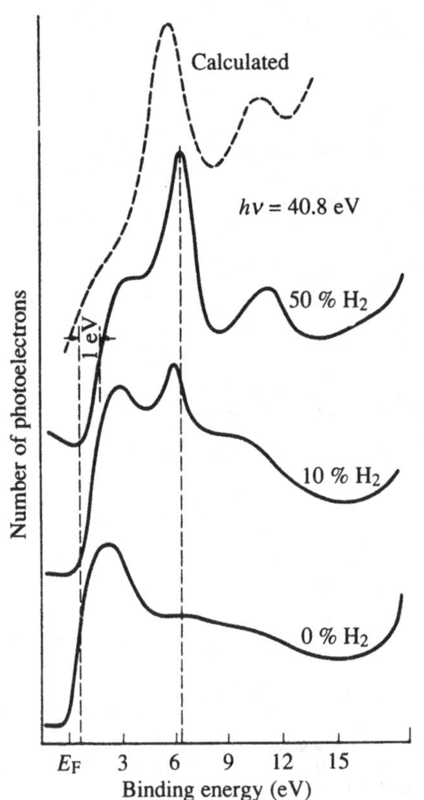

surface, it is quite possible that the structure at a depth of 50 Å is not properly representative of the bulk, and is a difficult problem with the interpretation of photoemission experiments.

Fig. 3.6 shows measured valence and conduction bands of a-Si:H, compared with crystalline silicon (Jackson *et al.* 1985). The valence band shape of crystalline silicon agrees well with the calculations in Fig. 3.3 when the different cross-sections for s and p states are taken into account and the three peaks predicted by theory are present in the data. The main difference between the a-Si:H and crystalline silicon spectra is that the two lower energy peaks merge into one broad band in a-Si:H, which also agrees with theory. The p states at the top of the valence band are hardly changed from the crystal. The similarity of the results for the two materials reflects the fact that the tetrahedral bonding is the same in both phases and that the disorder energy is fairly small. The conduction band also has a similar shape as in the crystal, except that the sharp features are broadened out.

The top of the valence band shifts to higher binding energy when the hydrogen content is increased (Ley 1984). The effect is ascribed to a replacement of Si–Si bonds with stronger Si–H bonds. The conduction band is less affected by hydrogen, so that the overall effect is to increase the band gap energy. Lower photon energy UPS spectra are used to observe the hydrogen bonding, and some examples are shown in Fig. 3.7 (von Roedern, Ley and Cardona 1977). Unlike the XPS data in Fig. 3.6, these spectra are dominated by a peak at 6 eV with a weaker peak at higher binding energies, particularly when the hydrogen content is

Fig. 3.8. Comparison of the spin–orbit split 2p core levels of (*a*) unhydrogenated and (*b*) hydrogenated amorphous silicon. The data for crystalline silicon are also shown in (*a*) for comparison. A-Si:H data comprise contributions from different SiH_n configurations, with n as indicated (Ley *et al.* 1982).

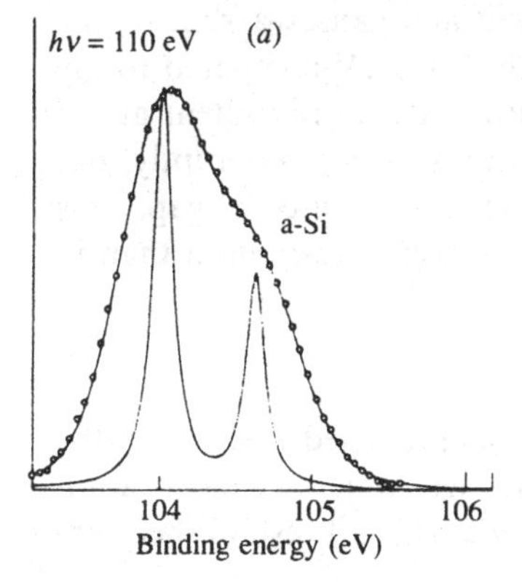

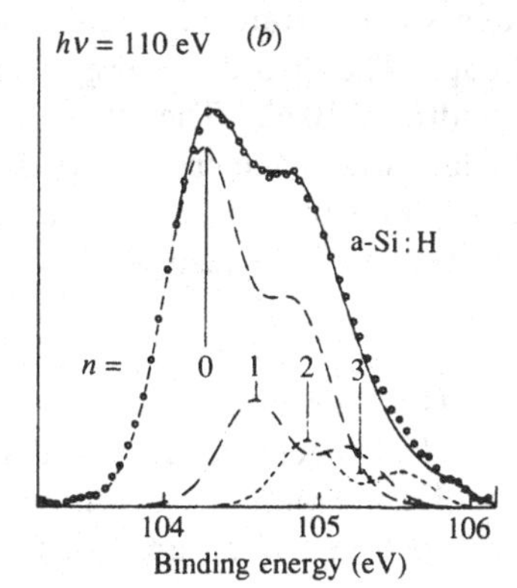

large. These features originate from hydrogen and are absent in unhydrogenated material. As might be expected, the shape and intensity of the hydrogen peaks depend on the plasma deposition conditions and reflect the different bonding structures, but the variations are too small to give a clear indication of the specific hydrogen bonds.

Photoemission also measures the energy and shape of the deeper core levels. The interaction between the bonding electrons, which broadens the valence and conduction bands, is absent in the core levels which are consequently much sharper. The 2p atomic levels in crystalline silicon consist of a pair of sharp lines split by 0.6 eV due to the spin–orbit interaction. Any change in the electron charge of the valence states around an atom alters the potential of the core electrons because of screening effects, resulting in a chemical shift of the measured core state energy. The obvious origins of a chemical shift of the silicon in a-Si:H are the difference between Si–Si and Si–H bonding and any changes in the bond charge due to the bonding disorder of Si–Si bonds (Ley, Reichart and Johnson 1982). The core levels of unhydrogenated amorphous silicon are broader than those of the crystal and there is a further asymmetrical broadening in a-Si:H as is shown in Fig. 3.8. The symmetrical broadening of about 0.2 eV is attributable to the bonding disorder and is indicative of fluctuations in the net bond charge, and the extra shift in a-Si:H is explained by the various Si–H bonding configurations. These measurements confirm that the disorder potential in a-Si:H is a small fraction of the bond strength.

The conclusions from the photoemission data are that the electronic bands in a-Si:H are much as predicted. The overall structure is determined by the short range order of the bonding, which is the same as in the crystal. Disorder effects are secondary, resulting in some broadening and mixing of the electronic states. Si–H bonding introduces extra states in the valence band and causes a shift of the band edge. The disorder energy is of order 0.1–1 eV compared to the band width of 10 eV. The effects of disorder are more evident at the band edges where the crystalline density of states drops abruptly, but the amorphous materials have band tails extending into the gap. This region needs to be investigated with a finer energy resolution than in Fig. 3.6.

3.2 The band tails

The existence of localized states was predicted early on in the studies of amorphous semiconductors by the Anderson localization theory (Section 1.2.5) and their presence is well established experimentally. The influence of localized states is apparent in electrical transport, doping, recombination, and many other measurements. Indeed, those electronic properties of amorphous semiconductors which differ markedly from the equivalent crystalline materials are almost invariably due to the band tail states. The question of how the localized band tail states merge into the extended conduction and valence band states is complicated and is discussed in Chapter 7. Where necessary for the interpretation of experiments, we assume that there is a well-defined mobility edge separating localized from delocalized states. It is shown in Chapter 7 that such an assumption is at best questionable, but does not significantly modify any of the conclusions made in this chapter.

The photoemission data in Fig. 3.6 provide some direct information about $N(E)$ in the band tails, but the results are limited by the low sensitivity and energy resolution of the experiment. The density of states can be determined only to about 5–10 % of the peak density in the band, with an energy resolution of 0.05–0.1 eV, after careful correction for the experimental resolution. Both band edges are found to have an approximately linear energy dependence of $N(E)$ over an energy range of about 0.5 eV from the band edges and down to a density of about 3×10^{21} cm^{-3} eV^{-1}. The density of states deep in the band gap of good quality a-Si:H is no more than 10^{15}–10^{16} cm^{-3} eV^{-1}, so that another 5–6 orders of magnitude of sensitivity are needed to describe the band tails properly. A different photoemission technique which measures the total electron yield has the extra sensitivity (Winer and Ley 1989). Photon energies of 5–10 eV are used in this experiment, just sufficient to excite electrons over the work function so that they are emitted into the vacuum. The number of electrons emitted is given by the joint density of states and the transition matrix element P,

$$Y(\hbar\omega) = \text{const.}\ P^2(\hbar\omega) \int_{E_{\mathrm{VAC}}}^{\infty} N_{\mathrm{V}}(E) N_{\mathrm{C}}(E+\hbar\omega)\, \mathrm{d}E \tag{3.2}$$

The probability of electron emission into the vacuum jumps abruptly at the work function energy, E_{VAC}, and is approximately constant above this energy, so that the valence band density of states is given by $\mathrm{d}Y(\hbar\omega)/\mathrm{d}\hbar\omega$.

An example of the valence band tail of a-Si:H obtained using this technique is shown in Fig. 3.9. The position of the mobility edge, E_{V}, is not obtained in this experiment, which does not distinguish localized from extended states, but is estimated to be at about 5.6 eV. There is a linear density of states near and above E_{V}, and an exponential band tail over several orders of magnitude of $N(E)$ below E_{V}. The slope of the exponent is about 50 meV and agrees well with other experimental results which are described shortly. The slope of the tail is often described by a characteristic temperature T_V, such that

$$N(E) = N_0 \exp(-E/kT_V) \tag{3.3}$$

where E is the energy from the band edge. Only the occupied states up to the Fermi energy can be obtained in this experiment, so that measurements of the conduction band must rely on other techniques.

3.2.1 *Dispersive trapping in a band tail*

The energy distribution of localized band tail states is also obtained from the rate of thermal excitation of trapped carriers during electronic transport. Conduction of electrons and holes occurs by frequent trapping in the tail states followed by excitation to the higher energy conducting states. The effective carrier mobility (called the drift mobility) is consequently lower than the actual mobility of the conducting states and is temperature-dependent because of the thermal activation. The fact that free carrier transport cannot be observed directly is one reason why the properties of the mobility edge have been difficult to resolve. The drift mobility provides a sensitive means of exploring the density of states distribution in the band tail. The drift mobility is also of special interest because it has an unusual time

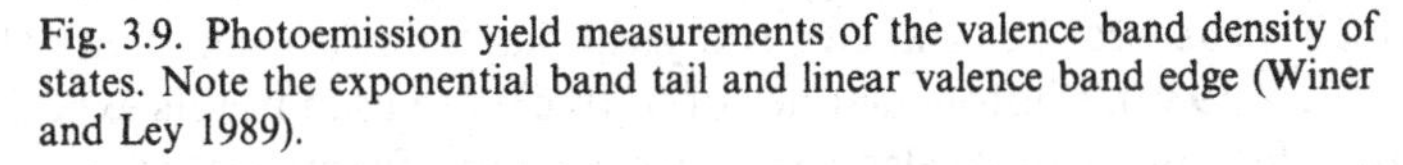
Fig. 3.9. Photoemission yield measurements of the valence band density of states. Note the exponential band tail and linear valence band edge (Winer and Ley 1989).

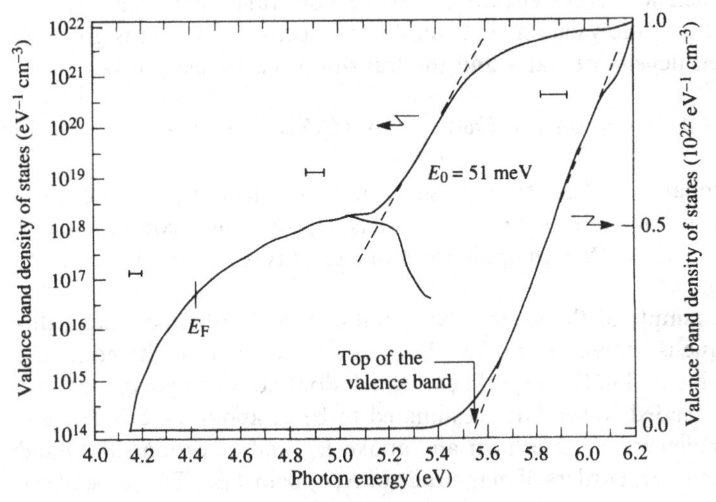

dependence which is a universal feature of amorphous semiconductors (Scher and Montroll 1975). This property, known as dispersive transport, is a consequence of a peculiar distribution of trap release times for the band tail. Many experiments in amorphous semiconductors depend for their interpretation on distributions of times, distances, etc. In fact, the properties of these distributions are responsible for many of the distinctive features of amorphous semiconductors and several examples are given in this book (e.g. Sect. 6.3.1 and 8.3.3).

To explain the trap-limited transport it is useful to consider the model of a single trapping level of density, N_T, at energy E_T below the conducting states, as illustrated in Fig. 3.10. The drift mobility is the free carrier mobility reduced by the fraction of time that the carrier spends in the traps, so that,

$$\mu_D = \mu_0 \tau_{free}/(\tau_{free} + \tau_{trap}) = \mu_0 N_C/[N_T \exp(E_T/kT) + N_C] \approx \mu_0 (N_C/N_T) \exp(-E_T/kT) \quad (3.4)$$

The approximate expression applies when $\mu_D \ll \mu_0$. Thus the drift mobility is thermally activated with the energy of the traps. When there is a distribution of trap energies, $N(E)$, the drift mobility reflects the average release time of the carriers, which is given by,

$$\tau_{av} = \frac{1}{N_T} \int_0^\infty N(E) \omega_0^{-1} \exp(E/kT)\, dE \quad (3.5)$$

Fig. 3.10. Illustration of trap-limited transport of carriers for a discrete or distributed trap level.

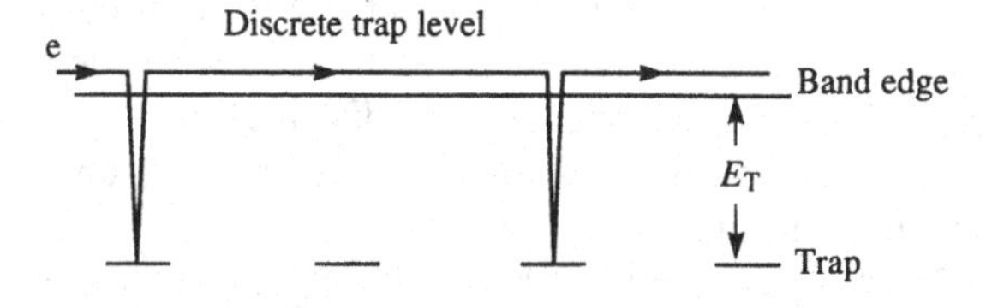

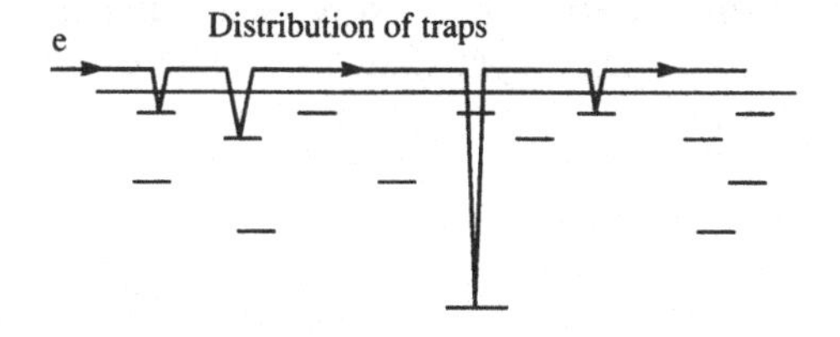

where N_T is the total density of traps. For the exponential distribution of Eq. (3.3), applied to a conduction band tail of slope, kT_C,

$$\tau_{av} = \omega_0^{-1} N_0 \int_0^\infty \exp\left(-\frac{E}{kT_C} + \frac{E}{kT}\right) dE \tag{3.6}$$

At temperatures larger than T_C, integration gives

$$\tau_{av} = \omega_0^{-1} N_0 \frac{kTT_C}{(T - T_C)} \tag{3.7}$$

However, when T is less than T_C, the exponent in Eq. (3.6) is positive, so that the integral diverges and the average trap release time becomes infinite. The drift mobility should consequently drop abruptly to zero. To see that this is not what actually happens, consider the median release time of the carriers, defined by

$$\tau_m = \omega_0^{-1} \exp(E_m/kT) \tag{3.8}$$

where E_m is the median energy of the traps, given by

$$\int_0^{E_m} N_0 \exp(-E/kT_C)\, dE = \tfrac{1}{2} N_0 kT \tag{3.9}$$

E_m is a well-defined finite energy, $E_m = kT_C \ln 2$, so that although the average release time diverges, the median time, τ_m does not. Its value is

$$\tau_m = \frac{2}{\omega_0} \exp\left(\frac{T_C}{T}\right) \tag{3.10}$$

which is of order ω_0^{-1}, provided T_C is not too different from T. Most of the trap release events occur very rapidly and there is no discontinuous change in τ_m when T decreases below T_C, as there is with the average release time, τ_{av}. The unusual release time distribution at $T < T_C$ is evident from a median release time of order 10^{-13} s and an average release time which is infinite.

The reason for the divergent τ_{av} but finite τ_m is that the average release time from the traps is dominated by the exponentially small density of states with very large trap energies. In any measurement that involves a finite number, n, of trapping and release events, the average carrier will not fall into a trap deeper than E_{max}, where

$$E_{max} = kT_C \ln(N_0/n) \tag{3.11}$$

The release time at this energy is,

$$\tau_{max} = \omega_0^{-1} (n/N_0)^{T/T_C} \tag{3.12}$$

τ_{max} is the largest contribution to the distribution of release times during the particular experiment. Therefore, although the average release time for an infinitely long experiment diverges, such is not the case for any measurement of finite duration, which has a finite average release time and so a measurable drift mobility. However, as the number of trapping and release events of the experiment is increased, the average release time increases roughly as n^{T/T_C}, giving an apparent time dependence to the drift mobility.

Before going further into the analysis, it is useful to describe the specific experimental measurement of the drift mobility for which the analysis applies. The time-of-flight technique was developed in the 1960s by Spear (1968) for application to low mobility insulators and is illustrated in Fig. 3.11. The sample consists of a capacitor structure with electrodes on either side of a thin film material. Carriers are generated near to one electrode, usually by a short pulse of highly absorbed light, resulting in mobile electrons and holes. The carriers separate under the action of the applied electric field, so that one sign of charge is collected by the near electrode and the other crosses the sample to the opposite electrode. As the charge, q, moves a distance x across the sample it induces on the electrode a charge,

$$Q_I = qx/d \tag{3.13}$$

where d is the sample thickness. The current that flows in the external circuit is

$$I = \mathrm{d}Q_I/\mathrm{d}t = (q/d)\mathrm{d}x/\mathrm{d}t = qv/d = q\mu_D F/d \tag{3.14}$$

Fig. 3.11. Schematic diagram of the time-of-flight apparatus for measuring drift mobility.

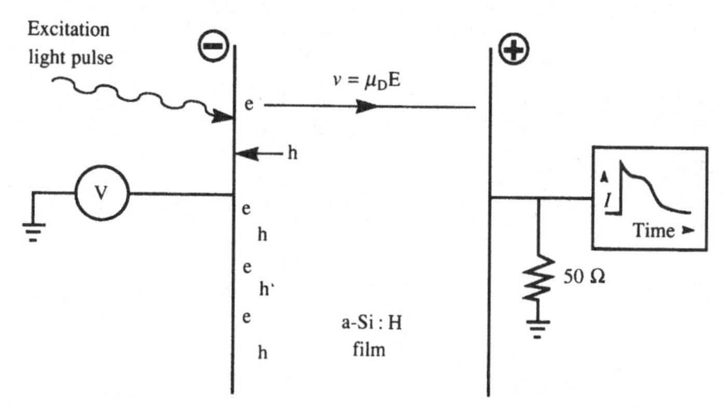

where $v = \mu_D F$ is the average carrier velocity in the conducting states and F is the electric field, assumed to be uniform. The current persists until the time when the carriers have crossed the sample, referred to as the transit time and given by

$$\tau_T = d/\mu_D F \tag{3.15}$$

The drift mobility is usually obtained from Eq. (3.15) rather than from Eq. (3.14), for which it is necessary to know the initial charge q.

Some examples of current transients for electrons and holes in a-Si:H are shown in Fig. 3.12 (Marshall, Street and Thompson 1984, 1986). Above room temperature the electron transport is described by the non-dispersive trapping model. There is a well-defined transit time, which is inversely proportional to the applied voltage, and a constant mobility. The rounding of the current pulse near the transit time is due to the normal broadening of the charge packet as it crosses the sample. The equivalent room temperature data for holes and the low temperature data for electrons are different, having neither a constant current before the transit nor a well-defined transit time. Instead the initial current decays as a power law in time, $t^{\alpha-1}$, where α is about 0.5 for holes at room temperature, and at longer times there is a change to a steeper decay, often but not always, of the form $t^{-\alpha-1}$. The change of slope is identified as the transit time.

The drift mobility in this dispersive regime has an unusual electric field and thickness dependence. Fig. 3.13 shows the field dependence of the electron and hole mobility at different temperatures (Marshall *et al.* 1986, Nebel, Bauer, Gorn and Lechner 1989). The electron drift

Fig. 3.12. Examples of the (*a*) electron and (*b*) hole transient current pulses at different temperatures, showing the increasingly dispersive behavior at low temperature.

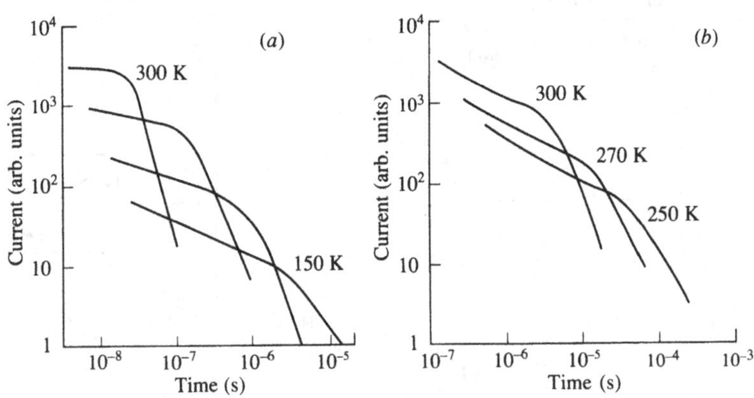

mobility is thermally activated, as expected for a trap-limited process and at room temperature there is no field dependence. However, in the low temperature dispersive regime, there is a large increase in μ_D with the applied field. This effect is widely studied in chalcogenide glasses, where the mobility has the form (Pfister and Scher 1977).

$$\mu_D \sim (F/d)^{1/\alpha - 1} \tag{3.16}$$

The parameter α characterizes the dispersive transport, describing the time, field and thickness dependence. The temperature dependence of α is shown in Fig. 3.14 and is approximately proportional to the temperature, $\alpha = T/T_C$, with a temperature T_C of 250–300 K for electrons and $T_V = 400$–450 K for holes (Tiedje, Cebulka, Morel and Abeles 1981).

With the experimental results in mind we return to the analysis of the trap-limited transport. The time-dependent decrease in the apparent mobility is obviously consistent with our earlier argument that the average trapping time will increase with the number of trapping events for an exponential band tail. Scher and Montroll (1975) were the first to point out this property of a very broad distribution of release times and to associate the effect with transport in disordered semiconductors. They analyzed the random walk of carriers with such a distribution and

Fig. 3.13. Temperature dependence of the (*a*) electron and (*b*) hole drift mobility at different applied fields ranging from 5×10^2 V cm^{-1} to 5×10^4 V cm^{-1}. The field dependence of μ_D is caused by the dispersion (Marshall *et al.* 1986, Nebel *et al.* 1989).

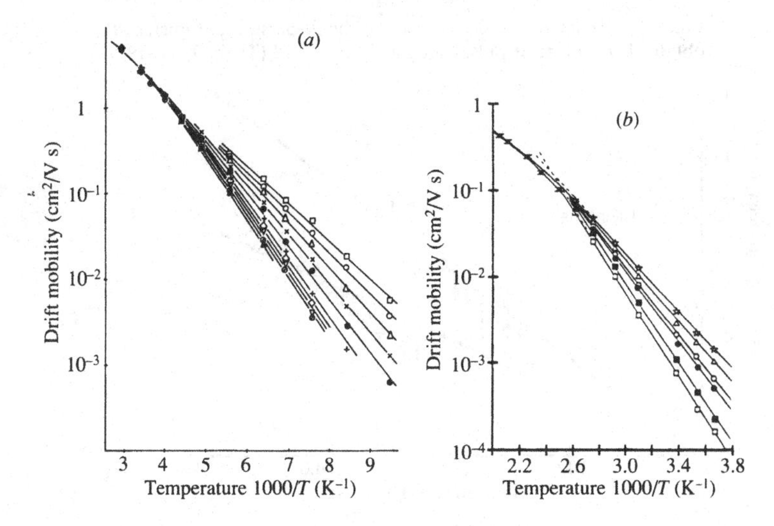

showed that a long time release probability of the form $t^{-\alpha-1}$ accounted for all the characteristics of the dispersive transport.

Of all the theoretical studies of dispersive transport by multiple trapping, the analysis of Tiedje and Rose (1980) (and similarly Orenstein and Kastner (1981)) is particularly instructive, because the physical mechanism is easy to understand. The model is specifically for an exponential density of states, $N_0 \exp(-E/kT_C)$. The main problem in analyzing the transport is to deal with the broad distribution of release times. The approach in this model is to define a demarcation energy E_D which varies with the time t after the start of the experiment,

$$E_D = kT \ln(\omega_0 t) \tag{3.17}$$

E_D is the energy at which the release time is just equal to the time t. Provided the temperature is less than T_C, electrons in traps which are shallower than E_D are excited to the mobility edge and trapped many times, but electrons in states deeper than E_D have a very low probability of release within the time t. Thus the states deeper than E_D are occupied in proportion to the density of states, which decreases as $\exp(-E/kT_C)$, but the states above E_D have had time to equilibrate and follow a Boltzmann distribution, $N(E)\exp[-(E-E_D)/kT]$. The electron distribution therefore has a peak at E_D and, from the definition of E_D, this peak moves to larger trapping energies as time progresses. In order to analyze the transport further, the approximation is made that *all* the electrons reside at E_D. The problem can then be treated as trapping at a single level, with the added property that the trap energy is time-

Fig. 3.14. Temperature dependence of the dispersion parameter, α, obtained from transit pulses such as in Fig. 3.12 (Tiedje *et al.* 1981).

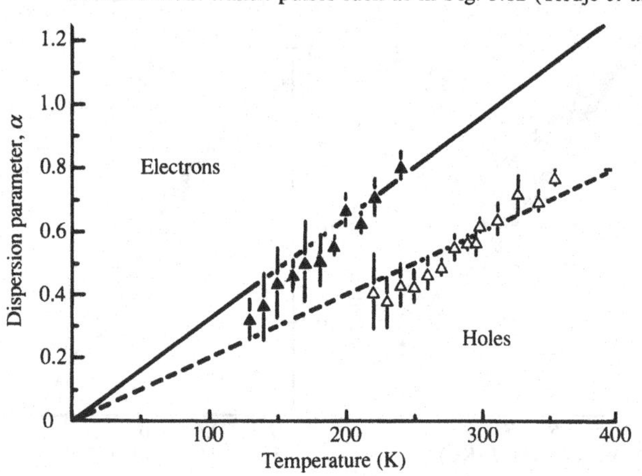

dependent. Substituting E_D from Eq. (3.17) into Eq. (3.4), with the assumed exponential density of states gives

$$\mu_D(t) = \mu_0\,\alpha(1-\alpha)\,(\omega_0\,t)^{\alpha-1} \tag{3.18}$$

where $\alpha = T/T_C$. It is assumed that the exponential density of states continues up to the mobility edge.

The drift mobility has the expected power law decrease with time. To obtain μ_D from the transit time in the time-of-flight experiment, the transit time is defined, somewhat arbitrarily, as the time when the average carrier is half way across the sample,

$$\int F\mu_D(t)\,\mathrm{d}t = d/2 \tag{3.19}$$

Substituting the time-dependent mobility from Eq. (3.18) gives an expression for the transit time from which the experimental mobility is obtained,

$$\mu_D^{\exp} = \omega_0\left[\frac{\omega_0}{2(1-\alpha)}\right]^{-\frac{1}{\alpha}}\mu_0^{\frac{1}{\alpha}}\left(\frac{F}{d}\right)^{\frac{1}{\alpha}-1} \tag{3.20}$$

The difference between the mobility described by Eqs. (3.18) and (3.20) is discussed shortly.

At longer times than the transit time, carriers are lost by reaching the electrode rather than being retrapped below E_D and for this reason the time-dependent decay of the current becomes stronger. The assumption that all the carriers excited from E_D are lost to the contact yields the following expression for the current,

$$I(t>\tau_T) = I_0\,\alpha(1-\alpha)\,(\omega_0\,\tau_T)^{2\alpha}\,(\omega_0 t)^{-\alpha-1} \tag{3.21}$$

Eqs. (3.18), (3.20) and (3.21) contain all the features of the dispersive transport, specifically the power law time decay which becomes steeper after the transit time, and the thickness and field dependence of the mobility. The results are confirmed by many other types of calculations, ranging from more rigorous analysis of the release time distribution (Michiel, Adriaenssens and Davis 1986), to Monte Carlo simulation of the transport (Marshall 1977).

Returning to the expressions for the mobility given by Eqs. (3.18) and (3.20), it must be stressed that the thickness and field dependences are artifacts of the time-of-flight experiment and are not intrinsic properties of the mobility, which is only time-dependent. The artifact arises because the transit time changes with the sample thickness or applied field and the measurement of the mobility $\mu_D(t)$ then applies to a different time. It is also of note that the measured mobility in Fig. 3.13

has a reasonably well-defined activation energy in the dispersive regime whereas neither Eq. (3.18) nor Eq. (3.20) contains any exponential terms. One way to understand this result is to consider the path of a carrier in a particular experiment. After many trapping and release events, the carriers have crossed the sample and near the end of their transit have a distribution centred at some demarcation energy, say E_{D1}. If the same experiment is performed at a different temperature, the path of an average carrier will be unchanged, because the choice of which trap it occupies is random and independent of temperature. The demarcation energy is therefore also at E_{D1}. The only difference is that the thermal release of carriers is slowed down, so that the transit takes longer. As the demarcation energy at the end of the experiment is independent of temperature, the drift mobility is proportional to $\exp(-E_{D1}/kT)$, and is thermally activated. In effect the temperature and time dependences of E_D exactly cancel out.

Fig. 3.15 illustrates the point, and shows the mobility as a function of temperature according to Eq. (3.18) for constant measurement time. The actual mobility in a time-of-flight experiment is measured at constant drift length (the sample thickness) and is indicated by the dashed intersecting line, because the time of the experiment increases

Fig. 3.15. The temperature dependence of the mobility for constant drift time (solid lines) and constant drift distance (dashed line).

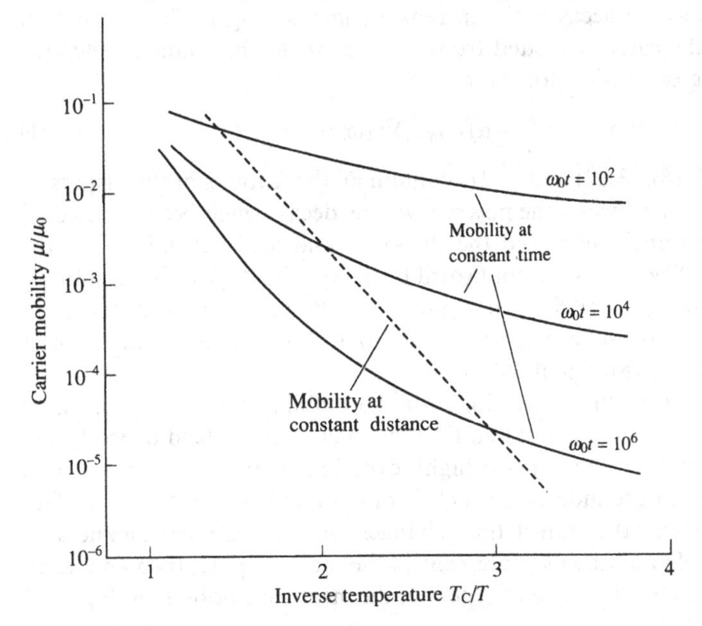

with decreasing temperature. A general property of dispersive diffusion is a difference between measurements performed under constant time and constant distance. This distinction also applies to the dispersive diffusion of hydrogen which is discussed in Section 6.4.1.

3.2.2 *The band tail density of states distribution*

The time-of-flight data for a-Si:H agree quite well with the predictions of the model for the exponential band tail. Slopes of 220–270 K for the conduction band tail and 400–450 K for the valence band tail are obtained from the temperature dependence of the dispersion parameter in Fig. 3.14. Electrons are non-dispersive at room temperature but holes are quite strongly dispersive. Although the results suggest that the band tails are exponential all the way to the mobility edge, in fact the experimental information is too restricted to make this conclusion. The range of energies which can be observed in the time-of-flight experiment is only that covered by the demarcation energy during the experiment. The logarithmic dependence of E_D on time means that a very wide time range is needed to explore the

Fig. 3.16. Experimentally derived conduction band and valence band density of states distribution in a-Si:H.

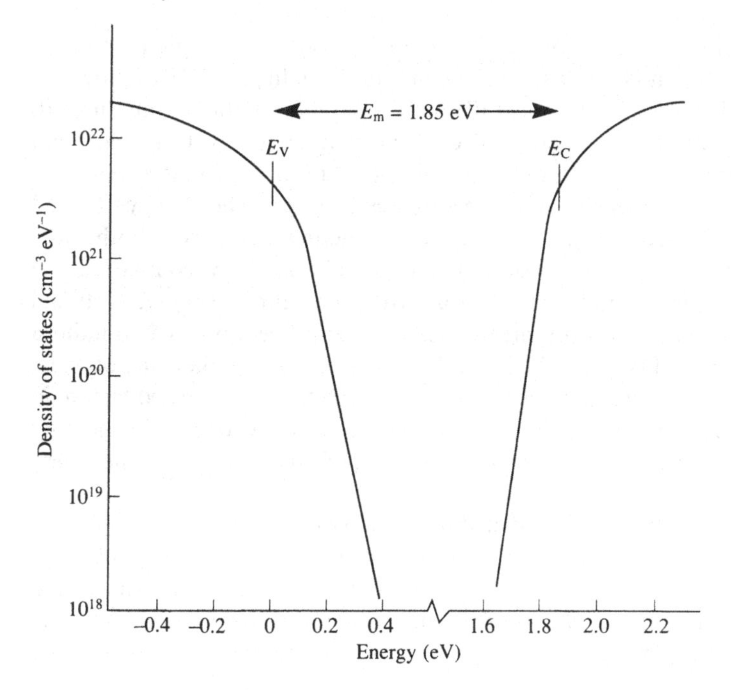

complete band tail. In practice, most experiments cover no more than 2–3 decades of time and less at the higher temperatures, so that the energy range is no more that 0.1–0.2 eV. Also, detailed measurements indicate that the tail is not strictly exponential as there are departures from the predicted results, with the value of α varying with time (Marshall *et al.* 1986). Pollak (1977) gives a general expression relating α to the shape of the density of states,

$$\alpha = kT\left[\frac{\mathrm{d}\ln N(E)}{\mathrm{d}t}\right]_{t=kT\ln(\omega_0 t)} \tag{3.22}$$

For the exponential tail this expression reduces to $\alpha = T/T_C$, but this is the only shape for which α does not depend on E_D and therefore also on time.

Interpreting the small differences between the data and the predictions of the exponential tail requires caution, because some of the basic assumptions of the model are not soundly based. For example, it is assumed that the transport occurs at a well-defined mobility edge, with a temperature-independent mobility, and both points are open to question. Other effects, such as the internal electric fields of the contacts and deep trapping, lead to distortions of the current pulse from its ideal form.

The information about the band tail distribution from photoemission and time-of-flight measurements on good quality a-Si:H is shown in Fig. 3.16. The energy range of the time-of-flight data is quite limited, while the photoemission yield data cover the whole valence band tail range and agree quite well with the other results, so that at least the shape of the valence band seems well established. The energies E_V and E_C are the deduced position of the conducting states from both time-of-flight and photoemission, the latter obtained by comparing the Fermi energy and the conductivity activation energy. Internal photoemission measurements described in Section 9.1.2 obtain a similar mobility gap of 1.9 eV. The question of whether or not these energies measure the mobility edge correctly is addressed again in Chapter 7. However, a point to note is that the exponential edge does not continue up to either E_C or E_V, which are instead in the linear region of the tail.

The density of states distribution in Fig. 3.16 represents the best current estimates for low defect density a-Si:H. The results are sufficiently accurate to use in calculations of the transport and recombination etc., but are continually being refined by new and more accurate experiments.

Although the experiments establish fairly convincingly that the band tails are at least approximately exponential, the underlying reason for this shape is less clear. A simple model for random disorder is expected to have a gaussian distribution of disorder energies and therefore gaussian band tails. Many attempts have been made to explain the exponential shape. Some models are specific to the optical absorption, which also has an exponential edge, and introduce excitonic effects (see Section 3.3.1). Other models attribute the broadening to an electron–phonon interaction or to the thermodynamic equilibrium occupancy of the different possible configurations, a point that is discussed further in Section 6.2.4. Despite these theories, the precise relation between the structural disorder and the band tail shape remains unclear, in part because there are probably different ways in which tail states can be created. The disorder energy may be confined to a single highly strained bond or to the combined effect of several less strained bonds. One model of a-Si:H envisages small volumes of low band gap material bounded by higher gap hydrogenated material (Brodsky 1980).

The difference in slope between the conduction and valence band tails is perhaps easier to understand, based on the model of Fig. 3.2. The conduction band edge is made up of s-like states which are spherically symmetrical and are not strongly influenced by bond angle disorder. On the other hand, the top of the valence band is made up of p-like bonding states and so the bond angle disorder has a greater effect, giving a large disorder energy.

3.3 Optical band-to-band transitions

Optical transitions between the valence and conduction bands are responsible for the main absorption band and are the primary measure of the band gap energy. The optical data are also used to extract information about the band tail density of states. However, the absorption coefficient depends on both conduction and valence band densities of states and the transition matrix elements and these cannot be separated by optical absorption measurements alone. The independent measurements of the conduction and valence state distributions described in Section 3.1.1 make it possible to extract the matrix elements and to explore the relation between $N(E)$ and the optical spectrum.

Optical transitions are defined in terms of the complex dielectric function $\varepsilon_1(\omega)+i\varepsilon_2(\omega)$. From this, the refractive index, n, and the absorption coefficient $\alpha(\omega)$ are obtained,

$$\alpha(\omega) = 2\pi\varepsilon_2(\omega)/n\lambda \tag{3.23}$$

The absorption occurs through a dipole interaction between the radiation field and the electrons in the medium, with the imaginary part of the dielectric function given by

$$\varepsilon_2(\omega) = 4\pi^2 e^2 \frac{1}{\Omega} \sum_{\mathrm{C,V}} \left[R_{\mathrm{CV}} \right]^2 \delta(E_{\mathrm{C}} - E_{\mathrm{V}} - \hbar\omega) \tag{3.24}$$

where Ω is the illuminated volume of the sample. R is the dipole matrix element connecting valence and conduction band states and the delta function expresses the conservation of energy. The sum is over all initial and final states. The matrix elements in a crystalline semiconductor are non-zero only for states of the same crystal momentum k. The optical transitions connect those energy levels which conserve both energy and momentum and are termed direct transitions. Indirect transitions to a state of different k can occur through the absorption or emission of a phonon to conserve the total momentum. Such transitions are generally weak and only observed at energies where the direct transitions are forbidden. In an amorphous semiconductor, momentum is not a good quantum number and the requirement of k-conservation is relaxed. Allowed transitions then occur between any two states for which energy conservation applies. The matrix elements therefore reduce to an average $R(\hbar\omega)$ over all pairs of states separated by energy $\hbar\omega$. This average value is still energy-dependent because the states which contribute to $R(\hbar\omega)$ change with energy. Eq. (3.24) becomes

$$\varepsilon_2(\omega) = 4\pi^2 e^2 a^3 R^2(\hbar\omega) \int N_{\mathrm{V}}(E) N_{\mathrm{C}}(E+\hbar\omega)\,\mathrm{d}E \tag{3.25}$$

The integral is the joint density of states for which the energy difference is constant. The shape of the absorption spectrum is the product of the energy dependence of this integral and the matrix element.

There are several differences between the absorption of amorphous and crystalline phases:

(1) Crystals exhibit sharp structure in $\varepsilon_2(\omega)$ (van Hove singularities), which corresponds to transitions at the band extremities and where two bands have parallel dispersion curves, and hence a large joint density of states. The lack of k-conservation and the disorder of the amorphous phase remove these sharp features.

(2) The distinction between a direct and an indirect gap is absent in a-Si:H because the momentum is not conserved in the transition.

(3) Some of the transitions near the band gap energy of the amorphous material involve localized states. The transition probability for localized-to-extended transitions is the same as for extended-to-extended transitions, but between two localized states separated by a large distance compared to the localization radius, the probability is reduced by the low wavefunction overlap $\exp(-R/R_0)$ where R_0 is the localization length (Davis and Mott 1970).

(4) Crystals exhibit excitonic effects near the band edges, in which the Coulomb interaction between an electron and a valence band hole results in absorption which does not follow the one-particle joint density of states in Eq. (3.25). Excitons produce an absorption peak just below the band gap energy and modify the absorption at higher energies. There is no exciton absorption peak observable in any amorphous semiconductor, because it is broadened out by the disorder. The Coulomb interaction is present in a-Si:H, but its significance in the optical absorption is unclear.

There are several techniques available to measure the optical absorption in a-Si:H. Optical transmission and reflectivity, and ellipsometry measure the high energy, high absorption transitions. In the weak absorption regime near the band edge, the measurement is made difficult by the limitation on sample thickness to a few microns. The transmitted light intensity, T, through a sample is approximately,

$$T = (1-R)^2 \exp(-\alpha d) \tag{3.26}$$

Where R is the reflectivity and d is the thickness. Very accurate values of R and T are needed when the absorptance, αd, is small. The technique of photothermal deflection spectroscopy (PDS) overcomes this problem by measuring the heat absorbed in the film, which is proportional to αd when $\alpha d \ll 1$. A laser beam passing just above the surface is deflected by the thermal change in refractive index of a liquid in which the sample is immersed. Another sensitive measurement of αd is from the spectral dependence of the photoconductivity. The constant photocurrent method (CPM) uses a background illumination to ensure that the recombination lifetime does not depend on the photon energy and intensity of the illumination. Both techniques are capable of measuring αd down to values of about 10^{-5} and provide a very sensitive measure of the absorption coefficient of thin films.

Fig. 3.17 compares the $\varepsilon_2(\omega)$ spectra for a-Si and crystalline silicon. As expected the van Hove singularities of the crystal are completely

Fig. 3.17. Comparison of the imaginary part of the dielectric function (ε_2) for a-Si and crystalline silicon ($\hbar\omega_g$ is the energy of the peak of ε_2) (Pierce and Spicer 1972).

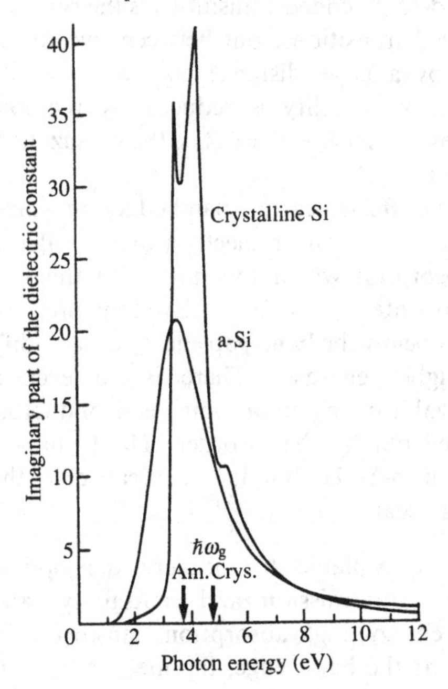

Fig. 3.18. The energy dependence of the dipole matrix element $R^2(\hbar\omega)$ for a-Si:H and crystalline silicon. The dashed line is an harmonic oscillator fit (Jackson *et al.* 1985).

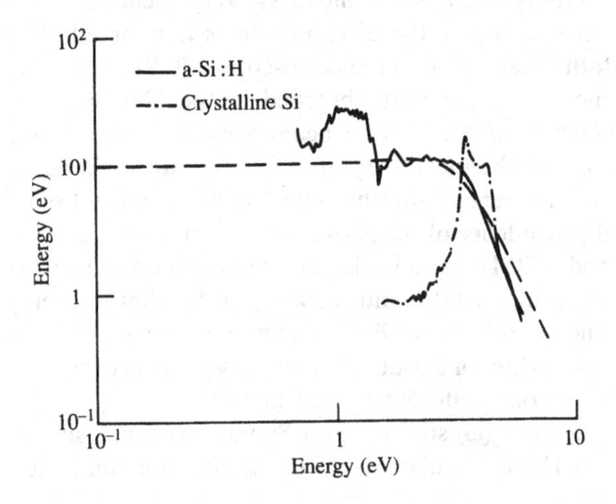

absent in the amorphous phase, but apart from this difference the overall shapes are similar, which is another consequence of the same short range order and density of band states in the two phases. $\varepsilon_2(\omega)$ is larger for a-Si than for crystalline silicon below 3 eV, even though the band gap of the crystal at 1.1 eV is smaller than that of a-Si:H. The crystalline band gap is indirect, so that the smaller $\varepsilon_2(\omega)$ at these energies is due to the different momentum selection rules. The direct gap of crystalline silicon is near 3 eV and $\varepsilon_2(\omega)$ increases rapidly above this energy. The energy dependence of the matrix element, $R^2(\hbar\omega)$, for a-Si:H is shown in Fig. 3.18 and is obtained from Eq. (3.25), by dividing $\varepsilon_2(\omega)$ by the joint density of states, computed from the measured conduction and valence bands (Jackson *et al.* 1985). $R^2(\hbar\omega)$ is constant up to about 3 eV and decreases rapidly at higher energy – the larger value below about 1.5 eV is in the sub-gap region. The peak in $\varepsilon_2(\omega)$ in Fig. 3.17 is primarily due to the shape of $R^2(\hbar\omega)$ because the joint density of states rises monotonically with energy. The form of the matrix element is quite close to that expected for a classical damped harmonic oscillator,

$$R^2(\hbar\omega) = \frac{1}{(\omega^2-\omega_0^2)^2-\Gamma^2\omega^2} \tag{3.27}$$

An energy $\hbar\omega_0 = 3.5$ eV with a width, Γ, of 4 eV fits the data well as is shown in Fig. 3.18.

The optical parameter of most interest is the band gap energy E_G. There is no precise location of the gap because the band tail density of states decays continuously with energy, so that E_G can only be defined in terms of an extrapolation of the bands. Taking this approach, Tauc, Grigorovici and Vancu (1966) introduced a simple model for the band gap. It is assumed that $N(E)$ increases as a power law of energy, $(E-E_{GC})^{n_C}$ from an extrapolated conduction band edge, E_{GC}, with a similar shape for the valence band. If the matrix elements are independent of energy, then the absorption near the gap is given from the joint density of states by

$$\hbar\omega\,\alpha(\hbar\omega) = A(\hbar\omega - E_G)^{(n_C+n_V+1)} \tag{3.28}$$

where E_G is the band gap defined by the separation of the two extrapolated band edges. For parabolic band edges, $n_C = n_V = \frac{1}{2}$,

$$[\hbar\omega\,\alpha(\hbar\omega)]^{\frac{1}{2}} = A'(\hbar\omega - E_G) \tag{3.29}$$

This relation provides a simple procedure to extract E_G from the measurement of $\alpha(\hbar\omega)$ and one example of the absorption plotted in this way is given in Fig. 3.19, from which a gap of about 1.7 eV is obtained. We have just seen, however, that the matrix elements are not exactly constant and that the band edges are more nearly linear than parabolic. However, it happens that these effects nearly cancel out so that the 'Tauc plot' of Eq. (3.29) does give quite a reasonable value for gap and is certainly useful in comparing the band gaps of a-Si:H prepared in different ways. It must be emphasized that this definition of the gap is simply an extrapolation of the density of states. The mobility gap – defined as the energy separation of the valence and conduction band mobility edges – generally as a larger energy.

3.3.1 *The Urbach edge*

The optical absorption has an exponential energy dependence in the vicinity of the band gap energy,

$$\alpha(\hbar\omega) = \alpha_0 \exp[(E - \hbar\omega)/E_0] \tag{3.30}$$

where E_0 is usually 50–100 meV – examples are shown in Fig. 3.20. This form of the absorption typically occurs for an absorption coefficient between 1 and 3000 cm^{-1} in a-Si:H with a low defect density. The exponential tail is called the Urbach edge after its first observation in alkali halide crystals (Urbach 1953) and is found in all amorphous semiconductors. The theoretical ideas of its origin have gone through

Fig. 3.19. The absorption coefficient plotted in the form of Eq. (3.29) showing an extrapolated band gap of 1.7 eV (Cody 1984).

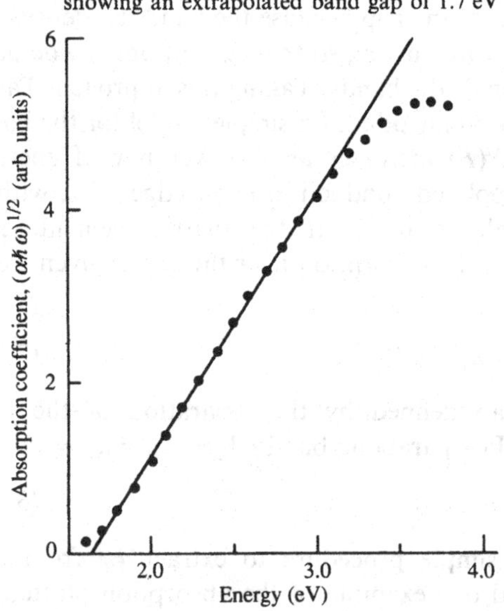

a complete transformation over the years. Originally, the possibility that the shape was given by the joint density of states was largely discounted in favor of explanations in terms of the energy dependence of the matrix elements (see Mott and Davis 1979). The reasons given for this view were that the similarity of the slope in all amorphous semiconductors seemed an unlikely coincidence if it represented the density of states and the observation of an Urbach edge in alkali halide crystals is obviously not a density of states effect. The most successful of the various models proposed was of an exciton transition broadened by random internal electric fields arising from the disorder (Dow and

Fig. 3.20. The optical absorption edge of a-Si:H measured at various temperatures, T_M, and measured at room temperature after annealing at the indicated temperatures T_H (Cody *et al.* 1981).

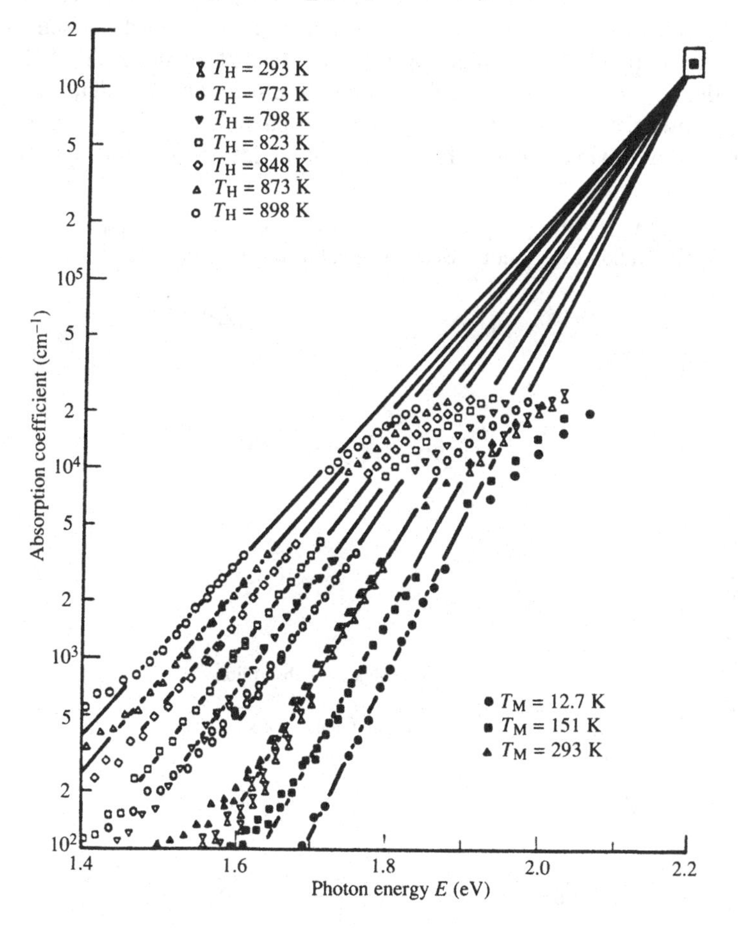

Redfield 1970). This model is able to account for the exponential slope with internal fields which are reasonably consistent with the disorder of the amorphous semiconductors.

Currently this explanation of the Urbach absorption edge has been largely discarded in favor of one in which the shape is simply given by the joint density of states and so reflects the disorder broadening of the bands. The valence band tail of low defect density a-Si:H has an exponential slope of about 45 meV. It is broader than the conduction band tail (see Fig. 3.16) and so dominates the joint density of states, and is indeed close to the slope of the Urbach edge. Fig. 3.21 compares the calculated density of states with the optical data and shows that the agreement is excellent. The same conclusion is also contained within the matrix element data of Fig. 3.18, since there is little energy dependence of $R^2(\hbar\omega)$ through the Urbach edge region. (The increase in $R^2(\hbar\omega)$ below 1.5 eV is in the region of defect absorption discussed in the next chapter.) This explanation of the absorption in terms of the joint density of states implies that excitonic effects are not significant since these are not described by the one-particle density of states distributions. There must be some modification of the optical

Fig. 3.21. Comparison of the measured room temperature optical absorption edge with the calculated joint density of states of Fig. 3.16.

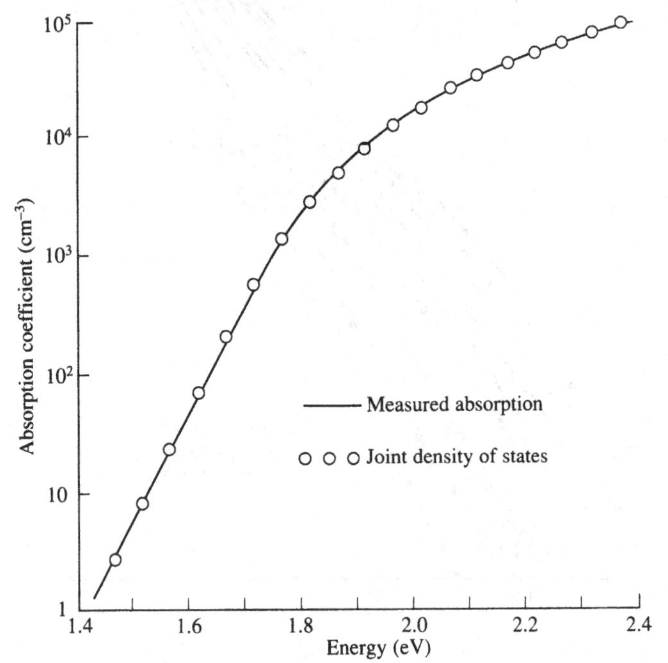

absorption by exciton effects, but their detection needs a more precise comparison of the absorption spectrum with the joint density of states than is presently possible.

Since the slope, E_0, of the Urbach absorption reflects the shape of the valence band tails, it follows that E_0 varies with the structural disorder. For example, one measure of the disorder is the average bond angle variation, which is measured from the width of the vibrational spectrum using Raman spectroscopy (Lannin 1984). Fig. 3.22 shows an increasing E_0 with bonding disorder, which is caused by changes in the deposition conditions and composition (Bustarret, Vaillant and Hepp 1988; also see Fig. 3.20). The defect density is another measure of the disorder and also increases with the band tail slope (Fig. 3.22). A detailed theory for the dependence of defect density on E_0 is given in Section 6.2.4.

3.3.2 *Thermal and static disorder*

Thermally-induced network vibrations broaden the absorption edge and shift the band gap of semiconductors. The thermal disorder couples to the optical transition through the deformation potential, which describes how the electronic energy varies with the displacement of the atoms. The bond strain in an amorphous material is also a displacement of atoms from their ideal position, and can be described by a similar approach. The description of static disorder in terms of 'frozen' phonons is a helpful concept which goes back 20 years. Amorphous materials, of course, also have the additional disordering of the real phonon vibrations.

Fig. 3.22. Dependence of the Urbach slope E_0 on (*a*) the average bond angle deviation (Bustarret *et al.* 1988) and (*b*) the defect density (Stutzmann 1989).

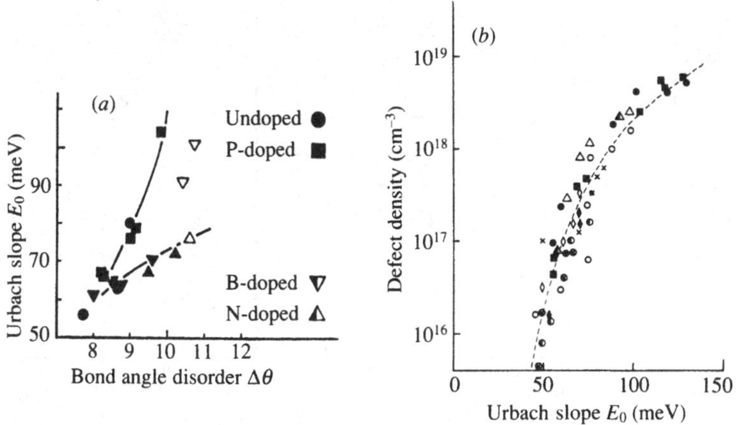

The frozen phonon approach relates the thermal and static disorder to the optical absorption spectra (Cody *et al.* 1981). The data in Fig. 3.20 shows a family of absorption curves obtained at different temperatures and for samples at different stages of thermal annealing, which changes the static disorder by releasing hydrogen from the a-Si:H network. The observation that a single family of curves describes the two types of disorder is evidence that both cause similar effects on the absorption. Increasing the disorder causes the band gap, E_G, to decrease and the slope of the Urbach edge, E_0, to increase. Fig. 3.23 shows the temperature dependence of both E_G and E_0 and that there is a common relation between these two parameters.

The temperature dependence of the optical band gap is given by

$$(\mathrm{d}E_G/\mathrm{d}T)_P = (\mathrm{d}E_G/\mathrm{d}T)_V - \beta/\kappa(\mathrm{d}E_G/\mathrm{d}T)_T \tag{3.31}$$

where β is the thermal expansion coefficient and κ is the compressibility. The first term on the right hand side is the explicit temperature dependence which originates from the electron–phonon interaction through the deformation potential, *B*, and the second term is due to the thermal expansion of the network. From the known values of β and κ, the second term contributes only about 2% to the measured temperature dependence and can be neglected. The electron–phonon

Fig. 3.23. Temperature dependence of (*a*) the slope, E_0, of the Urbach edge, and (*b*) the band gap energy; and (*c*) the correlation between the band gap and the band tail slope (Cody *et al.* 1981).

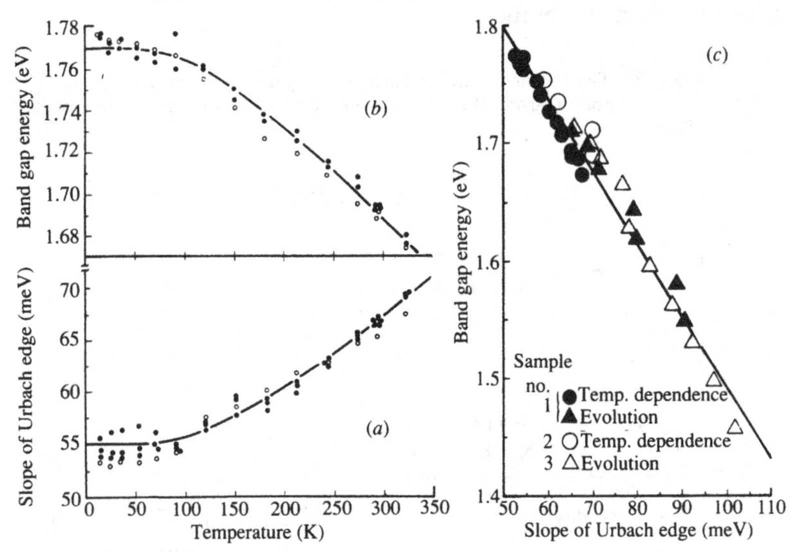

interaction on the temperature dependence of the band gap is written as

$$E_G(T) = E_G(0) - B[r(T) - r(0)] \tag{3.32}$$

where $r(T)$ is the mean square displacement of the atoms due to the thermal vibrations, and $r(0)$ represents the zero point motion.

According to the theory of the Urbach absorption edge in crystals, the slope E_0 is proportional to the thermal displacement of atoms $r(T)$. The frozen phonon model assumes that an amorphous semiconductor has an additional temperature independent term, r_D, representing the displacements which originate from the static disorder, so that

$$E_0(T, r_D) = K[r(T) + r_D] \tag{3.33}$$

where K is a constant. Similar reasoning suggests that the temperature dependence of the gap should also contain the extra term, r_D, and is therefore given by

$$E_G(T, r_D) = E_G(0, 0) - B[r(T) + r_D - r(0)] \tag{3.34}$$

Combining Eqs. (3.33) and (3.34) gives

$$E_G(T, r_D) = E_G(0, 0) - Br(0)[E_0(T, r_D)/E_0(0, 0) - 1] \tag{3.35}$$

Eq. (3.35) relates the broadening of the Urbach edge and the shift of the band gap, both of which originate from the thermal and bonding disorder.

The temperature dependence of the gap is next calculated, based on an Einstein model of the lattice vibrations, described by the Einstein temperature θ which is $\frac{3}{4}$ of the Debye temperature, θ_D,

$$r(T)/r(0) = (\hbar^2/2\theta kTM)[\coth(\theta/2T) - 1] \tag{3.36}$$

where M is the atomic mass. The gap is obtained by substituting Eq. (3.36) into Eq. (3.34). A fit to the temperature dependence data of Fig. 3.23 is obtained with a Debye temperature of 536 K, which is satisfactorily close to the value of 625 K in crystalline silicon. The same model gives for the temperature dependence of $E_0(T, r_D)$

$$E_0(T, r_D) = (\theta/2)[\coth(\theta/T) + r_D/r(0)] \tag{3.37}$$

where the constant K has been eliminated by requiring that at sufficiently high temperature $E_0 \rightarrow kT$. The solid line in Fig. 3.23(*a*) shows that the same choice of Debye temperature also fits the temperature dependence of the Urbach slope. Lastly the linear relation between E_G and E_0 predicted by Eq. (3.35) is observed in the data of Fig. 3.23(*c*). Of particular note is that exactly the same relation between E_G and E_0 applies to the data for thermal and bonding disorder.

The Urbach edge represents the joint density of states, but is dominated by the slope of the valence band, which has the wider band tail. Expression (3.37) for E_0 is therefore also an approximate description of the thermal broadening of the valence band tail. It is worth noting that the slope is quite strongly temperature-dependent above 200 K. This may have a significant impact on the analysis of dispersive hole transport, in which the temperature dependence of the slope is generally ignored.

The above model provides a consistent picture of the effects of static and thermal disorder on the band gap energy and the slope of the Urbach tail. One consequence of the results in Fig. 3.23 is that the band gap of a-Si:H apparently has no explicit dependence on the concentration of hydrogen, but only the implicit effect in which the hydrogen concentration influences the disorder. While this conclusion is evidently valid for the set of data shown, it cannot be generally true. Samples of a-Si:H can be made with a high hydrogen content, in which the band gap is substantially larger than these samples, but with an Urbach slope that is not correspondingly reduced. These alloying effects only become apparent with much larger hydrogen concentrations than is present under the optimum deposition conditions.

4 Defects and their electronic states

Electronic defects reduce the photosensitivity, suppress doping and impair the device performance of a-Si:H. Their high density in pure amorphous silicon makes this material of lesser interest and is the reason for the attention on the hydrogenated material, in which the defect density is greatly reduced. The remaining defects in a-Si:H control many electronic properties and are centrally involved in the substitutional doping process. The phenomena of metastability, which are described in Chapter 6, are caused by the defect reactions.

Defects are described by three general properties. First is the set of electronic energy levels of their different charge states. Those defects with states within the band gap are naturally of the greatest interest in understanding the electronic properties because of their role as traps and recombination centers. Second is the atomic structure and bonding of the defect, which determine the electronic states. Third are the defect reactions which describe how the defect density depends on the growth and on the treatment after growth. This chapter is mostly concerned with the first two properties, and the defect reactions are discussed in Chapters 5 and 6.

The defects are one of the more controversial aspects of a-Si:H. Many models have been proposed and there is active debate about the interpretation of several of the important experiments. This chapter concentrates on what is currently the most widely accepted model, but also compares the different possibilities and discusses the underlying reasons for disagreement.

4.1 Defects in amorphous semiconductors

The first question to address is the definition of a defect in an amorphous material. In a crystal any departure from the perfect crystalline lattice is a defect, which could be a point defect, such as a vacancy or interstitial, an extended defect, such as a dislocation or stacking fault, or an impurity. A different definition is required in an amorphous material because there is no perfect lattice. The inevitable disorder of the random network is an integral part of the amorphous material and it is not helpful to think of this as a collection of many defects. By analogy with the crystal one can define a defect as a departure from the ideal amorphous network which is a continuous random network in which all the bonds are satisfied. In ideal a-Si:H all the silicon is four-fold coordinated and all the hydrogen singly coordinated.

This approach immediately leads to the idea of a coordination defect in which an atom has a distinctly different bonding state from the ideal. All the electrons in the ideal network are paired in bonding or non-bonding states. If, however, the local coordination of an atom is one greater or less than the ideal, then the neutral state of the atom has an unpaired electron. The addition of an electron results in paired electrons but a net charge. This type of defect therefore has the distinguishing characteristic of either a paramagnetic spin or an electric charge, which sets it apart from the electronic states of the ideal network. The three-fold coordinated silicon dangling bond is one example.

Other types of defects are possible in an amorphous network, but have less well-defined characters. For example, a hydrogenated void in a-Si:H is certainly a defect in the sense that it has a local structure different from the rest of the amorphous network. However, from the atomic coordination point of view, it is difficult to distinguish a void from the ideal network. If the atoms around the void have a similar distribution of bonding disorder, then the electronic states probably fall within the ensemble of bulk states and are indistinguishable from those of the bulk. Localized states of particularly large binding energy may result when the local disorder is greater than in the ideal network. Unlike the coordination defect, this difference is quantitative rather than qualitative.

In one respect the defects of an amorphous solid are easier to deal with than those of a crystal. Any small deviation in the local structure of the defect in a crystal results in an identifiably different state, resulting in many possible defect structures. More than 50 point defects are known in crystalline silicon and there is probably an even larger diversity of extended defects. In the amorphous material, small differences in local structure which fall within the disorder of the ideal network cannot be resolved meaningfully. Thus one expects fewer separate classes of defects, but with their energy levels broadened out by the disorder, as illustrated schematically in Fig. 4.1.

Similar arguments apply to impurity states. Any impurity which is bonded with its optimum valency is expected to form a part of the ideal network and contribute only to the conduction and valence bands. Oxygen, nitrogen, carbon, and germanium all behave in this way, forming alloys with a-Si:H. Most of the phosphorus and boron atoms which are added as dopants, are in three-fold coordinated inactive sites with states that merge into the bands. The four-fold donor and acceptor states, on the other hand, are similar to the coordination defects, having either a spin or a charge and resulting in distinctive gap states.

4.1.1 *Lattice relaxation at defects*

Defect states in the gap influence the electronic properties because their electron occupancy is easily changed by doping or by trapping and excitation of carriers. The energy level of the defect depends on the local atomic structure and therefore on the bonding electrons of the neighboring atoms. An electron–phonon coupling between the bonding electrons and the gap state electrons causes the defect structure and energy to depend on the charge state. The lattice relaxation is generally described by a configurational coordinate diagram of the type shown in Fig. 4.2. This semiclassical model describes the combined effect of the electronic excitation and the local distortions of the bonding. In order to be specific, consider the trapping of an electron from the conduction band onto a defect. The potential energy of the upper state consists of two terms, one of which is the energy, E_C, of the electron at the bottom of the conduction band, and the second term is the additional energy of lattice vibrations. These are

Fig. 4.1. Comparison of the general form of the defect gap state distribution in amorphous and crystalline semiconductors.

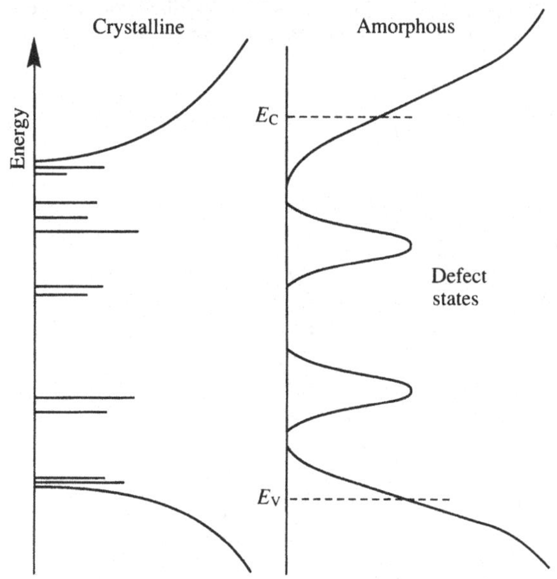

considered to be a single vibrational mode described by a configurational coordinate, q, which for simple harmonic oscillator approximation gives an energy,

$$E(q) = E_C + Aq^2 \tag{4.1}$$

A defines the strength of the network vibrations and the equilibrium state is at $q = 0$. The quantum mechanical solution to the harmonic oscillator gives the phonon energies $(n+\frac{1}{2})\hbar\omega$ with the frequency ω given by $(2A/m)^{\frac{1}{2}}$. The model is readily extended to include more than one vibrational mode.

The trapping of an electron by the defect releases an electronic energy E_T, and without any electron–phonon interaction, the energy of the defect state is E_T below the upper energy state. The interaction introduces an additional term which couples the energy to the configuration q. A first order, linear coupling results in an energy

$$E_{\text{defect}}(q) = E_C - E_T + Aq^2 - Bq \tag{4.2}$$

where B is the deformation potential.

The defect state has a potential minimum at $q = B/2A$, rather than at $q = 0$, and the minimum energy of the defect state is $E_C - E_T - W$, where,

$$W = B^2/4A \tag{4.3}$$

Fig. 4.2. Configurational coordinate diagram describing the capture and release of an electron from the conduction band into a defect state.

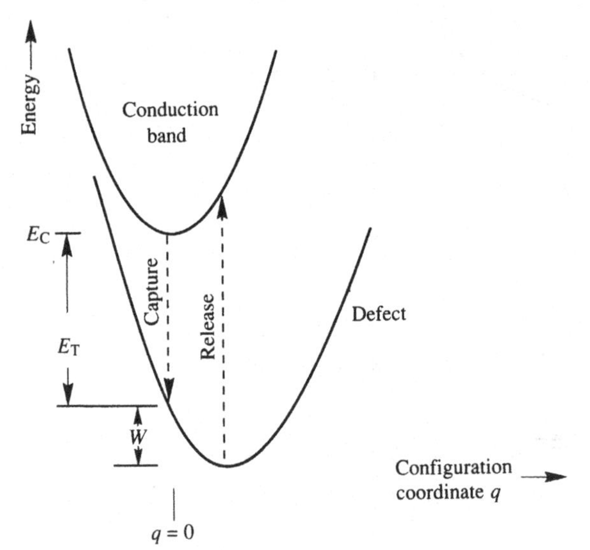

The potential well is indicated in Fig. 4.2. The electron–phonon coupling therefore causes the network to relax to a new equilibrium configuration at a lower energy.

The main consequence of the lattice relaxation is that different types of excitation between the lower and upper levels measure apparently different energies. Thermal excitation of the electron requires an energy $E_T + W$, which is the energy difference of the potential minima in Fig. 4.2. However, direct optical transitions between the two potential minima have low probability, because the involvement of many phonons is needed to change the configuration and the matrix element is small. Instead, the highest transition probability is that of the vertical transition in Fig. 4.2, in which there is no change in configuration, but which leaves the upper state vibrationally excited. The electron then relaxes to the equilibrium state by emitting phonons. The dominant transition from the upper to lower states is also vertical and leaves the lower state vibrationally excited. Thus, defect electronic transitions are characterized by three energies, E_T, $E_T + W$, and $E_T + 2W$, of which the first and third correspond to optical transitions, and the second to thermal excitation. The optical transitions are discussed further in Chapter 8.

4.1.2 *Correlation energies*

The Pauli exclusion principle states that each quantum level of the defect may be occupied by up to two electrons, so that a defect with a single level can exist in three charge states depending on the position of the Fermi energy, as illustrated in Fig. 4.3. For example, the dangling bond defect is neutral when singly occupied, and has a charge $+e$, 0 and $-e$ when occupied with zero, one or two electrons.

Fig. 4.3. Illustration of the one-electron and two-electron energy levels of a defect and the four possible transitions to the conduction and valence bands. The charge state is indicated when the Fermi energy lies in the different energy ranges. The defect is assumed to be neutral when singly occupied, with a positive correlation energy, U.

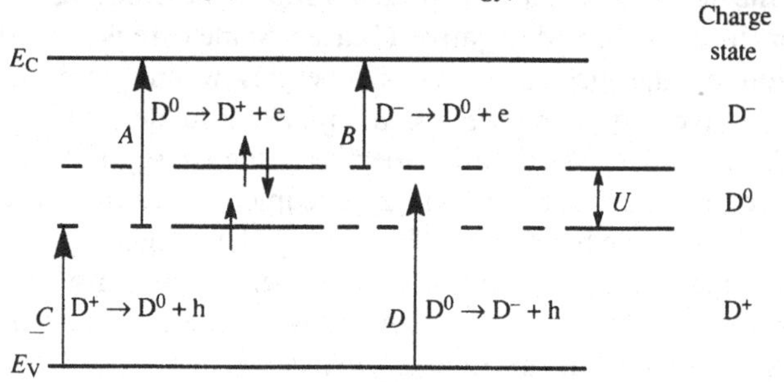

The electronic energies of the first and second electrons are not the same because of electron–electron interactions. The two electrons repel each other with a Coulomb interaction which is absent in the singly occupied state. The energy levels are split by the correlation energy,

$$U_C = e^2/4\pi\varepsilon\varepsilon_0 r \tag{4.4}$$

where r is the effective separation of the two electrons and so is roughly the localization length of the defect wavefunction. The energy is difficult to calculate exactly, because the wavefunction is not accurately known and because the state is generally too localized for the dielectric screening to be given correctly by the static dielectric constant. As a rough estimate of the Coulomb interaction, U_C is 130 meV, when $r =$ 10 Å and the static dielectric constant of 12 is assumed and is about 10 times larger if the high frequency dielectric constant is used instead. The correlation energy is therefore a significant fraction of the band gap energy.

Another contribution to the correlation energy arises from the lattice relaxation at the defect. The previous section showed that the addition of an electron to a localized state may cause a change in the bonding, which lowers the electronic energy by the amount W given in Eq. (4.3). The total correlation energy is a combination of the Coulomb and relaxation energies,

$$U = e^2/4\pi\varepsilon\varepsilon_0 r - W \tag{4.5}$$

The second term on the right hand side raises the interesting possibility of a defect that has a negative total correlation energy. This concept was introduced by Anderson (1975) and was first observed in the defect structure of chalcogenide glasses (Street and Mott 1975).

Positive and negative correlation energy defects have characteristically different electronic properties. Consider a coordination defect denoted D (for example, the dangling bond), which is neutral, D^0, when it contains one electron and its other two charge states are the empty D^+, and the doubly occupied D^-. Fig. 4.4 shows the ordering of the gap states for the positive and negative U cases. Some care is needed in the description of the electronic states' energies when more than one electron is involved. Here we use a one-electron description of the states; the 1e and 2e states are respectively the energy needed to add one electron to the empty and singly occupied states, measured with respect to the valence band edge. These energy levels therefore represent the actual measured energies of any one-electron transition. Note, however, that the 2e state is only defined when the 1e state is occupied and does not exist at an unoccupied defect.

The transfer of charge between two neutral defects is described by the reaction

$$2D^0 \leftrightarrows D^+ + D^- \tag{4.6}$$

According to the energy definitions given in Fig. 4.4, the energy of the left hand side of the reaction is $2E_{d1}$ and that of the right hand side, $2E_{d1}+U$, the difference being just the correlation energy U needed to put the two electrons on one site. If U is positive, the left hand side is the low energy state and all the defects are singly occupied (assuming that the temperature is such that $U \gg kT$). On the other hand, a negative U results in an equilibrium state which comprises an equal density of D^+ and D^- defects with no singly occupied states. One immediate observable consequence is that negative U coordination defects show no paramagnetism, whereas positive U defects generally do.

Negative U defects pin the Fermi energy. Since the upper 1e level is empty, and the lower 2e level is filled, E_F must obviously lie between these two energy levels. The pinning of the Fermi energy position is demonstrated by assuming that the N defects contain a variable density of electrons, n, where $0 < n < 2N$. The law of mass action (see Section 6.2.1) applied to Eq. (4.6) gives

$$N^2_{D^0} = N_{D^+} N_{D^-} \exp(U/kT) \tag{4.7}$$

where N_{D^0}, N_{D^+} and N_{D^-} are the densities of the different charge states. The density N_{D^0} is also given in terms of the Fermi energy by

$$N_{D^0} = N_{D^+} \exp[-(E_{d1}-E_F)/kT] \tag{4.8}$$

At sufficiently low temperatures such that $|U| \gg kT$, N_{D^0} is very small so that the approximations can be made

$$n = 2N_{D^-} \text{ and } N = N_{D^+} + N_{D^-} \tag{4.9}$$

Fig. 4.4. Ordering of the 1e and 2e energy levels for positive and negative correlation energy defects. The diagram indicates the charge state (D^0, D^+, D^-) when the Fermi energy is in the different energy regions. Note that there is never a large concentration of singly occupied negative U defects.

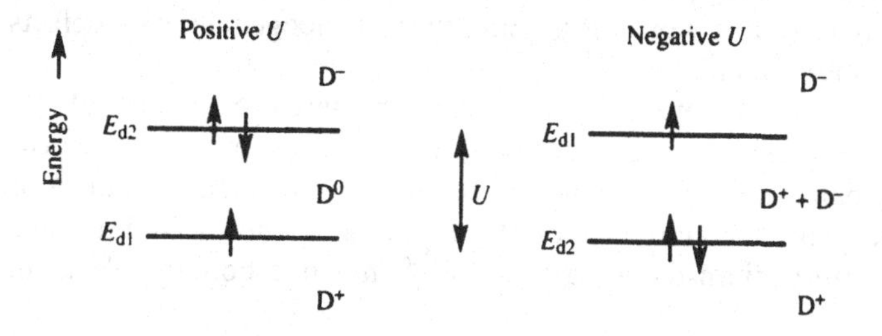

The solution for E_F is then

$$E_F = E_{d1} + U/2 - (kT/2)\ln(2N/n - 1) \tag{4.10}$$

Adler and Yoffa (1976) obtained the same result from a statistical mechanics analysis of an ensemble of defect states.

Under conditions of charge neutrality, $n = N$, the Fermi energy is midway between the 1e and 2e states and as n varies it departs very little from this position. For example, when $n = 0.1N$, the shift is less than $2kT$, so that the Fermi energy is strongly pinned. In contrast, when the defect has a positive U, the Fermi energy moves rapidly from the lower 1e level to the upper 2e level near $n = N$. The different behavior of E_F is shown in Fig. 4.5.

The properties of the two types of defect are summarized as follows: the singly occupied paramagnetic state is the lowest energy state of a positive U defect. The Fermi energy is unpinned and the defect usually has little lattice relaxation, so that the thermal and optical energies are nearly the same. Negative U defects are diamagnetic with low occupancy of the singly occupied state. The Fermi energy is strongly pinned between the energy levels and a large lattice relaxation is a necessary condition for the negative correlation energy.

4.1.3 *Valence alternation pairs – the example of selenium*

Lattice relaxation is an essential property of negative U defects because the Coulomb contribution to the correlation energy in Eq. (4.5) is always positive and must be offset by a large W. The electron–phonon coupling which allows the relaxation implies that there is a change in the configuration of the bonding electrons in the different charge states of the defect. In the limit of strong coupling, the change in electron occupancy may completely break one of the bonds. Coordination defects in chalcogenide glasses such as selenium have this property, which takes the form of valence alternation pairs (Kastner, Adler and Fritzsche 1976). Such defects are characterized by a different atomic coordination in the different charge states. The positive defects are overcoordinated and the negative defects are undercoordinated. Since the doping properties of a-Si:H have many similarities to valence alternation, it is worth describing these chalcogenide glass defects in a little more detail.

Selenium is normally two-fold coordinated, Se_2^0, and of its four valence p electrons, two are bonding electrons and the other two are lone pair states which make up the top of the valence band. A simple molecular orbital model of the bonding is shown in Fig. 4.6. The neutral selenium dangling bond, Se_1^0, has one bonding electron, and

Fig. 4.5. The dependence of the Fermi energy position on defect occupancy, n/N, for (*a*) positive and (*b*) negative U defects (Adler and Joffa 1976).

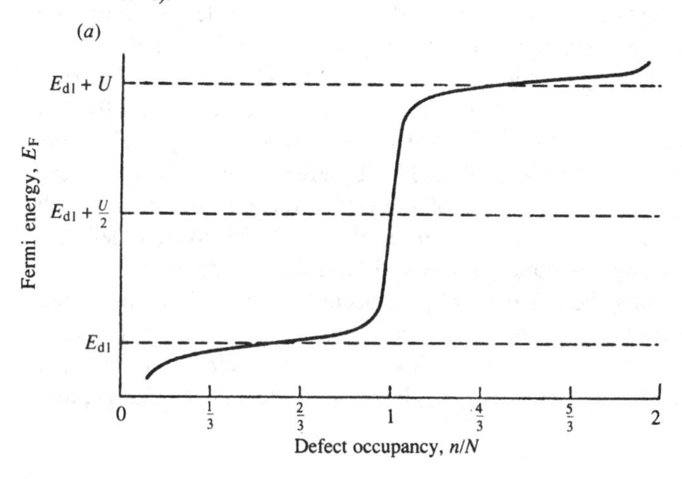

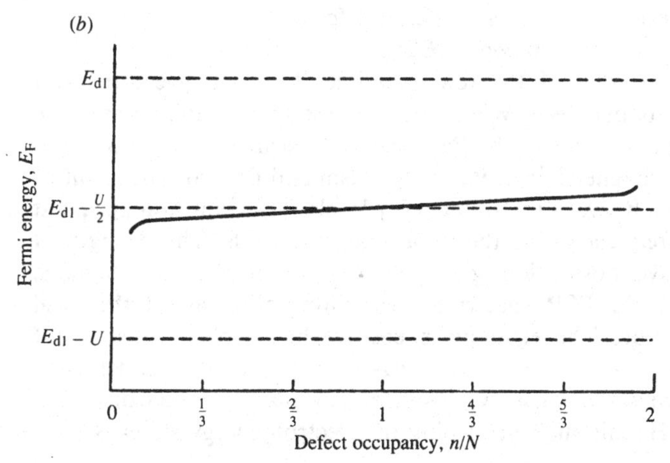

Fig. 4.6. A molecular orbital model for the electronic structure of different bonding states of selenium, illustrating the valence alternation defects, Se_1^- and Se_3^+.

	Se_2^0	Se_1^-	Se_3^+	p states
E_σ^*				Anti-bonding
E_{LP}	↑↓	↑↓ ↑↓		Lone pair
E_σ	↑↓	↑	↑↑↑	Bonding

three other electrons filling the two remaining lone pair states. The addition of an extra electron making Se_1^- fills the lone pair states, so that all the valence states are satisfied. The positive selenium dangling bond, Se_1^+, has two electrons missing from the lone pair states. Considerable energy is gained by forming a bond with a neighboring selenium atom so that the two remaining lone pair electrons form a strong bond. The energy gained by this bond is $2(E_{LP}-E_{\sigma})$ – the energies are defined in Fig. 4.6. The bonding interaction with the neighboring atom is the source of the lattice relaxation energy which causes the negative correlation energy. Note that the positive defect is no longer a dangling bond, but is a three-fold coordinated selenium atom, Se_3^+. Thus the negative U pair occurs as an undercoordinated negative state and an overcoordinated positive state. These defects are also the states predicted by the $8-N$ rule, which anticipates an increase in the coordination when there is a decrease in the valence charge, and vice versa.

4.2 Experimental measurements of defects

4.2.1 *Electron spin resonance (ESR)*

ESR is one of the few experiments which give structural information about defects. When a quantum state is occupied by a single electron, the two states of the Pauli pair are normally degenerate, but are split by a magnetic field. Paramagnetism and ESR are the result of the transitions between the split energy levels. The transition occurs at microwave frequencies for the usual magnetic fields. The strength of the microwave absorption gives the density of the paramagnetic electrons and the ESR spectrum gives information about the local bonding structure. The technique is sensitive, being able to detect about 10^{11} spins and is well suited for the measurement of low defect densities.

When there is a low density of isolated paramagnetic electrons, as in a-Si:H, the Hamiltonian describing the electron energy states is

$$H = \mu_B \mathbf{H}_0 \cdot \mathbf{g} \cdot \mathbf{S} + \sum_i I_i \cdot A_i \cdot S \qquad (4.11)$$

The first term on the right hand side is the Zeeman interaction energy, where μ_B is the Bohr magneton, $\mathbf{H}_0$ is the externally applied magnetic field, $\mathbf{g}$ is the gyromagnetic tensor, and $\mathbf{S}$ is the electron spin. The second term is the nuclear hyperfine interaction in which the magnetization of adjacent nuclei influences the magnetic field at the electron and changes the resonance energy. In Eq. (4.11), $\boldsymbol{I}_i$ and $\mathbf{A}_i$ are the nuclear spin and hyperfine tensors and the sum is over all the contributing nuclei near the electron. The structural information about the local configuration of the defect is contained in the interaction terms **g** and $\mathbf{A}_i$. The hyperfine interaction is usually less than 10% of the Zeeman splitting, which therefore dominates the resonance energy. A schematic diagram of the energy level splitting is shown in Fig. 4.7.

The Zeeman term is considered first and the hyperfine interaction is discussed in Section 4.2.2. The free electron g-value of $g_0 = 2.0023$ is shifted when the electron is surrounded by material, because of the spin–orbit coupling to the other electron states. The shift is

$$[\Delta g]_{ij} = g_{ij} - g_0 = -2 \sum_{n \neq d} \frac{\langle d|L_i|n\rangle \langle n|L_j|d\rangle}{E_n - E_d} \tag{4.12}$$

where d refers to the ground state of the paramagnetic defect and the sum is over all the excited electronic states of the material. The L_i are the components of the orbital momentum operator. The g-shift, $\Delta g = g - g_0$ contains detailed information about the local bonding structure of the defect, particularly the symmetry of the wavefunction, but unfortunately much of this information is lost by the orientational averaging and bonding disorder in an amorphous material.

The example of the silicon dangling bond is illustrated in Fig. 4.8. The wavefunction has approximate axial symmetry, so that the g-tensor contains terms from two interactions, corresponding to the direction along the bond, $\Delta g_{\parallel}$, and perpendicular to the bond $\Delta g_{\perp}$. The matrix elements for the spin–orbit interaction in the axial direction are zero by symmetry, so that $\Delta g_{\parallel} \sim 0$, and the main contribution to the g-shift is from $\Delta g_{\perp}$. The sign of Δg is given by the energy denominator in Eq. (4.12). The spin–orbit coupling to the p-like valence band is greater than to the s-like conduction band and so the coupling is dominated by states for which $E_n < E_d$, resulting in a positive g-shift. Thus defect states lying nearest to the valence band tend to have the largest g-shifts.

A single dangling bond therefore has a small Δg when oriented along

Fig. 4.7. Illustration of the Zeeman and nuclear hyperfine magnetic interactions.

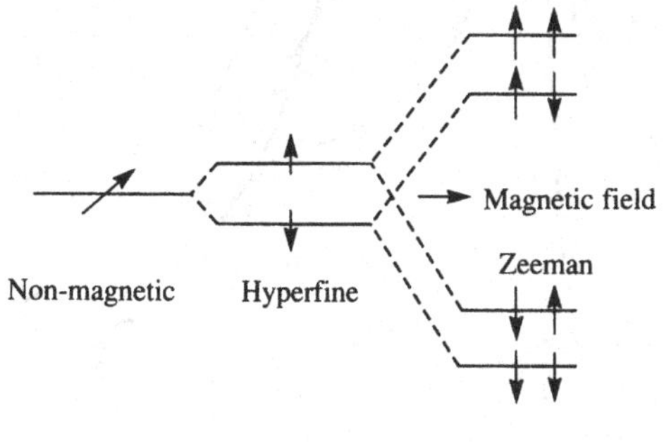

the magnetic field, but a larger positive shift when perpendicular. Any arbitrary orientation of the defect results in a g-shift which is intermediate between $\Delta g_{\parallel}$ and $\Delta g_{\perp}$. An ensemble of dangling bonds with random orientation exhibits a 'powder pattern' spectrum of g-values. The ideal powder pattern for the axial defect has a peak at the

Fig. 4.8. Principal components and angular dependence of the g-shift for a silicon dangling bond.

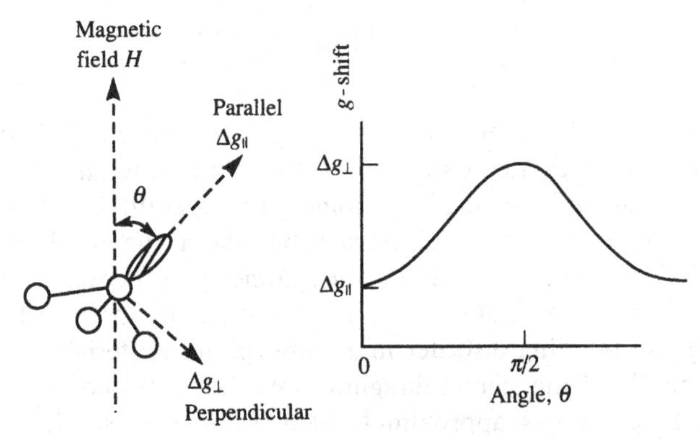

Fig. 4.9. Calculated powder pattern ESR absorption spectra for an axial defect such as the dangling bond, showing the effect of disorder broadening (increasing values of $g_d/\Delta g$). The spectra are calculated using the g-values obtained from the crystalline Si–SiO_2 interface (Street and Biegelsen 1984).

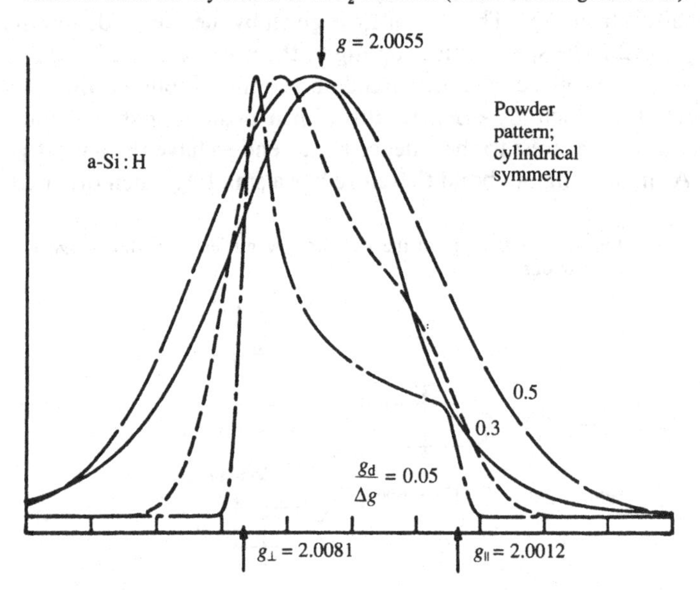

$\Delta g_\perp$ position and a shoulder at $\Delta g_\parallel$, but when an additional disorder broadening is added, only a slightly asymmetrical single peak remains. Fig. 4.9 illustrates the powder pattern spectrum for an axial defect, together with the effects of additional disorder broadening. The orientational averaging replaces the g-tensor by a scalar g-value corresponding to the peak of the resonance.

All forms of undoped amorphous silicon, including crystalline silicon made amorphous by ion bombardment, pure amorphous silicon deposited by evaporation or sputtering, as well as all forms of a-Si:H have a paramagnetic defect with its resonance at a g-value of 2.0055. Fig. 4.10 gives an example of a typical ESR spectrum, which is the first derivative of the absorption spectrum, measured by keeping the microwave frequency fixed and varying the magnetic field. The g-value corresponds to the zero crossing and the line width, 7.5 G, at the X-band frequency of 9 GHz, is given by the separation of the upper and lower wings. There is experimental confirmation that the line width is given by the g-value distribution rather than by a homogeneous broadening mechanism from a small relaxation time.

When the 2.0055 line was first observed (Brodsky and Title 1969), the suggested defect was the silicon dangling bond, although in the context of a microcrystalline structure model. The dangling bond remains the preferred interpretation, although not the only possibility, as is discussed in Section 4.3. Any model must, of course, be consistent with the measured g-value distribution. The powder pattern average for the dangling bond can be predicted from the values of $\Delta g_\perp$ and $\Delta g_\parallel$ obtained from the crystalline Si–SiO_2 interface dangling bond g-values (Caplan, Poindexter, Deal and Razouk 1979). These values are used to model the powder pattern in Fig. 4.9 and predict a broad absorption peak very close to $g = 2.0055$ and with about the correct width (Biegelsen 1981). The comparison is good evidence for the dangling bond interpretation.

Fig. 4.10. Derivative ESR spectrum of the paramagnetic states in undoped a-Si:H, showing the $g = 2.0055$ defect, with line width 7.5 G.

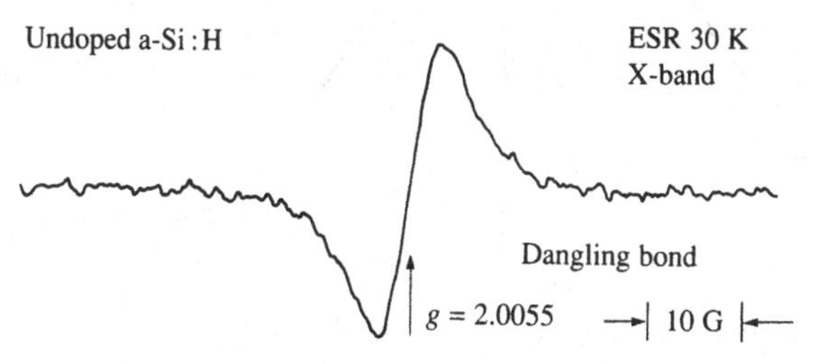

The spin density of the 2.0055 defect varies greatly with the deposition conditions of a-Si:H and was used early on as a measure of film quality in optimizing the film growth conditions. Fig. 4.11 shows that the lowest spin density results from low rf plasma power, undiluted silane, and a growth temperature of 200–300 °C, these, of course, being the conditions for CVD-like growth (see Section 2.1.1). Samples with a columnar microstructure, deposited at high power and with diluted silane, typically have spin densities of about 10^{18} cm^{-3}. Samples deposited at around room temperature, but without columnar structure have a similarly high defect density. Annealing to 250 °C greatly reduces the spin density of the room temperature material, but not that of the columnar material. In the optimum films, the spin density is below 10^{16} cm^{-3} and is close to the limit of the ESR sensitivity, at least for films of a few microns thickness. An accurate measure of the bulk defect density in these films is complicated by paramagnetic surface defects.

The spin density of the $g = 2.0055$ defects decreases in doped a-Si:H. However, the reduction is not because the defect density decreases – in fact, it increases rapidly with doping (see Section 5.2.1). Instead the decrease in spin density is due to a change in the charge state of the defects. The movement of the Fermi energy by doping causes the defects to be doubly occupied by electrons (D^-) in n-type material, and empty (D^+) in p-type, as indicated by Fig. 4.3. Neither of these states

Fig. 4.11. Dependence of the ESR spin density in undoped a-Si:H on deposition parameters, specifically rf power, silane dilution in argon and substrate temperature (Knights 1979).

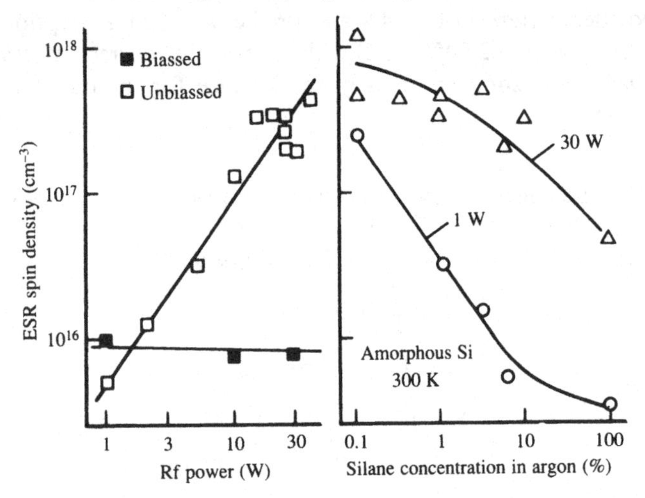

is paramagnetic. The equilibrium spin density gives a correct measure of the defect density only in undoped a-Si:H.

4.2.2 *ESR hyperfine interactions*

The nuclear hyperfine interaction splits the paramagnetic states of an electron when it is close to a nucleus with a magnetic moment. For a random orientation of spins and nuclei, the tensor quantities in Eq. (4.11) are replaced by scalar distributions, and the resonance magnetic field is shifted from the Zeeman field H_0 by

$$H = H_0 - \sum m_i A_i/g \tag{4.13}$$

where the m_i are the nuclear spin states of the quantum level I. When a single nucleus dominates the hyperfine splitting, there is a separate resonance line for each m_i and the total number of lines is $(2I+1)$. Thus there are two hyperfine lines for a spin $\frac{1}{2}$ atom and three for spin 1, etc. In undoped a-Si:H, the atoms of interest are obviously silicon and hydrogen. Hydrogen has a nuclear spin of $\frac{1}{2}$ but deuterium has spin 1. ^{28}Si is the dominant isotope of silicon and also has zero spin, but ^{29}Si has spin $\frac{1}{2}$ and a natural abundancy of 4.7%. The 2.0055 has no obvious hyperfine splitting, which is consistent with the central nucleus being a silicon atom and implies that the overlap of the wavefunction with neighboring hydrogen atoms cannot be large. However, a ^{29}Si hyperfine splitting is observed in careful measurements and is greatly enhanced in material which is isotopically enriched with ^{29}Si (Biegelsen and Stutzmann 1986). Examples of the ESR spectra for both types of material are shown in Fig. 4.12. A pair of hyperfine lines is observed with a splitting of about 70 G and with the expected dependence on the ^{29}Si abundance.

The hyperfine ESR data are valuable because they are the best measure of the electron wavefunction at the defect. The form of the hyperfine spectrum, which contains two broad lines, implies that the defect state is highly localized on a single silicon atom. Further analysis makes use of an approach that is successful in analyzing the hyperfine interaction in crystalline materials and describes the defect wavefunction, Ψ, by a linear combination of atomic orbitals. The wavefunction of a single silicon valence electron is written in terms of s and p orbitals as

$$\Psi = c[(\alpha s) + (\beta p)]: \qquad \alpha^2 + \beta^2 = 1 \tag{4.14}$$

where c is the extent to which the wavefunction is localized at a single site.

The hyperfine interaction for a dilute system of localized states contains the direct overlap of the electron wavefunction with the nucleus which depends only on the s-component of the wavefunction, and a dipole interaction term which depends only on the p-component. The total hyperfine splitting is,

$$\Delta H = c^2\alpha^2\Delta H(\mathrm{s}) + c^2\beta^2(3\cos^2\theta - 1)\,\Delta H(\mathrm{p}) \qquad (4.15)$$

where θ is the angle between the magnetic field and the p orbital direction. $\Delta H(\mathrm{s})$ and $\Delta H(\mathrm{p})$ are known atomic hyperfine constants. The

Fig. 4.12. ESR spectra of (*a*) normal and (*b*) ^{29}Si-enriched a-Si:H, showing the silicon hyperfine interaction in the $g = 2.0055$ defect resonance (Biegelsen and Stutzmann 1986).

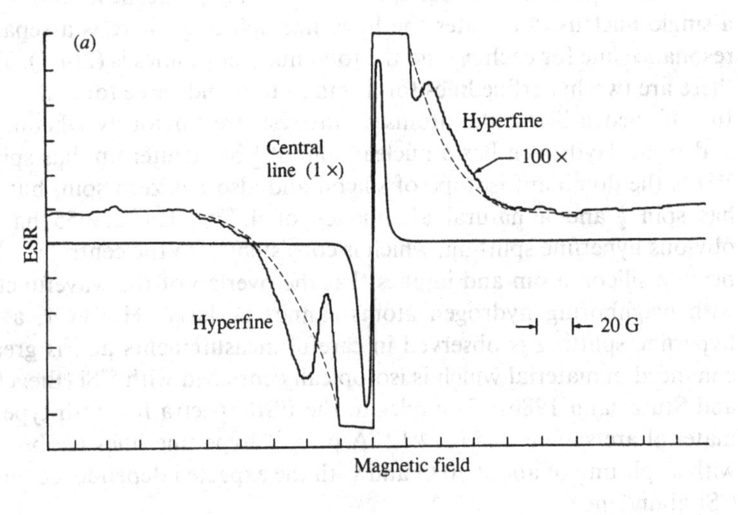

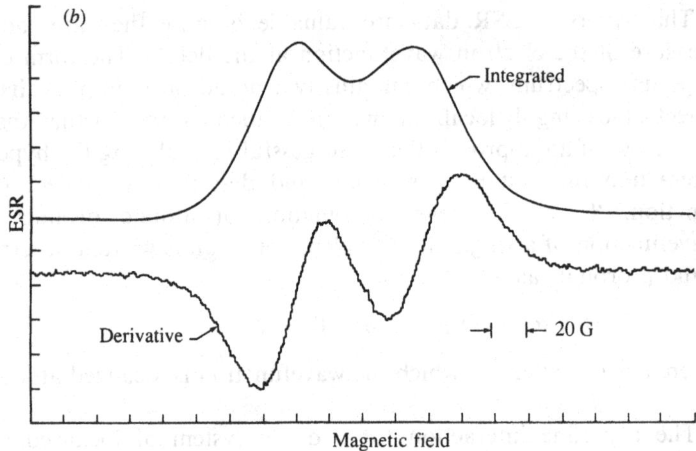

sum over different orientations in a random material results in no contribution to the hyperfine splitting from the second term on the right hand side of Eq. (4.15), because the angular-dependent factor averages to zero. The splitting is therefore given only by the isotropic s-component of the interaction, while the anisotropic p-component contributes to the broadening of the hyperfine lines. Of course, any site-to-site disorder of the s orbital also causes a broadening of the lines, as does the g-value distribution of the Zeeman term.

Based on this model, the hyperfine spectra for the defect can be related to the s- and p- components of the wavefunction (Stutzmann and Biegelsen, 1988). Table 4.1 shows the results and compares them with the silicon defects which are known to be of the dangling bond type in other materials. An sp^3 dangling bond has $\frac{1}{4}$ s-like and $\frac{3}{4}$ p-like character, so should have $\alpha = 0.5$ and $\beta = 0.87$. In practice, all the defects in Table 4.1 have a slightly smaller s-character and larger p-character and also incomplete localization, compared to the sp^3 dangling bond model.

Both the g-value distribution and the hyperfine splittings of the $g = 2.0055$ defect are consistent with the expected properties of dangling bonds. Consistency, however, does not constitute proof of the structure, and other possibilities have been proposed, which are discussed below. The ESR parameters do provide quantitative constraints that must be met by alternative models and, at present, are the only specific experimental information that we have about the defect wavefunctions.

Although the silicon hyperfine splitting of the 2.0055 defect is large, the hydrogen hyperfine interaction is too small to be observed as a splitting of the central line (Stutzmann and Biegelsen 1986). The small effect is puzzling in view of the large hydrogen concentration, but may occur because the defect wavefunction is strongly localized on a single silicon atom. A weak hyperfine interaction is observed by the technique of electron nuclear double resonance (ENDOR). In this experiment, ESR is measured under conditions of partial saturation by the

Table 4.1. *Wavefunction coefficients in Eq. (4.14) for silicon dangling bonds in different environments*

Dangling bond	α	β	c
sp^3	0.5	0.87	1
a-Si:H	0.32 ± 0.04	0.95 ± 0.01	0.8 ± 0.1
Si/SiO_2	0.35	0.94	0.89
SiH_3 in Xe	0.39	0.92	0.97

microwave power, so that the signal strength depends on the spin-lattice relaxation process. An additional relaxation channel is provided for the electron when a neighboring nuclear spin is brought into resonance, and the absorption signal increases. The effect therefore provides a measure of the coupling between the electron spin and the nucleus. Fig. 4.13 shows an ENDOR spectrum obtained by varying the nuclear resonance frequency. The sharp peaks at 2.8, 4.4 and 13.9 MHz are all due to hydrogen, the lower frequency peaks being the odd harmonics of the 13.9 MHz resonance. The origin of the ENDOR interaction is either the overlap of the defect wavefunction with a nearby hydrogen nucleus or a dipolar coupling. In principle, more details about the electronic wavefunction can be obtained from the shape of the peaks in Fig. 4.13, but, in practice, microwave coherence effects tend to dominate the line shape so that all that can be concluded is that the defect is sufficiently localized that the overlap with hydrogen is very small.

The final ESR property to discuss is the spin-lattice relaxation time T_1, which is the time taken for the spin excitation to return to the ground state, dissipating its energy into the thermal bath of network vibrations. Fig. 4.14 shows the temperature dependence of T_1 for the $g = 2.0055$

Fig. 4.13. ENDOR spectrum of a-Si:H showing the hydrogen signal at 13.9 MHz. The lines at 4.5 and 2.8 MHz are overtones (Stutzmann and Biegelsen 1986).

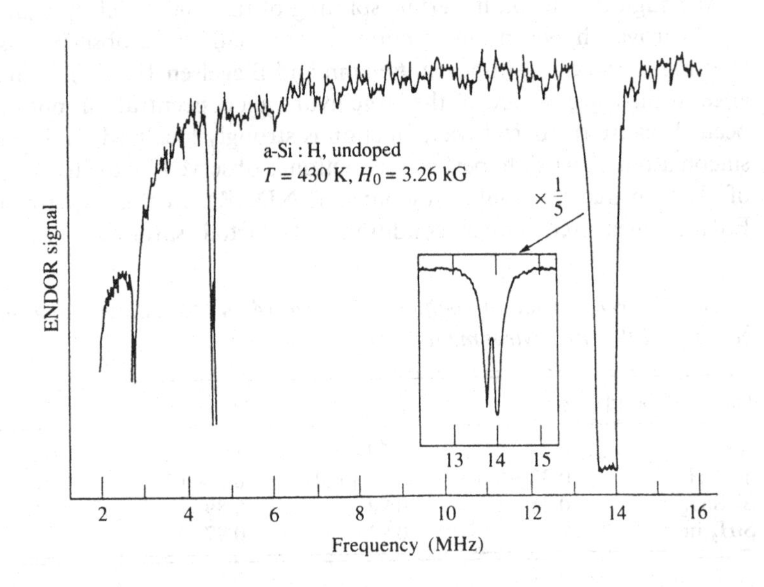

defect and the result of annealing the sample to remove the hydrogen (Stutzmann and Biegelsen 1983). Above 50 K there is no effect of annealing and T_1 varies with temperature as T^{-2}, but at lower temperature there is a smaller T_1 in the annealed samples.

Spin relaxation occurs by transferring the excitation energy to lattice vibrations. The majority of network phonons have an energy of order $k\theta_D$, where θ_D is the Debye temperature of about 600 K; $k\theta_D$ is much greater than the Zeeman energy, $k\theta_Z$, since $\theta_Z \sim 0.5$ K. The most efficient relaxation mechanism is usually a two-phonon Raman process in which one phonon of high energy is absorbed and another of slightly lower energy is emitted, the difference being $k\theta_Z$. The predicted temperature dependence of this mechanism has the form

$$T_1^{-1}(T) \approx \text{const. } T^2 \int n(E)\, n(E - k\theta_Z)\, \mathrm{d}E \qquad (4.16)$$

where the integral is the average probability of thermal excitations of the phonons, and the T^2 term is from the matrix elements of the

Fig. 4.14. The temperature dependence of the spin-lattice relaxation time T_1 of the $g = 2.0055$ defect and its variation with annealing temperature, T_A (Stutzmann and Biegelsen 1983).

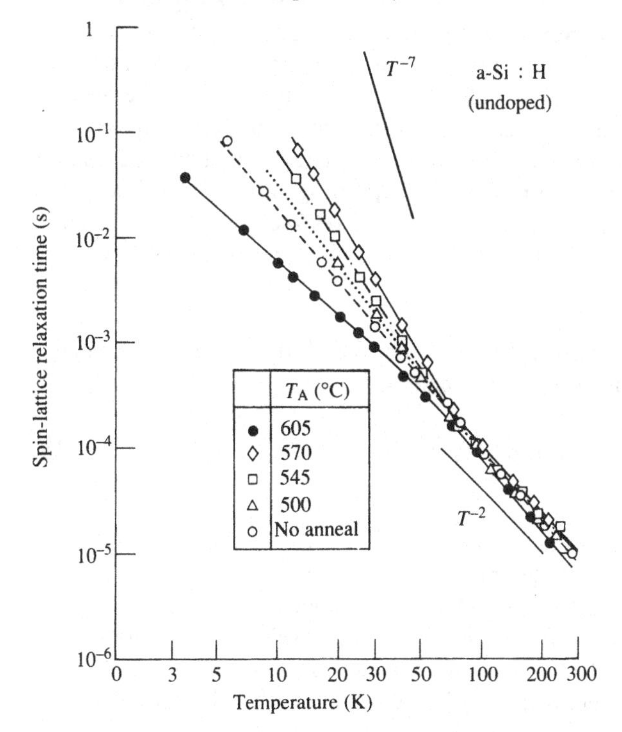

transition. All the phonons can be excited at high temperatures, so that the relaxation time varies as T^{-2}, but at low temperature, the usual shape of the phonon density of states results in a much faster dependence of T^{-7}. The T^{-2} behavior in the unannealed a-Si:H samples indicates that the two-phonon mechanism operates, but this continues down to 4 K rather than changing to the T^{-7} mechanism. The result implies that there is an additional relaxation channel at low energy and is interpreted in terms of the same tunneling modes that are seen in the low temperature specific heat (see Section 2.2.3). A model which assumes that the density of these modes is nearly independent of energy, as is suggested by the specific heat data, predicts a value of T_1 which varies as T^{-2}, at least for a relaxation mechanism that involves one phonon and one tunneling mode (Stutzmann and Biegelsen 1983). The changes that take place with annealing therefore seem to imply a reduction in the tunneling mode density as hydrogen is evolved from the material. Conceivably this could be the result of the network reconstruction that takes place as hydrogen leaves, but as yet there is no detailed structural model for the tunneling modes.

4.2.3 *Defect level spectroscopy – thermal emission energies*

In measurements of the defect energies, it is essential to distinguish between the thermal emission and optical transition energies, to account properly for lattice relaxation effects (see Section 4.1.1). Thermal emission measurements are generally based on electrical transport and in most cases involve the effects of a trapped space charge, Q. The two main ways of extracting information about the density of states distribution are from the release time τ_R of the charge from the traps

$$\tau_R = \omega_0^{-1} \exp(E_T/kT) \tag{4.17}$$

where E_T is the trap depth, or from the shift of the Fermi energy resulting from the excess space charge,

$$Q = \int_{E_F}^{E_F+\Delta E_F} N(E)\mathrm{d}E; \ \Delta E_F \simeq Q/N(E_F) \tag{4.18}$$

(1) *Capacitance measurements*

The first of these approaches is used in the technique of deep level transient spectroscopy (DLTS), which is perhaps the most common experiment for measuring deep levels in crystalline semiconductors (Lang 1974). The DLTS experiment is the measurement of the transient capacitance of a Schottky contact to the sample and is best understood by considering first the steady state capacitance. Fig. 4.15 illustrates the depletion layer of a Schottky barrier on an n-type semiconductor under zero and reverse bias. The depletion layer of width $W(V)$ constitutes an insulating layer bounded by the metal contact on one side and the conducting semiconductor on the other, and so behaves as a parallel plate capacitor, with capacitance $\varepsilon\varepsilon_0/W$ per unit area. The width of the depletion layer is given by Poisson's equation.

$$d^2V/dx^2 = -\rho(x)/\varepsilon\varepsilon_0 \tag{4.19}$$

where $\rho(x)$ is the charge density at position x.

The solution to Poisson's equation for the depletion layer is discussed further in Chapter 9. The hatched region in Fig. 4.15 represents the gap states which change their charge state in depletion and so contribute to $\rho(x)$. When there is a continuous distribution of gap states, $\rho(x)$ is a spatially varying quantity. For the simpler case of a shallow donor-like level, the space charge equals the donor density N_D and the solution for the dependence of capacitance on applied bias, V_A, is (see Section 9.1.1)

$$1/C^2 = 2(V_A + V_B)/eN_D\varepsilon\varepsilon_0 A^2 \tag{4.20}$$

where A is the sample area and the zero bias Schottky barrier potential is V_B. A plot of $1/C^2$ versus V_A is predicted to be a line with a slope given by the density of states. This method is commonly used to obtain the doping density in crystalline semiconductors.

There are several complications in using this technique for a-Si:H, first of which is the frequency dependence. The capacitance is measured by the response to a small alternating applied electric field. The depletion layer capacitance is obtained only when the free carriers within the bulk of the semiconductor can respond at the frequency of the applied field, ω_m. The appropriate condition is that ω_m must be smaller than the inverse of the dielectric relaxation time,

$$\omega_m < 1/\varepsilon\varepsilon_0\sigma \tag{4.21}$$

where σ is the bulk conductivity. The room temperature conductivity of undoped a-Si:H is about $10^{-9}\ \Omega^{-1}\ \mathrm{cm}^{-1}$, so that the upper limit on the frequency is approximately 10 Hz. Such low frequency measurements are possible, but are complicated by the high impedence of the junction, $1/\omega_m C$, which accentuates any effects of resistive loss, for example, due to leakage at the contacts. The bulk conductivity increases rapidly at elevated temperatures, so that higher frequencies may be used.

The capacitance technique is more readily applicable to doped a-Si:H, where σ is larger and higher frequencies can be used. Fig. 4.16 shows an example of capacitance–voltage data for n-type a-Si:H (Hack, Street and Shur 1987). The $1/C^2$ plot does not obey Eq. (4.20), so that no single value of the density of states can be extracted. The curvature is caused by the presence of deep defect levels in addition to the shallow donors. The relative contributions of the different states to the space charge change with applied voltage, with the result that the $1/C^2$ slope is dominated by the shallow states at low V_A, with the deep states contributing increasingly at high V_A. There is not enough information to invert the capacitance–voltage data to obtain the density of states distribution. However, the capacitance–voltage data can be fitted to an assumed distribution. This is done for the data in Fig. 4.16, assuming a broad defect level 0.8 eV below the conduction band edge and the band tail distribution of Fig. 3.16.

The analysis of capacitance–voltage data is made difficult by the lack of specific energy information. The DLTS experiment provides this information and consequently is a better technique to obtain the density of states. DLTS distinguishes between deep and shallow states

Fig. 4.15. Illustration of the depletion layer at zero bias and at a reverse bias of V_A. The shaded region represents the gap states depleted of charge by the bias.

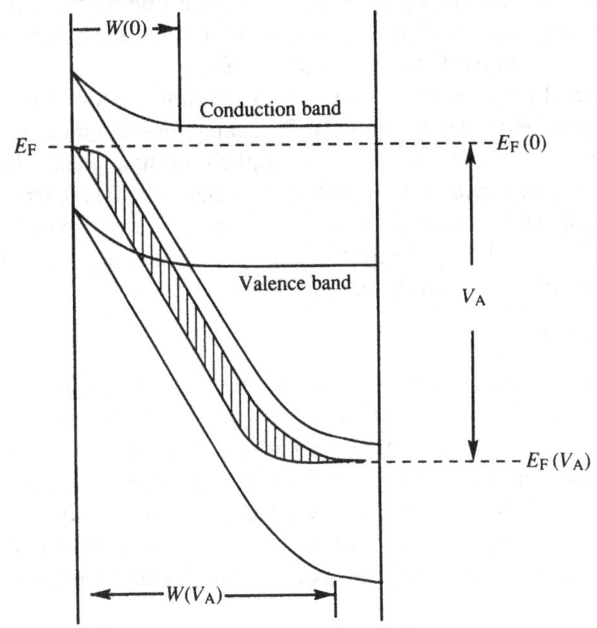

by the time-dependent release of the electrons, based on Eq. (4.17). A forward bias is applied to the Schottky barrier to collapse the depletion layer and fill all the traps. When a reverse bias is subsequently applied, the depletion layer is initially wide, but decreases with time to its steady state value as carriers are released from traps. The change in the depletion layer at time t reflects the number of traps of binding energy $kT\ln(\omega_0 t)$ within the depletion layer. The density of states is therefore derived from the transient capacitance. The calculation of $N(E)$ from the data is straightforward, although somewhat involved, as it must account for the distribution of states and the voltage dependence of the space charge layer (Lang, Cohen and Harbison 1982a). Deep states are detected only down to the middle of the gap, because deeper states are filled by electrons from the valence band faster than they are excited to the conduction band.

DLTS experiments are performed either by measuring the temperature dependence of the capacitance at a fixed time after the application of the reverse bias or the time dependence is obtained at constant temperature (Johnson 1983). The result is an equivalent measure of $N(E)$, but the exponential dependence of the release time on energy in Eq. (4.17) means that a wide range of times are needed to measure the full energy range. Figs. 4.17 and 4.18 show examples of the density of states distribution for n-type a-Si:H obtained using both techniques. A broad defect band 0.8–0.9 eV below E_C is observed.

The energy scale for the deep traps, given by Eq. (4.17), depends on having the correct value for the attempt-to-escape frequency ω_0, and here there has been considerable controversy. Theoretical estimates of

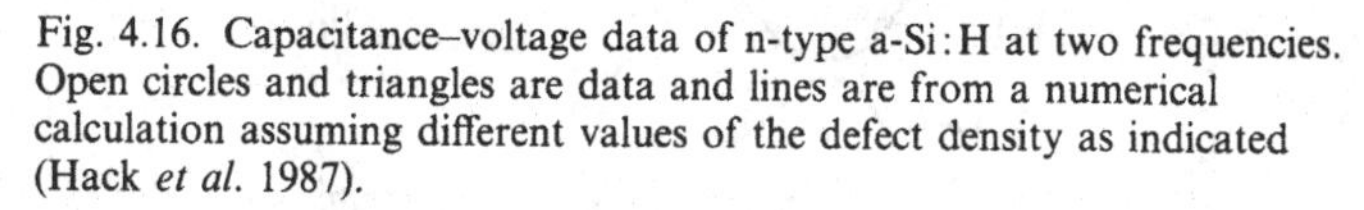
Fig. 4.16. Capacitance–voltage data of n-type a-Si:H at two frequencies. Open circles and triangles are data and lines are from a numerical calculation assuming different values of the defect density as indicated (Hack *et al.* 1987).

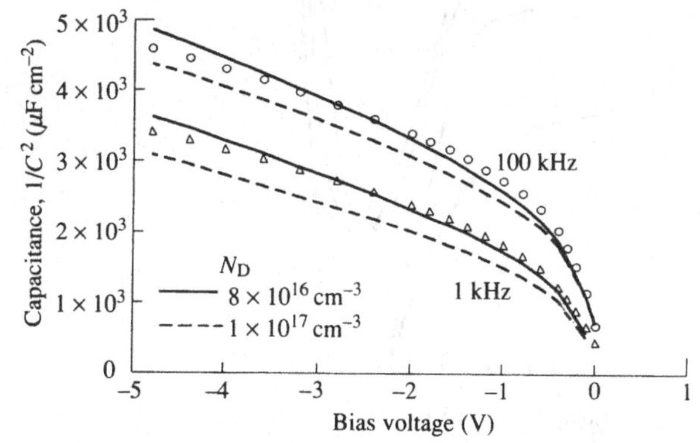

ω_0 are 10^{-12}–10^{-13} s^{-1}, approximately equal to the phonon frequency (see Section 8.1.3). The magnitude of ω_0 is measured from the DLTS experiment by varying the emission time window. Lang *et al.* (1982a) find that this approach does indeed give $\omega_0 \sim 10^{13}$ s^{-1}. However, a different method of extracting ω_0 used by Okushi *et al.* (1982) obtained a value of only 10^8 s^{-1}. The resulting difference in energy is almost a factor of 2 for an emission time of 10^{-2}–1 s, so that quite different density of states distributions are deduced from the same experimental data. Okushi's method has been criticized and it now appears that the larger value is correct (Lang *et al.* 1984). As further evidence, the value of ω_0 is related by detailed balance to the capture cross-section, σ_c, for trapping from the conduction band into the deep level,

$$\omega_0 = kTN(E_C)\, v_C \sigma_c \tag{4.22}$$

where v_C is the free carrier velocity of the mobile electrons. The measured cross-sections for deep trapping yield a value for $v_C \sigma_c$ of

Fig. 4.17. Some examples of the density of states distribution of n-type a-Si:H obtained from DLTS. Samples with the largest phosphorus concentration have the greatest density of states at the middle of the gap. The Fermi energies of the different samples are indicated (Lang *et al.* 1982a).

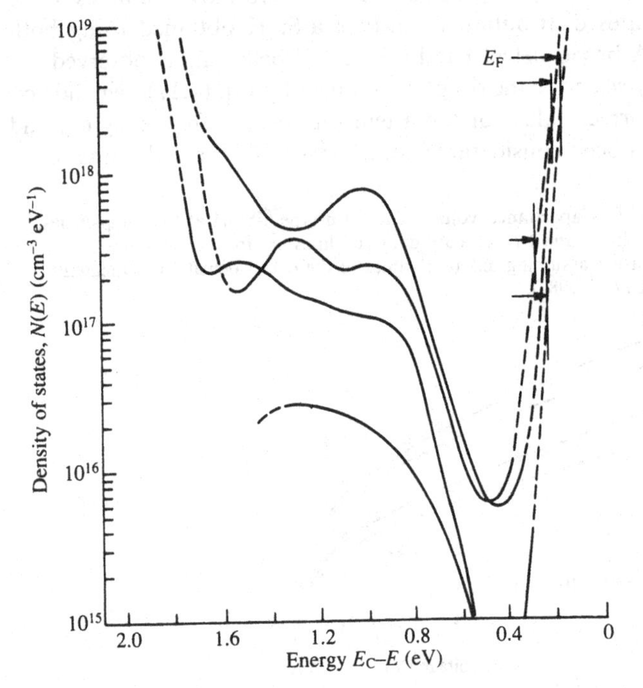

about 4×10^{-8} cm^3 s^{-1} (Section 8.4.2). Taken with a density of states at E_C of $(2\text{–}4) \times 10^{21}$ cm^{-3} eV^{-1}, this results in ω_0 between 10^{12} and 10^{13} s^{-1}. The evidence of DLTS is therefore that the main deep trap in n-type a-Si:H is 0.8–0.9 eV below the conduction band and is a peak with a width of about 0.1–0.15 eV. The DLTS data also show that there is a deep minimum in the density of states between the trap and the conduction band edge.

It is obviously of interest to know whether the deep trap observed in DLTS is the same as the $g = 2.0055$ ESR defect and two different experiments have confirmed this. Both use the DLTS approach, but measure ESR in combination with the capacitance (Cohen, Harbison and Wecht 1982, Johnson and Biegelsen 1985). When a reverse bias is applied to n-type a-Si:H there is a transient increase in the spin density of the 2.0055 resonance. The temperature dependence of the transient spin signal is identical to that found in the usual DLTS experiment. The interpretation is that a 2.0055 spin occurs whenever an electron is released from the deep trap, so that the trap can be identified with the doubly occupied charge state of the defect. The other experiment made

Fig. 4.18. Density of states in n-type a-Si:H measured by isothermal DLTS. The inset shows the data and the numerical fit (Johnson 1983).

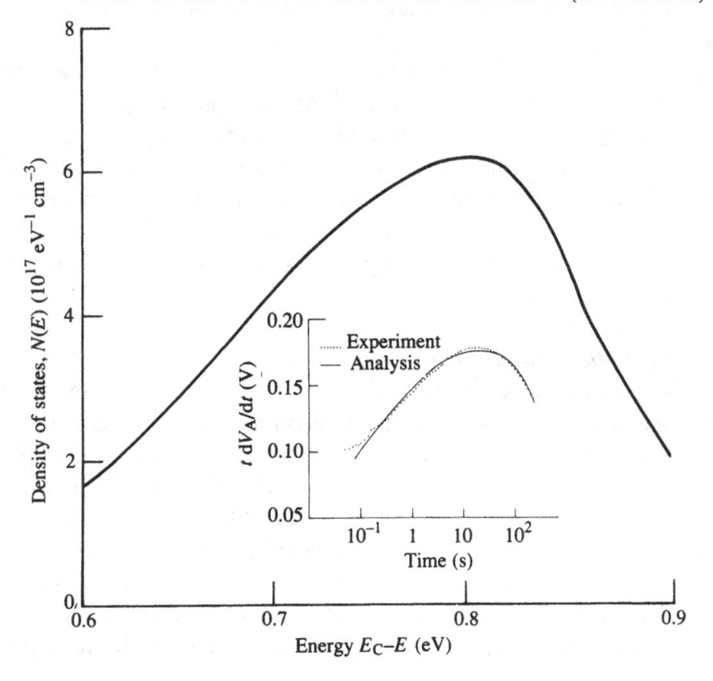

use of the optical transitions in a photocapacitance experiment (see Section 4.2.4). A simultaneous measurement of the capacitance and the ESR was again able to show that one $g = 2.0055$ spin resulted for every electron released from the trap into the conduction band. It should be noted here that the defect density in Figs. 4.17 and 4.18 is about 10^{17} cm^{-3}, which is much larger than the defect density in undoped a-Si:H. The increase in defect density is one of the characteristic features of doping in a-Si:H and is described in the next chapter.

The DLTS measurements so far describe only electronic states in the upper half of the band gap. The lower half of the gap is harder to study because of the difficulty of making a stable junction to p-type a-Si:H. One technique uses light illumination of n-type samples to probe the lower half of the gap (Lang *et al.* 1984). The optical absorption populates the defect states with holes which are removed by thermal excitation to the valence band. This experiment measures the thermal emission energy because the information comes from the thermal emission step rather than from the initial optical excitation. The data for the lower half of the gap in Fig. 4.17 are derived from this type of experiment. The results are consistent with the usual DLTS where the two results overlap, but there are various new peaks seen in the spectra, with no obvious correlation with the sample growth properties. The addition of the illumination makes the analysis much harder and it is difficult to judge whether all the extra structure is real.

DLTS gives good results for defects in the upper half of the gap in n-type material, but in other situations the defect spectroscopy measurements are limited. DLTS is difficult in undoped a-Si:H because of the need for low frequencies and its application to p-type material has been limited by the difficulty of making good Schottky contacts. There are many variations of transient capacitance and conductivity which are applied to the study of deep states (e.g. Cohen 1989, Crandall 1981, Nebel *et al.* 1989).

(2) *Field effect measurements*

The limitations of capacitance measurements in undoped a-Si:H have resulted in a greater emphasis on measurement techniques which use the shift of E_F with a trapped space charge (Eq. (4.18)). An example is the field effect experiment, which is of special interest because it was the first technique used to obtain $N(E)$ in a-Si:H (Madan, LeComber and Spear 1976). The experimental configuration is shown in Fig. 4.19. A voltage V_A across the dielectric layer induces a space charge $Q = C_d V_A$ in the a-Si:H film, where C_d is the capacitance of the dielectric. The Fermi energy in the a-Si:H near the interface moves towards the band edge and the lateral conductivity $\sigma(V_A)$ increases. A low density of gap states results in a large shift of the Fermi energy and a large change in the conductivity. In principle, the density of states can be deduced from $\sigma(V_A)$ by solving Poisson's equation.

Fig. 4.19. Schematic diagram of the field effect. The applied voltage V_A induces charge carriers near the interface giving an excess conductance in the plane of the interface.

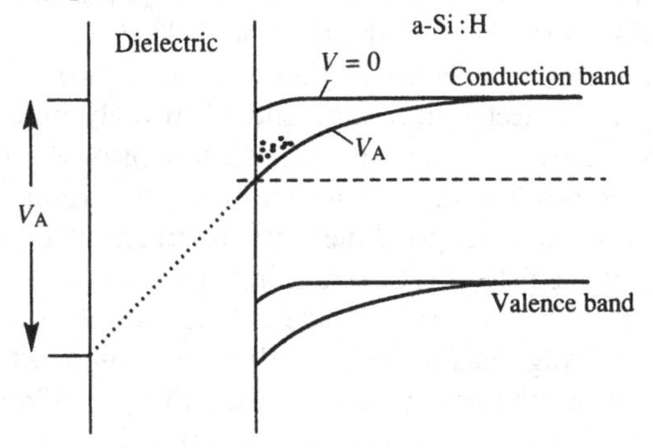

Fig. 4.20. The density of states distribution obtained from space charge limited current measurements on undoped a-Si:H. The various symbols represent different samples and deposition conditions. Field effect (FE) and DLTS data are shown for comparison. T_D is the substrate temperature during growth (Mackenzie, LeComber and Spear 1982).

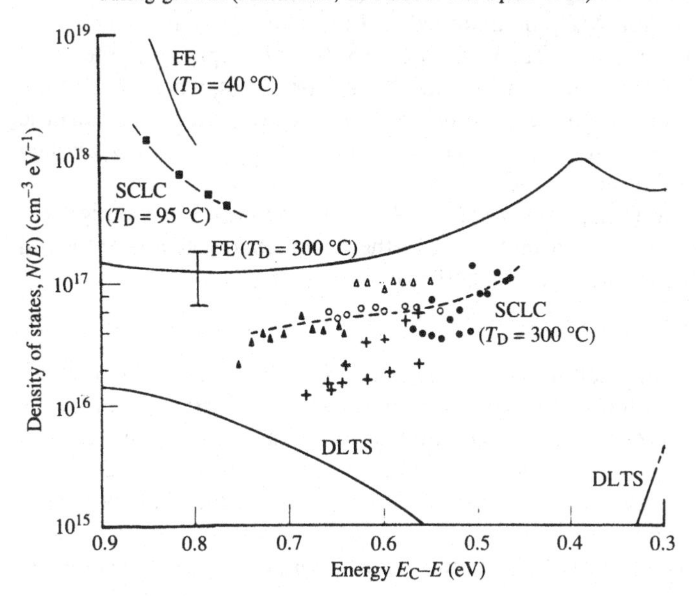

There are, however, some problems in the analysis. Poisson's equation (Eq. (4.19)) relates the space charge to the second derivation of the voltage, so that very precise data are needed to get the detailed shape of the density of states, particularly for deep states of low density, although the approximate magnitude is obtained more easily. The early experiments found two peaks in the density of states; the one in the upper half of the gap is reproduced in Fig. 4.20. However, analysis of subsequent data with essentially the same form of $\sigma(V_A)$ was unable to reproduce those peaks (Goodman and Fritzsche 1980).

Another problem with the field effect is that most of the space charge is within 100 Å of the dielectric interface and is strongly influenced by the presence of interface states or near-surface growth related defects. With the advent of field effect transistor technology, much has been learned about how to make good dielectric interfaces with silicon nitride; an example of the field effect current in such a device is shown in Fig. 10.8. Analysis of the current–voltage characteristics now finds a monotonically increasing density towards the band edges, which is completely different from the density of states found by DLTS in the n-type samples and also the early field effect data. The important question of whether the defects have the same density of states distribution in doped and undoped a-Si:H is discussed in Section 6.2.5.

(3) *Space charge limited currents*

Space charge limited current (SCLC) measurements are also used to obtain $N(E)$ in undoped a-Si:H and again the results are derived from the shift of E_F due to trapped space charge. The measurement is typically made on n^+–i–n^+ structures, where the heavily doped contact layers allow carriers to move freely into the intrinsic layer. The current–voltage relation, $J(V)$, is ohmic at low applied voltages, followed by a SCLC regime at higher voltage characterized by a more rapid voltage dependence. When the voltage is not too high, the space charge is insufficient to move the Fermi energy significantly and the current is given by (Rose 1955)

$$J(V) = 1.13 f_n \varepsilon\varepsilon_0 \mu_D V^2/d^3 \tag{4.23}$$

where f_n is the fraction of electrons above E_F which are mobile. Further increase in voltage gives a more rapid increase in current, because the quasi-Fermi energy is shifted towards the band edge by the trapped space charge. The first SCLC regime allows a measure of $N(E_F)$, and, in principle, the second regime gives $N(E)$ over the range of energies in which E_F moves.

Some results of SCLC experiments in undoped a-Si:H are shown in

Fig. 4.20 (Mackenzie *et al.* 1982). The deduced density of states distribution is fairly constant in material deposited at 300 °C, increasing slightly towards the band edge. The results are similar to the density of states found in early field effect data, but have a different shape from the DLTS data and a much larger density than is measured by ESR on good quality undoped a-Si:H. A larger defect density is found for low deposition temperatures in agreement with the ESR results. The SCLC analysis suffers from the same difficulties as the field effect, in that the overall density should be given correctly but the structure in $N(E)$ is very sensitive to the exact shape of $J(V)$. It also seems to the author that an important aspect of the experimental situation has been overlooked in the analysis. A basic assumption is that the injected space charge forms an equilibrium with all the localized states and so can be described by a quasi-Fermi energy. In low defect density a-Si:H, however, an electron in the band tail has a mean free path for deep trapping which is much larger than the sample thickness. Time-of-flight measurements of the drift mobility would not be possible otherwise. This implies that there cannot be complete equilibration of the excess electrons and the band tail charge should be described by a different quasi-Fermi energy from the charge in deep traps. This effect should alter the analysis, influencing both the calculation of the trap density and the shape of $N(E)$.

The thermal emission experiments, therefore, do not give a completely clear picture of the gap state distribution because of the conflicting results. A remaining question is whether the differences between the DLTS results in the doped samples and the field effect and SCLC data on undoped a-Si:H are due to an effect of doping on the shape of the density of states, or whether they are due to experimental artifacts.

4.2.4 *Defect level spectroscopy – optical transition energies*

The optical absorption arising from the defect transitions is weak because of the low defect densities and in a thin film cannot be measured by optical transmission. The techniques of PDS, CPM and photoemission yield, described in Section 3.3, have sufficient sensitivity. Photocapacitance, which measures the light-induced change in the depletion layer capacitance, is similarly sensitive to weak absorption (Johnson and Biegelsen 1985). PDS measures the heat absorbed in the sample and detects all of the possible optical transitions. At room temperature virtually all the recombination is non-radiative and generates heat by phonon emission. CPM detects photocarriers and so is primarily sensitive to the optical transitions which excite electrons to the conduction band, because electrons dominate the photocurrent. Photocapacitance also primarily detects electrons and, in addition, requires doped samples in order to measure the junction capacitance correctly. Photoemission yield measures the distribution of electrons below the Fermi energy and within about 50 Å of the surface. Apart from these slight differences, each of the techniques gives similar results when applied to the same type of sample, and is apparently much less affected by the differences that are found in the thermal emission experiments.

Examples of PDS measurements on undoped a-Si:H samples are shown in Fig. 4.21. The Urbach edge of the band-to-band transitions is present above about 1.5 eV and at lower energy there is a distinct, broad, absorption band extending to about 0.5 eV due to transitions from deep defect states. The series of spectra in Fig. 4.21 corresponds to samples deposited at different rf power in the plasma which causes a range of defect densities when measured by the $g = 2.0055$ ESR line. Fig. 4.22 shows that the strength of the low energy absorption is directly proportional to the spin density, which obviously suggests that exactly the same defects are being measured. The relation between the defect density and the integrated absorption is (Jackson and Amer 1982)

$$N_D = \frac{cnm_e}{2\pi^2\hbar^2}\left[\frac{(1+2n^2)^2}{9n^2e^2f_{oj}}\right]\int_0^{E_{max}} \alpha_{meas}(E)\,dE \tag{4.24}$$

$$= 7.9\times10^{15}\int_0^{E_{max}} \alpha_{meas}(E)\,dE\ (\mathrm{cm^{-3}}) \tag{4.25}$$

where N_D is the defect density, α_{meas} is the measured defect absorption, c is the velocity of light, n is the refractive index, m_e is the electron mass and the integral is continued up to an energy, E_{max}, of about 1.6 eV, after subtracting the Urbach edge contribution. The constant of proportionality is obtained by assuming unity oscillator strength, f_{oj}, and a factor of 2 local field correction. Eq. (4.25) gives an excellent quantitative fit to the data in Fig. 4.22 and is further evidence of the correspondence between the defects observed optically and in ESR.

Any defect with two states in the gap has four possible optical transitions which are denoted *A–D* in Fig. 4.3. The actual contribution to an optical absorption experiment depends largely on the position of the Fermi energy. Only transitions *A* and *D* in Fig. 4.3 contribute to the absorption when E_F lies between the two levels and the defects are singly occupied, but when the defect states are either doubly occupied or empty, the only possible transitions are *B* and *C* respectively. (The defect is assumed to have a positive correlation energy.) For the dangling bond defect, these three possible charge states of the defect apply respectively to undoped, n-type and p-type material. Although the broadness of the absorption bands makes it difficult to be precise,

Fig. 4.21. PDS measurements of the optical absorption edge and defect transitions of a-Si:H samples deposited at different rf power, as indicated (Jackson and Amer 1982).

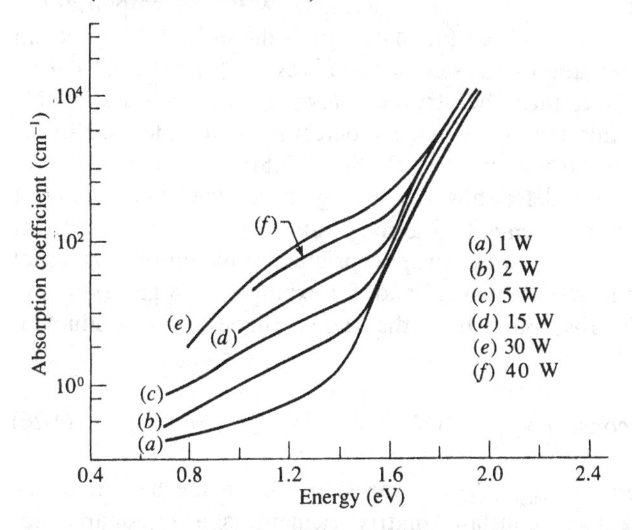

Fig. 4.22. Plot showing the proportionality between the strength of the defect absorption from the data in Fig. 4.21 and the $g = 2.0055$ ESR spin density. The solid line is given by Eq. (4.25) (Jackson and Amer 1982).

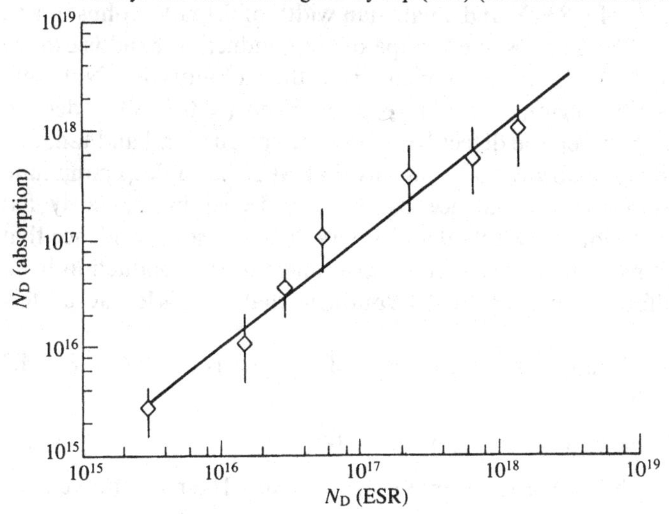

the n-type samples have their absorption at lower energy than the undoped or p-type material. This leads to the conclusion that the average energy of the defect levels is slightly above midgap, which is in agreement with the n-type conductivity of undoped a-Si:H. It is therefore expected that the low energy part of the spectrum in undoped films is more dominated by transition A than D. The results of the CPM experiment, which only measures transition A and yet has the same shaped spectrum, agree with this conclusion. With this simplification, the correlation energy, U (see Fig. 4.4), equals the shift of the spectrum between undoped and n-type samples and is estimated to be 0.2–0.4 eV (Jackson 1982). It must be stressed, however, that a questionable assumption of this analysis is that the defect states are identical in the doped and undoped material (see Section 6.2.5).

The shape of the defect absorption is given by the joint density of states and the matrix elements, and the position and width of the defect band can only be deduced by the appropriate deconvolution. The usual approach is to model the defect band, for example, by a gaussian, and to calculate the absorption from the known shape of the conduction band,

$$\alpha_D = \text{const.} \int N_{\text{defect}}(E)\, N(E+\hbar\omega)\, \mathrm{d}E \tag{4.26}$$

The parameters of $N_{\text{defect}}(E)$ are varied to obtain the best fit to the absorption data. A constant matrix element is a reasonable approximation for the limited energy range involved. Fig. 4.23 gives an example of the fit to the photocapacitance data of n-type a-Si:H, for a defect energy of 0.83 eV and a gaussian width of 0.15 eV (Johnson and Biegelsen 1985). The assumed shape of the conduction band is exactly that of Fig. 3.15 with the zero of energy at the mobility edge. Note that the low energy region of the absorption band (< 0.8 eV) is derived from states between the defect band and the conduction band tail, and the steep drop in absorption confirms that $N(E)$ has a deep minimum. The states nearer the valence band are reflected in the fairly flat absorption region between 1.0 and 1.5 eV. There is some evidence that $N(E)$ remains high here, but the precise shape of the conduction band strongly affects the calculated absorption, so that $N(E)$ is less accurately obtained.

The optical transition energies, E_{opt}, of transitions B and C in Fig. 4.3 are related by

$$E_M = E_{\text{opt}}(B) + E_{\text{opt}}(C) + U - 2W \tag{4.27}$$

where E_M is the mobility gap (not the Tauc gap). The relaxation energy

$2W$ enters because of the different energies for optical and thermal transitions due to lattice relaxation. $E_{opt}(B)$ is found to be 0.83 eV from Fig. 4.23. There has not been a similar deconvolution of the defect absorption in p-type material using the measured valence band density of states, but $E_{opt}(C)$ is estimated to be about 0.9 eV. The mobility gap is about 1.85 eV (see Fig. 3.16), so that $U-2W \approx 0.1$ eV with an uncertainty of about 0.2 eV. This result is consistent with a correlation energy of 0.2 eV and a small relaxation energy. However, other investigators, with essentially the same data but different procedures for extracting the energies, have obtained negative values of the

Fig. 4.23. Photocapacitance data for n-type a-Si:H corresponding to the excitation of an electron from the doubly occupied defect to the conduction band. The solid line is the calculated fit with the parameters shown (Johnson and Biegelsen 1985).

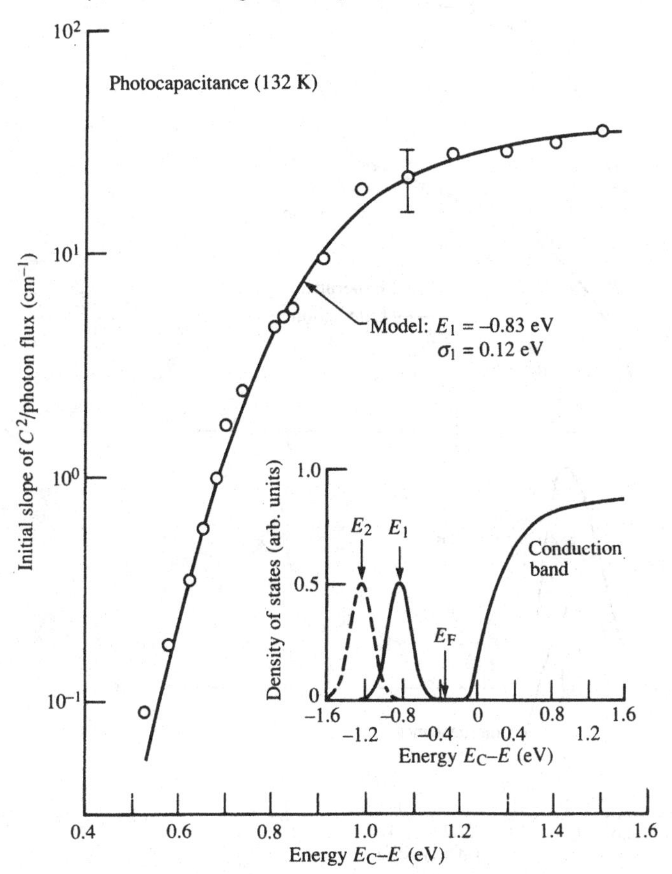

correlation energy, which suggests that the defects may not be at the same energy in doped and undoped samples (e.g. Kocka, Vanacek and Schauer 1987). At the time of writing, this point is not clearly resolved.

The defect energy levels are also obtained from optical emission transitions. Measurements of luminescence in a-Si:H are described in more detail in Chapter 8. Transitions to defects are observed as weak luminescence bands. The transition energies are about 0.8 eV and 0.9 eV in n-type and p-type material respectively and are identified with the inverse of the transitions *B* and *C* of Fig. 4.3.

Fig. 4.24. A compilation of different spectroscopic techniques used to obtain the energy of the doubly occupied defect state in n-type a-Si:H (Street *et al.* 1985).

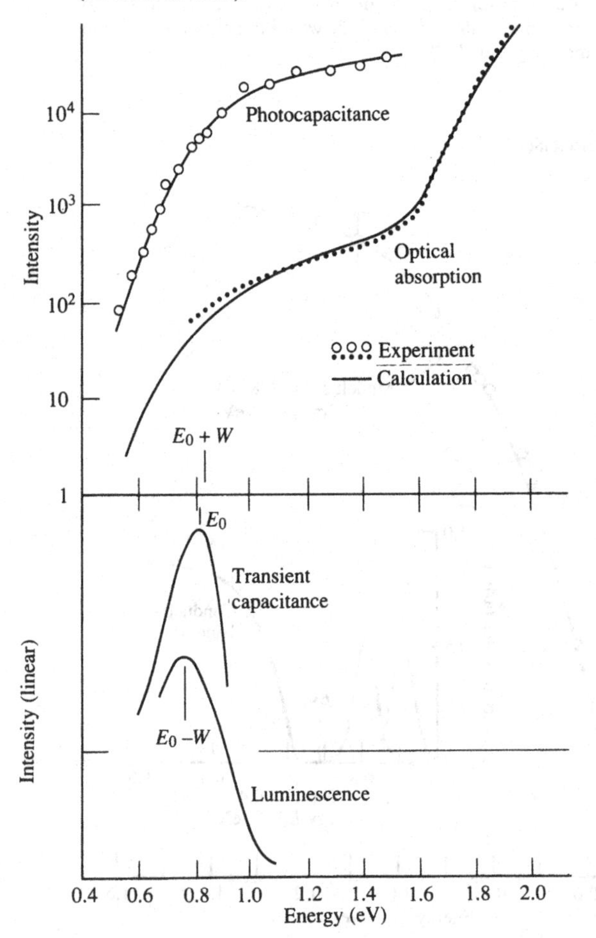

4.2.5 *Summary*

Perhaps the best way of summarizing the experimental information about the defect energy levels is to attempt to answer the following questions.

Is the 2.0055 ESR defect the dominant deep trap in a-Si:H?

The bulk of the evidence suggests that the answer is yes. No other ESR signature of a deep state has been found (although there is some indication of oxygen-related deep defects in alloys, and possibly a spinless defect in ion-bombarded a-Si:H). The defect absorption is proportional to the ESR spin density with the expected transition probability. The DLTS defects are definitely shown to be the same as the 2.0055 defect and further support for the conclusion is found from deep trapping and luminescence data described in Chapter 8. Virtually all the experiments for normal a-Si:H can be satisfactorily explained by the ESR-active defect.

How large is the lattice relaxation energy at the defects?

The best measurement of the lattice relaxation energy is the comparison of the optical and thermal transition energies of the same defect state. This comparison is made for n-type samples in Fig. 4.24, which shows absorption, thermal emission, and luminescence data attributed to the same transition. All the energies agree within less than 100 meV, so that there is no evidence of a large Stokes shift. Taking into account the experimental uncertainties, the lattice relaxation energy cannot be larger than 100 meV.

Fig. 4.25. Illustration of some of the possible coordination defects and their expected gap state structure.

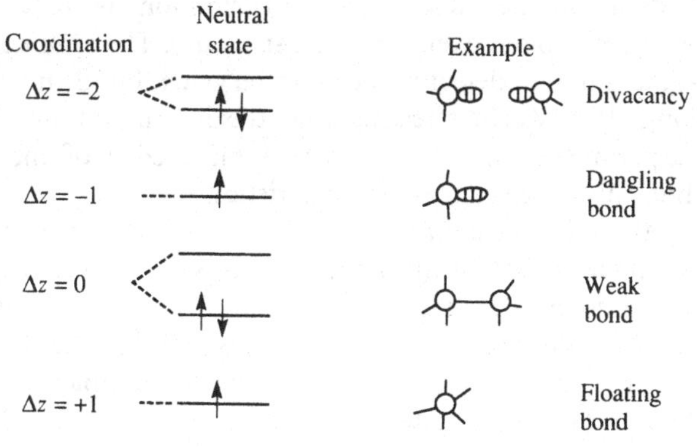

Coordination	Neutral state	Example
$\Delta z = -2$		Divacancy
$\Delta z = -1$		Dangling bond
$\Delta z = 0$		Weak bond
$\Delta z = +1$		Floating bond

What is the sign and magnitude of the correlation energy?

The 2.0055 defect is a positive U state, as can be immediately deduced from the observation of paramagnetism. The positive U is also consistent with the small lattice relaxation energy, since U must be positive in the absence of lattice relaxation. It is possible that U has a distribution of values, so that a small fraction of defects may have negative U. The magnitude of U is as yet imprecisely obtained, with experimental estimates varying from 0–0.5 eV. Recent calculations find $U \sim 0.4$ eV (Northrup 1989), although earlier calculations found a lower value (Bar-Yam and Joannopoulos 1986).

Are the defect energies the same in doped and undoped a-Si:H?

At the time of writing the evidence is controversial, with some experiments indicating little difference in the energy levels, and other data suggesting a shift as large as 1 eV. The gap state energy of the negative defect in n-type a-Si:H is 0.8–0.9 eV below E_C but is the only level for which there is reasonably clear evidence (see Fig. 4.24). The subgap absorption data on undoped and p-type a-Si:H are consistent with defect levels which do not vary with doping, but the correlation energy is not known accurately enough to be sure. The evidence of other experiments, including field effect and SCLC seem to indicate that the defects are at different energies, but there are significant uncertainties about the analysis. The author doubts that the shift can be as large as 0.5–1 eV, but 0.1–0.2 eV is consistent with the data. The shift is of great interest because it is predicted by the thermal equilibrium description of the doping process developed in the next two chapters (see Section 6.2.5).

4.3 Defect models

It is hardly surprising that the controversy over the experimental data of the defect state distribution is reflected in considerable disagreement over the defect structure. The arguments for and against the different defect models can be given, but definite proof is still lacking. It is worth stressing that the interpretation of most electronic experiments does not require a knowledge of the defect structure, but only its electronic characteristics.

Following the ideas of coordination defects introduced in Section 4.1, it is convenient to discuss the defects in terms of four generic types, based on the difference in coordination, Δz, from that of the ideal network. Fig. 4.25 illustrates the defects and a simple molecular orbital description of the energy levels, together with the occupancy of each type of defect in its neutral state.

$\Delta z = -2$: The divacancy or two-fold coordinated silicon atom

The divacancy is a stable defect in crystalline silicon. The six dangling bonds left by the removal of two adjacent atoms reconstruct to leave two strong bonds and a pair of weakly interacting dangling bonds which comprise the gap states. The defect has four electronic levels in the gap and its neutral state has paired electrons and is diamagnetic. The divacancy model was first proposed by Spear (1974) to explain the early field effect results. A two-fold coordinated silicon atom also has $\Delta z = -2$ and so is of the same type. Adler (1978) concluded from molecular orbital arguments that this defect has the lowest formation energy of any silicon coordination defect.

$\Delta z = -1$: The dangling bond

The properties of this defect have already been discussed in some detail. It has two states in the gap and is paramagnetic when neutral.

$\Delta z = 0$: The hydrogenated void or weak Si–Si bond

Hydrogenated void defects certainly exist in large concentrations in a-Si:H and are expected to have all their bonding requirements satisfied. An estimate that at least half the hydrogen is at the surface of small voids and that each void has about ten hydrogen atoms leads to a void density of approximately 2×10^{20} cm^{-3}. Gap states would only result if bonding distortions were large enough to move the bonding or anti-bonding states out of the band tails. Similarly, a sufficiently distorted Si–Si bond results in a defect state in the gap which is diamagnetic when neutral. The band tail states are attributed to distorted bonds, but it is not known how much the hydrogenated voids contribute to the tails. Deep states only result from particularly large local distortions.

$\Delta z = 1$: The floating bond

Until recently the possibility of five-fold coordinated silicon atoms was discounted. However, the analogy of the dangling bond and floating bond in the amorphous phase, with the vacancy and interstitial in crystalline silicon, suggests that this defect may have a comparable energy to the dangling bond (Pantelides 1986). The term floating bond is based on the predicted ability of the defect to migrate rapidly through the network by a bond switching mechanism (Pantelides 1987). The floating bond has two states in the gap and its neutral state is paramagnetic.

Earlier in this chapter it is argued that the 2.0055 defect is the dominant and perhaps the only deep state in a-Si:H. Both the divacancy and void types of defect have even values of Δz and so have paired electrons when neutral. Defects of this type will show no paramagnetism when undoped, but n-type or p-type doping will uncover two different paramagnetic states. The 2.0055 defect has exactly the opposite property, of an unpaired spin in the neutral state which becomes diamagnetic when the Fermi energy is moved by doping. On the other hand the different ESR resonances in doped material do have the character of the even Δz type and, in fact, are attributed to band tail states with $\Delta z = 0$ (see Section 5.2.2).

The ESR data clearly indicate that the 2.0055 defect has an odd value of Δz and so the discussion reduces to the choice of dangling bonds or floating bonds. The following arguments have been given for and against the two possibilities:

(1) *The wavefunction and g-value*

The ESR data give the most specific and detailed information about the defect, and it is from these results that the identity of the defect will probably be finally settled. The wavefunction of the dangling bond is located primarily on the broken bond and should be close to the ideal sp^3 orbital, having 75% p-character and 25% s-character. On the other hand, the floating bond is partially delocalized on the other bonds and is estimated to have only about 30% p-character. The analysis of the hyperfine data discussed in Section 4.2.2 found that the wavefunction is close to that expected of the dangling bond (Biegelsen and Stutzmann 1986). However, the analysis has been challenged with the counter claim that the wavefunction is nearer to that expected for the floating bond (Stathis and Pantelides 1988). At present it is unclear whether the different results reflect different assumptions in the modeling or whether one or other analysis is incorrect.

The *g*-value spectrum is consistent with the dangling bond interpretation, as is shown by Fig. 4.9 in which the surface dangling bond on crystalline silicon averages to the observed *g*-value. There is, however, no equivalent estimate for the floating bond which might also have roughly the same *g*-value. The *g*-value can, in principle, be calculated from the wavefunction, using the perturbation expansion given in Eq. (4.12), but the calculation is not easy and requires an accurate wavefunction.

(2) *Formation energy and topology*

It is expected, although not certain, that the defect with the lowest formation energy is the most abundant, provided that it is not further constrained by the local topology. At present not enough is known about the formation energies to decide either way. Pantelides (1986) argues that the energies of the vacancy and interstitial in crystalline silicon are about the same and that these defects can be considered to be four dangling bonds and four floating bonds respectively. The implication is that the two defects in a-Si:H have similar energies and that there is no preference for one or other based on energy.

The topological constraints return us to the analysis, presented in Section 2.2.1, that a four-fold coordinated network is overconstrained compared to the ideal coordination of 2.5. It is then argued that the presence of dangling bonds or of Si–H bonds are alternative ways of reducing the network coordination towards the ideal value. Thus it is difficult to understand the five-fold coordinated floating bond, which increases the overall coordination. The topology arguments therefore favor the dangling bond.

(3) *The correlation energy and gap states*

One calculation places the dangling bond level close to the middle of the gap, and the floating bond level near the conduction band (Fedders and Carlsson 1988, 1989). Comparison with experiment favors the dangling bond. Furthermore, if the five-fold floating bond can exist without excessive topological constraints, then a valence alternation pair is formed with the dangling bond. Much is known about such pairs from the studies of defects in chalcogenide glasses (see Section 4.1.3). In particular, valence alternation pairs have the property of a negative correlation energy. Indeed the pair should be viewed not as two separate defects, but as the different charge states of a single defect. Thus the positive defect would be three-fold coordinated with no electrons in its broken bond orbital, and the negative defect would be the five-fold state, with two electrons to form the fifth bond. The

Fig. 4.26. Illustration of the migration of a floating bond by a bond switching mechanism (Pantelides 1987).

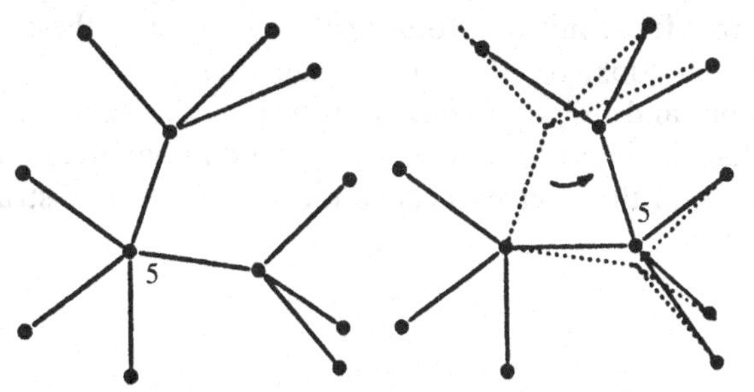

neutral defect would be expected to have some intermediate configuration. The energy gained by forming the fifth bond results in a large lattice relaxation energy and so a negative U. This approach suggests that if the floating bond can exist, it confers a negative correlation energy on the basic coordination defect, but if the five-fold state is prevented by a high formation energy or topological constraints, the dangling bond has very limited lattice relaxation, and has a positive U. The experimental evidence shows fairly clearly that the defect has a positive U, so that the dangling bond seems to be favored.

(4) *Defect migration*

In the floating bond model, the floating bond is attributed a high diffusivity through a bond switching mechanism illustrated in Fig. 4.26. Bond switching requires a substantial flexibility of the lattice because, in general, a bond length distortion of only about 20% is sufficient to break the bond completely. The diffusion mechanism in bond switching is one in which the bond transfers from a nearest neighbor atom to a second neighbor. Therefore in order to have a high diffusion, with a low energy barrier, the bond switching must occur when the two distances are not too different. The average second neighbor distance is 1.6 times the bond length, so that local distortions of 60% from the average are required for this diffusion mechanism. A soft flexible lattice is indicated, which seems to be at odds with the observation that the silicon network is rather rigid. Such defect migration should also relax the internal strain of the network.

Measurements of defect diffusion have been reported (Jackson 1989). The experiment uses hydrogen passivation to induce a concentration gradient of defects which is measured for different anneal temperatures. No defect diffusion is detected and the diffusion coefficient is deduced to be less than that of hydrogen (see Fig. 2.20).

Taking the different arguments together, it is the author's opinion that the dangling bond model remains the more plausible explanation of the 2.0055 defect. Perhaps within a short time, further studies of the hyperfine interaction or calculations of the defect energy levels, etc. will be able to provide definitive proof one way or the other. In the remainder of this book, for the sake of definiteness, we refer to the 2.0055 ESR spin and the associated deep trap as the dangling bond, recognizing that the interpretation of electrical data involves only the gap state levels and the electron occupancy, not the atomic structure.

5 Substitutional doping

In 1975 Spear and LeComber reported that a-Si:H could be doped by the addition of boron or phosphorus; their conductivity data are reproduced in Fig. 5.1. This first observation of electronic doping in an amorphous semiconductor set the stage for the subsequent development of a-Si:H electronic technology. The addition of small quantities of phosphine or diborane to the deposition gas results in changes in the room temperature conductivity by more than a factor 10^8. The activation energy decreases from 0.7–0.8 eV in undoped material to about 0.15 eV with phosphorus doping and 0.3 eV for boron. Subsequent experiments confirmed that the conductivity change was due to a shift of the Fermi energy, and that n-type and p-type conduction was occurring (Spear and LeComber 1977). The explanation of the results in terms of substitutional doping has never been doubted.

Examples of the conductivity temperature dependence $\sigma(T)$ of n-type and p-type a-Si:H are shown in Fig. 5.2 (Beyer and Overhof 1984). The thermally activated $\sigma(T)$ implies that the Fermi energy always remains in localized states and there is never metallic conductivity. E_F is prevented from reaching the conducting states above the mobility edge by the high density of band tail localized states and also by a low doping efficiency. The conductivity is lower in p-type samples than n-type, primarily because the wider valence band tail keeps E_F farther from the mobility edge. The high temperature extrapolation of $\sigma(T)$ is an indication that the free carrier conductivity is of order 100 Ω^{-1} cm^{-1}. However, careful analysis is needed to extract this information – the electronic transport mechanisms are discussed in Chapter 7. Although the conductivity is thermally activated, there is a change of slope near 100 °C, most noticeable in the n-type material (see Fig. 5.2). This effect originates from thermally induced changes in the material structure, which are described in more detail in Chapter 6.

The conductivity increases with doping concentrations up to about 1 %. Higher doping levels cause a large decrease in the conductivity and a corresponding increase in the activation energy (Tsai 1979, Leidich *et al.* 1983). At these concentrations, the material is more properly described as an alloy.

Substitutional doping is unsurprising to anyone familiar with

Fig. 5.1. The variation of the room temperature dc conductivity of a-Si:H films doped by the introduction of phosphine and diborane into the deposition gas (Spear and LeComber 1975).

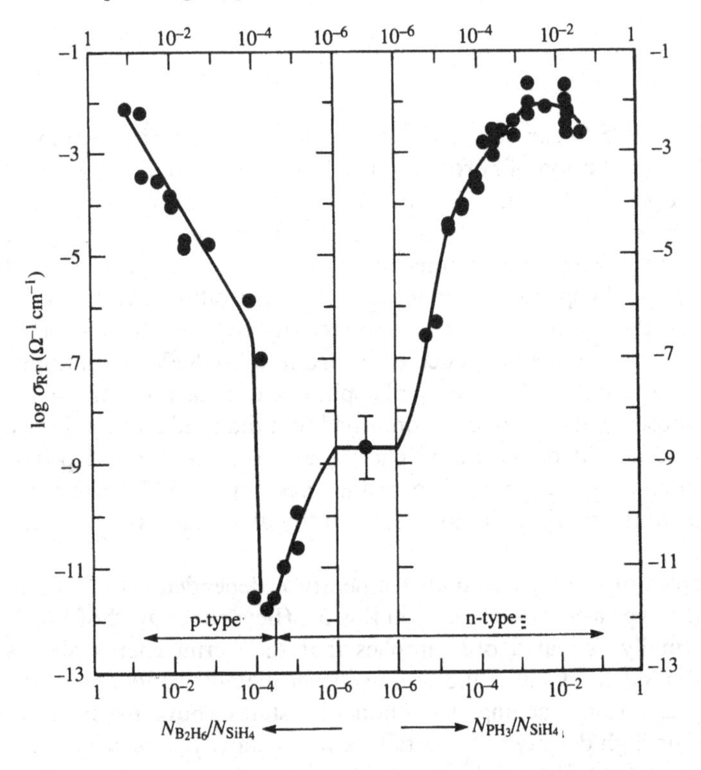

Fig. 5.2. Examples of the temperature dependence of the conductivity of (*a*) n-type and (*b*) p-type a-Si:H. The data correspond to gas phase doping levels ranging from 10^{-6} to 2×10^{-2} (Beyer and Overhof 1984).

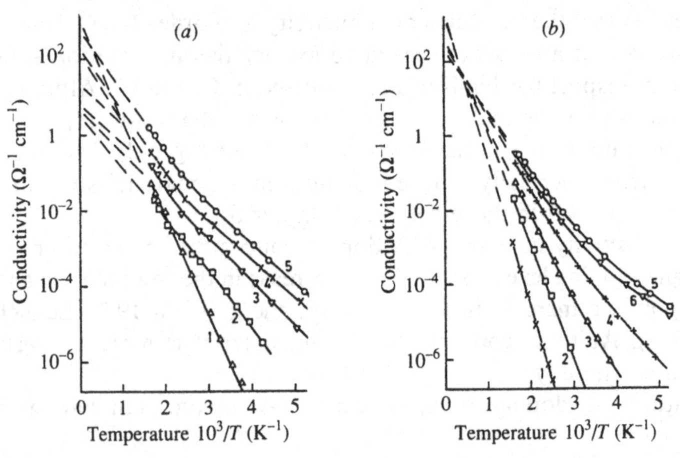

crystalline semiconductors. The observation is significant in a-Si:H because of the theoretical prediction that substitutional doping of an amorphous semiconductor is impossible. The argument follows from the $8-N$ rule for the chemical bonding (see Section 1.2.2), which states that all the elements in a random amorphous network bond with their optimal valency, irrespective of the neighboring atoms. According to this principle, each atom adopts a coordination which results in fully occupied bonding and lone pair states, but has no occupied anti-bonding states. Accordingly, both boron and phosphorus are expected to be three-fold coordinated. However, substitutional n-type doping requires that the dopant coordination is such that one electron is promoted to the anti-bonding state and becomes a free carrier in the conduction band. Thus doping seems incompatible with the expected bonding of the atoms in the network. In crystalline silicon, the periodic lattice of atoms constrains the impurity to have the same coordination as the silicon, in which case the phosphorus or boron atoms become dopant states. A different coordination of the impurity can only occur in the presence of a defect or impurity in the crystalline structure. It is, therefore, the topological constraint of the crystalline lattice that leads to the substitutional doping and this constraint is absent in the amorphous material.

The observation of substitutional doping in a-Si:H apparently contradicts this bonding argument. However, one of the main conclusions of this chapter and the next is that a modified $8-N$ rule

Fig. 5.3. Schematic diagram of the density of states and the electron distribution of n-type a-Si:H showing deep and shallow states.

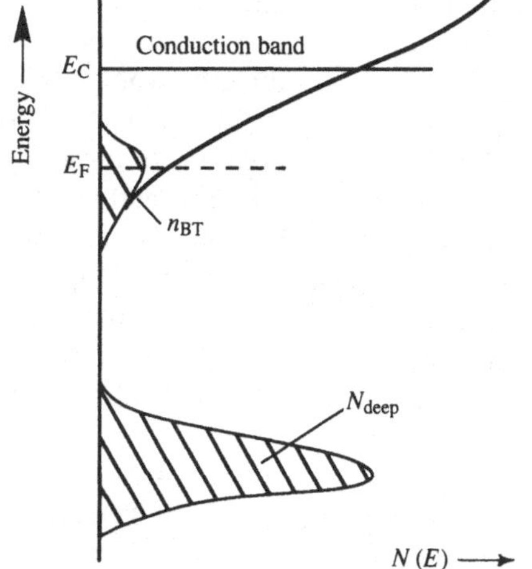

does apply. The doping process in a-Si:H is particularly interesting and quite different from that in crystalline silicon, with some unexpected properties. This chapter concentrates on the experimental information about the doping, particularly the changes in the localized state distribution. Chapter 6 develops the doping mechanisms further, in the context of a thermal equilibrium description of the electronic states and Chapter 7 discusses the transport mechanisms of both doped and undoped a-Si:H.

Given the possibility of both three-fold inactive impurity states and substitutional four-fold dopants, it is important to know the doping efficiency, η, which is defined as the fraction of impurities which are active dopants.

$$\eta = N_4/(N_3+N_4) = N_4/N \qquad (5.1)$$

where N_3 and N_4 are the densities of three-fold and four-fold states and N is the total impurity concentration. Alternatively, η can be defined from the electronic properties. Each donor adds an electron, so that the doping efficiency is obtained by counting the number of excess carriers,

$$\eta = N_4/N = (n_{BT}+N_{deep})/N \qquad (5.2)$$

Here N_{deep} is the density of deep states which can take an extra electron from the donor and n_{BT} is the density of electrons occupying shallow states near the band edge. Fig. 5.3 shows a schematic diagram of the occupancy of states by donor electrons. Thus the doping efficiency may be obtained from either structural or electronic information.

5.1 Growth and structure of doped a-Si:H

Doping of plasma deposited a-Si:H is most commonly achieved by adding phosphine, arsine, or diborane to the plasma gas. Other techniques include ion implantation (Müller, Kalbitzer, Spear and LeComber 1977), neutron transmutation doping (Hamanaka *et al.* 1984), and the generation of dopant radicals directly from a solid source in the reactor (Street, Johnson, Walker and Winer 1989). Interstitial dopants, such as lithium, are diffused in from a film deposited on the surface (Beyer and Fischer 1977). In general it does not matter how the impurities are introduced, although damage from ion implantation must be annealed to activate the doping. Doping is observed with all of the different deposition techniques used to grow a-Si:H, provided that the intrinsic defect density is not too high. This book concentrates on the most common and convenient mode, namely gas phase doping of plasma deposited films.

The dopant molecules are dissociated by the plasma, just as is silane, forming radicals or ions that bond to the growing surface. The relative deposition rates of the various species may be different, and are described by an impurity distribution coefficient, d_I,

$$d_I = C_s/C_g \tag{5.3}$$

where C_g is the molecular concentration in the plasma gas (e.g. $[PH_3]/[SiH_4+PH_3]$), and C_s is the atomic concentration in the solid (e.g. $P/(Si+P)$). The doping efficiency can be defined in terms of either the gas phase or the solid phase concentration and it is important to distinguish between these alternative descriptions.

The distribution coefficient is usually greater than unity and is sometimes very large. The rf power dependence of the incorporation of phosphorus and arsenic is compared in Fig. 5.4, in both cases for films deposited near 250 °C. The phosphorus distribution coefficient is about 5 at low concentrations and moderate rf power and decreases slowly

Fig. 5.4. The distribution coefficients of arsenic and phosphorus dopants as a function of the rf power in the plasma at different gas phase concentrations (Winer and Street 1989).

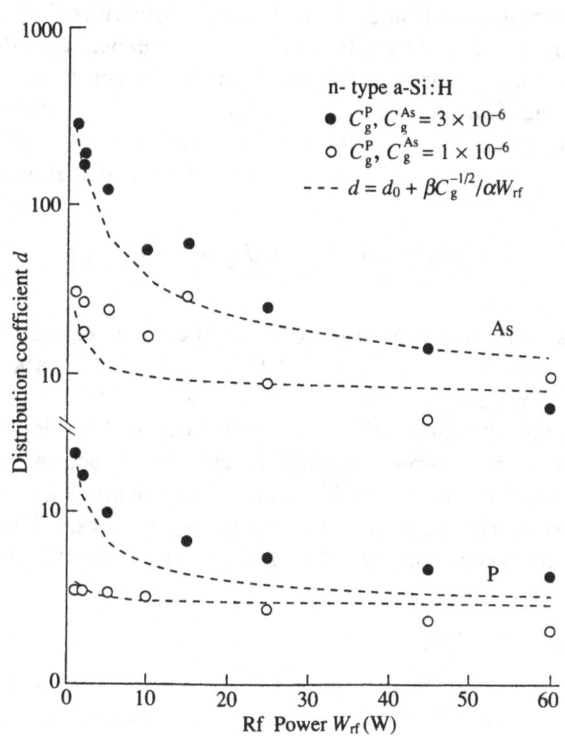

with increasing phosphorus concentration (Leidich *et al.* 1983). The slope of $d_I(C_g)$ is slightly less than unity. Boron has a similar behavior, with a distribution coefficient of 3–5 at low doping levels and a similar slightly sublinear doping dependence. On the other hand, arsenic has a large value of d_I, which reaches over 300 at low doping levels and rf power and decreases with doping roughly as (Winer and Street 1989),

$$d_{As}(C_g) \approx C_g^{\frac{1}{2}} \tag{5.4}$$

The arsenic distribution coefficient varies strongly with rf power (W_{rf}), decreasing approximately as $1/W_{rf}$, as is evident in Fig. 5.4.

The dopants also have a catalytic influence on the deposition of the silicon, with the largest effect being induced by boron doping. Under CVD growth conditions, the deposition rate of the silicon increases by about 40% at a boron gas phase concentration of 1%. Phosphorus doping suppresses the growth rate, but by a smaller amount. Fig. 5.5 shows the growth rates of compensated a-Si:H and illustrates both effects.

Neither the dopant distribution coefficient nor the catalytic effect on growth rate is understood in detail. It is reasonable to expect that the dissociation rate of an impurity molecule might be higher than for SiH_4, enhancing the concentration in the film. When the plasma contains molecular species α and β and the dissociation rate is higher for β by a factor d_{reg}, then a simple model predicts a distribution coefficient of

$$d_\beta = d_{reg}(C_g^\alpha + C_g^\beta)/(C_g^\alpha + d_{reg}C_g^\beta) = d_{reg} \text{ when } C_g^\beta \ll C_g^\alpha \tag{5.5}$$

This type of dopant incorporation has a linear concentration dependence, only changing when $C_g^\beta/(C_g^\alpha + C_g^\beta) \geqslant 0.1$, which is above the normal doping range. The data, however, show a doping- and power-dependent distribution coefficient, particularly at low doping levels, which must reflect some unidentified chemical reaction not described by the simple model. The values of d_I which are much greater than unity seem to be associated with CVD growth conditions. Thus, the arsenic distribution coefficient is reduced by increasing the rf power and is approximately,

$$d_{As} = d_{reg} + d_{irr}\, C_g^{-\frac{1}{2}} W_{rf}^{-1} \tag{5.6}$$

where W_{rf} is the rf power in the plasma and d_{reg} and d_{irr} are constants. The first term on the right hand side in Eq. (5.6) represents a regular deposition process, described by Eq. (5.5). The second term is irregular, varying with power and non-linear in the arsenic concentration. This type of deposition is also present in phosphorus doped material, but with a smaller value of d_{irr}. The analysis of the distribution coefficients in terms of chemical equilibrium growth processes is discussed in Section 6.2.8.

The anomalies in the growth rates also reflect some unknown chemical reactions either at the growing surface or in the gas phase. The impurity radicals in the gas might promote the formation of silicon radicals or, alternatively, boron on the growing surface may increase the rate of attachment of the silicon. Another possible mechanism to explain the enhanced growth rate in p-type material might be the faster release of hydrogen from the surface because it is known that hydrogen diffuses rapidly in boron doped a-Si:H (see Chapter 2).

The structural determination of the doping efficiency has proved to be difficult. One of the first experiments used EXAFS (extended X-ray absorption fine structure) to measure the coordination of the arsenic atoms in a-Si:As:H (Knights, Hayes and Mikkelson 1977). In the alloy regime of arsenic concentrations above 1 %, the coordination is close to the expected value of 3. The coordination increases as the arsenic concentration is reduced to doping levels and a doping efficiency of about 20 % is deduced from the data. However, EXAFS is insensitive to the hydrogen bonding and so a large correction term has to be applied and the doping efficiency can only be extracted on the questionable assumption that the hydrogen bonding is independent of the arsenic concentration. A few attempts have been made to obtain the atomic coordination using NMR, based on the expected shift of the resonance frequency with local bonding (Reimer and Duncan 1983,

Fig. 5.5. The change in deposition rate with the addition of dopants (diborane and phosphine) in compensated a-Si:H (Street *et al.* 1981).

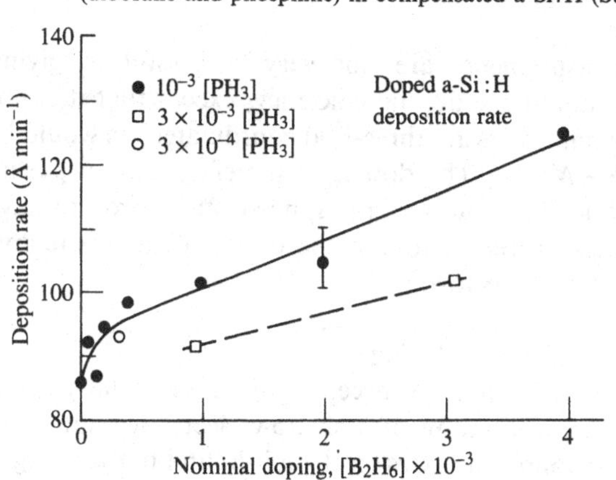

Greenbaum, Carlos and Taylor 1984, Boyce, Ready and Tsai 1987). The experiment is only possible at high doping levels because of the low sensitivity of the experiment. Only three-fold coordinated boron was observed in p-type material, but some evidence of four-fold phosphorus was obtained in n-type, with an estimated doping efficiency of about 20%. It is apparent in the next section that the electronic data give much smaller values of the doping efficiency. It may be that the NMR is observing some other configuration of phosphorus that is not contributing to the doping. The structural experiments confirm that most of the impurity atoms are three-fold coordinated.

Compensation is achieved by codoping with both boron and phosphorus. There are, again, indications of complex gas or surface reactions because the distribution coefficients for boron and phosphorus depend on the concentration of both species. One report finds that (Street, Biegelson and Knights 1981),

$$d_B > d_P \text{ when } C_g^P > C_g^B \tag{5.7}$$

but

$$d_B < d_P \text{ when } C_g^P < C_g^B \tag{5.8}$$

The deposition rate of the species with the lower concentration is enhanced, with the result that the impurity concentrations in the film are more nearly equal than in the gas. The effect suggests pairing of the boron and phosphorus atoms; indeed the chemical environment of phosphorus in compensated material is different from the singly doped material. Pairing is confirmed by NMR double resonance experiments in which the spin relaxation of one atom is modified by a resonance induced at its neighbor (Boyce *et al.* 1987). These experiments find that pairing accounts for about half of the boron and phosphorus atoms at high doping levels.

The structural measurements are not very successful in giving detailed information about the doping efficiency, except that it is not high and most of the impurities are three-fold coordinated, as would be expected from the $8-N$ rule. The doping is therefore only a partial deviation from the rule. The ability of phosphorus and boron to have a coordination of either three or four is the origin of the distinctive doping properties described below.

5.2 The electronic effects of doping

A substitutional donor or acceptor in a crystalline semiconductor results in the formation of a shallow electronic state. For the specific case of the donor, the extra electron is bound to the charged ion core by the Coulomb interaction. Provided that the electronic wavefunction is spread out over a large volume with low amplitude at the impurity core, then the donor behaves as a point charge, and the wavefunction is described by

$$\Phi(k) = F(r)\Psi(k) \tag{5.9}$$

where $F(r)$ represents a slowly varying envelope function and $\Psi(k)$ are the conduction band states of the pure semiconductor. The electron has bound states equivalent to those of the hydrogen atom screened by the dielectric constant. The donor binding energy is

$$E_{\mathrm{DNO}} = e^4 m^*/32\pi^2\hbar^2\varepsilon^2\varepsilon_0{}^2 = (m^*/\varepsilon^2)R \tag{5.10}$$

and the localization length of the wavefunction is

$$a = 4\pi\hbar^2\varepsilon\varepsilon_0/e^3 m^* = (\varepsilon/m^*)a_{\mathrm{H}} \tag{5.11}$$

where R and a_{H} are the Rydberg energy (13 eV) and the hydrogen atom Bohr radius (0.5 Å) and m^* is the effective mass of the electron, defined from the curvature, $\mathrm{d}^2E/\mathrm{d}k^2$, of the energy dispersion bands. The high dielectric constant and low effective mass of a semiconductor result in a small donor binding energy and a large localization length.

A more precise effective mass theory for a crystalline semiconductor takes into account the details of the band structure in a way that is not applicable to an amorphous semiconductor. The theory gives a donor binding energy in crystalline silicon of 30 meV. In practice, the measured values are larger and depend on the impurity, because there is a significant fraction of the wavefunction at the donor atom and the impurity atom contributes to the donor energy. This is the central cell correction, E_{cc}, and the total donor binding energy is

$$E_{\mathrm{DN}} = E_{\mathrm{DNO}} + E_{\mathrm{cc}} \tag{5.12}$$

The central cell correction results in a binding energy of 40 meV for phosphorus and 45 meV for arsenic in crystalline silicon and the boron acceptor energy is 45 meV.

The Coulomb interaction between the electron and the donor core is, of course, present in an amorphous semiconductor and binds an electron in much the same way, so the shallow donor state is preserved. The effective mass theory for dopants cannot be applied directly to amorphous semiconductors, because it is formulated in terms of the momentum–space wavefunctions of the crystal. It is not immediately obvious that the effective mass has any meaning in an amorphous material in which momentum is not conserved. Kivelsen and Gelatt (1979) argue, however, that the effective mass remains a valid concept and express m^* in terms of the average gap, E_0, and the minimum band gap energy, E_G,

$$m^* = m_e(7E_0/4E_G - 1)^{-1} \tag{5.13}$$

where m_e is the electron mass. According to this definition, m^* for a-Si:H is about 0.4 m_e, when $E_G = 1.7$ eV and E_0 is set equal to the peak in the imaginary part of the dielectric function (see Fig. 3.16).

The donor binding energy in the crystal is measured from the bottom of the conduction band, but the equivalent energy reference level in a-Si:H is less obvious. The donor wavefunction is made up of the low energy states of the a-Si:H conduction band lying within the donor radius and these might be extended states or localized tail states, depending on the local configuration. The disorder potential of the random network acts on the donor state to further localize its wavefunction and to create a broad distribution of binding energies. The localization radius of the donor should not be larger than that of a band tail state in a similar environment of disorder potentials. The additional localization increases the donor binding energy from the effective mass value and also increases the central cell correction. The donors should therefore overlap the band tail and have a broadening which approximately follows the shape of the band tail of the equivalent undoped material. A schematic diagram comparing the crystalline and amorphous material is shown in Fig. 5.6.

The expectation that substitutional impurities behave as effective mass donors and acceptors in a-Si:H is borne out by experiment. However, doping also induces deep states in a-Si:H and these are discussed first.

Fig. 5.6. Comparison of the band edge and dopant states in (*a*) crystalline and (*b*) amorphous silicon.

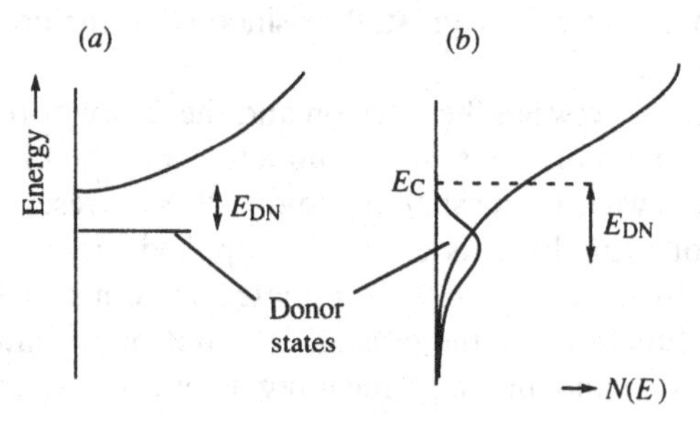

5.2.1 *Defects induced by doping*

When doping was first observed in a-Si:H, it appeared that the behavior was the same as in crystalline silicon, except perhaps for a lower doping efficiency. One of the first indications of different properties came from the light-induced ESR (LESR) experiments illustrated in Fig. 5.7. The ESR signal is very weak in the absence of illumination, with a spin density of about 10^{16} cm^{-3} for a doping level of $C_g = 10^{-4}$. The dark ESR resonance line shape is different in n-type and p-type a-Si:H and is attributed to band tail electrons and holes, as discussed in Section 5.2.2. When the sample is illuminated with absorbing light at low temperature, there is a much larger ESR signal. The resonance in boron-doped samples is composed of the 2.0055 dangling bond signal and an approximately equal density of band tail holes. The resonance in n-type material is less obviously resolved into components, but careful analysis shows that it contains dangling bonds and band tail electrons.

The LESR data are explained by the presence of a large defect density and by the shift in position of the Fermi energy. The movement of the Fermi energy from mid-gap to the band edge by the doping changes the equilibrium electron occupancy of the defects. These gain an electron and become negatively charged in n-type a-Si:H, and lose an electron to become positively charged in p-type. Neither configura-

Fig. 5.7. Light-induced ESR spectra of p-type and n-type a-Si:H, showing the induced $g = 2.0055$ resonance which is absent in the dark (Street and Biegelsen 1982).

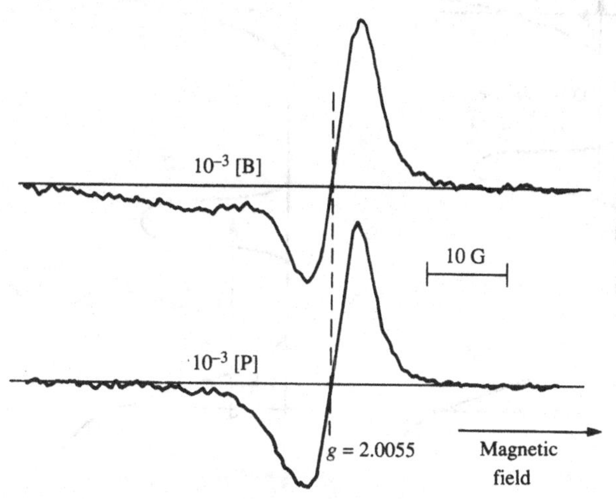

tion is paramagnetic, so that there is no equilibrium spin signal from the defects. The ground state is illustrated in Fig. 5.8. Illumination of n-type material excites an electron out of the doubly occupied D^- state, leaving it in the paramagnetic D^0 state (see Fig. 5.8). The electron is trapped in the band tail and recombination back to the defect is inhibited by the low temperature which renders the band tail electrons immobile. The LESR resonance contains both types of states because the band tail electrons are also paramagnetic. The density of light induced states, N_{DL}, is approximately,

$$N_{DL} = \frac{N_D \beta G\tau}{1+\beta G\tau} \leqslant N_D \tag{5.14}$$

where G is the illumination flux, β is the transition rate, and τ is the recombination time. (In practice there is a distribution of recombination times.) The LESR spin density is therefore only a lower limit on the total defect density. Some recombination does take place (see Chapter 8), and the LESR signal increases slowly with the illumination density. The LESR data demonstrate that the defect density is quite high in doped material, being about 10^{17} cm^{-3} at a doping level of 10^{-5}, compared to a density of less than 10^{16} cm^{-3} in similarly prepared undoped a-Si:H.

Several other experiments confirm the high defect density in doped

Fig. 5.8. Diagram showing how illumination changes the charge state of the defects and induced band tail carriers: (*a*) n-type and (*b*) p-type a-Si:H.

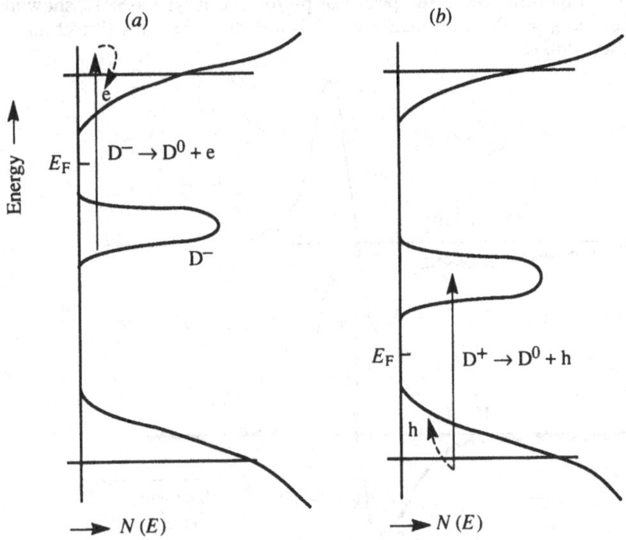

a-Si:H. PDS measurements find that the subgap absorption in doped a-Si:H is much stronger than in undoped material (Jackson and Amer 1982). The inferred defect density increases with doping and is roughly the same in both n-type and p-type material at the same doping level. The concentrations are found by applying the optical cross-section that is used for the neutral dangling bond undoped samples, and which is given by Eq. (4.26). DLTS data on deep defect states in n-type doped a-Si:H are described in Section 4.2.3, together with the evidence that the DLTS defect is the negative charge state of the dangling bond. The defect density and the charge state are inferred from time-of-flight and luminescence experiments – these are discussed in Chapter 8.

The variation of the defect density with doping level, as measured by the different experiments, is shown for phosphorus doping in Fig. 5.9. The defect density increases with the square root of the phosphorus concentration,

$$N_D = 3 \times 10^{19} C_g^{\frac{1}{2}} \text{ cm}^{-3} \tag{5.15}$$

The same relation applies for the dopants boron and arsenic and, indeed, the total defect density is approximately the same for all three dopants. The dopant concentration in Fig. 5.9 is that of the gas phase impurity concentration, C_g, not that of the solid. The equivalent plot against C_s, is changed only slightly for boron and phosphorus, but the arsenic data are substantially different, because of the large and non-linear distribution coefficient shown in Fig. 5.4. A challenge for any model of doping is to explain both the square root dependence of the defect density and why this applies to the gas concentrations and not to those of the grown film.

5.2.2 *Shallow electronic states*

The estimated dopant binding energies of about 100 meV indicate that the donor and acceptor levels overlap the intrinsic band tail states. Fortunately, ESR measurements allow a separate identification of both types of states. From the knowledge gained in the previous section that doping interacts with the silicon structure to create extra defects, one might anticipate that doping could also modify other aspects of the density of states distribution. Such effects, however, appear to be small and are significant only at very high doping levels. A boron doping of $C_g = 10^{-2}$ reduces the band gap by about 0.1 eV and the equivalent phosphorus doping has a smaller effect on the band gap. The electron and hole drift mobilities decrease slowly with doping, such that the band tail slope is increased by only about 10% at $C_g = 10^{-4}$. Even these changes may be more due to the random Coulomb potential introduced by charged defect and dopant states, than structural rearrangements of the silicon network (see Section 7.4.5). The broadening effects can be neglected in most situations and the density of states distribution near the band edges is approximately the sum of contributions from the band tails of undoped a-Si:H and the additional donor or acceptor states.

(1) *ESR measurements*

The dark ESR spectra of doped a-Si:H in Fig. 5.10 show resonances near $g = 2$, with different line shapes and g-values from those of the dangling bond (Stuke 1977). These lines are attributed to band tail states because they are observed when the Fermi energy is moved up to the band tails by doping and also in the low temperature LESR spectra of undoped a-Si:H, when electrons or holes are optically excited into the band tails. The larger g-shift for the valence band tail states than for the conduction band states is expected from Eq. (4.12).

Fig. 5.9. The gas phase doping dependence of the defect density and occupied band tail density, obtained in different experiments (Street *et al.* 1985).

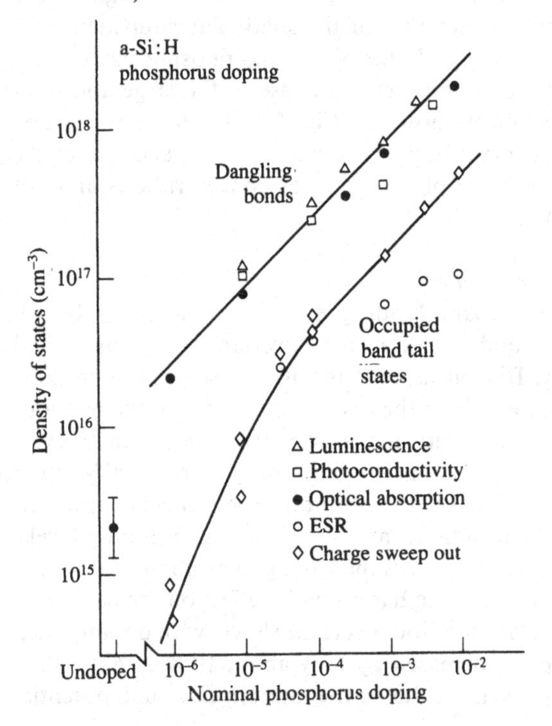

More sensitive ESR measurements over a wider magnetic field range find additional resonances in phosphorus- and arsenic-doped material, examples of which are shown in Fig. 5.11. The extra lines (two for phosphorus and four for arsenic) are due to the hyperfine interaction of the electron bound to the donor (Stutzmann and Street 1985). The ESR spectra have exactly the number of lines and relative intensities expected from the nuclear spins of $\frac{1}{2}$ and $\frac{3}{2}$ for phosphorus and arsenic atoms. The splitting of the lines, ΔH_{da}, is proportional to the electron density at the nucleus and is a measure of the localization length, r_{da}, of the donor.

$$\Delta H_{da} \approx \Delta H_{dx}(r_{dx}/r_{da})^3 \tag{5.16}$$

where ΔH_{dx} and r_{dx} are the hyperfine splitting and donor radius in crystalline silicon. ΔH_{dx} is smaller than ΔH_{da} by a factor of about 5, indicating a larger radius donor state, which is known from calculation to be 17 Å. By scaling the splitting according to Eq. (5.16), the localization length for phosphorus in a-Si:H is deduced to be 10 ± 1 Å, and that for arsenic 9 ± 1 Å. A simple model predicts that the localization length varies inversely as the square root of the donor binding energy (Eq. (1.10)). The extra localization of the states indicates a donor binding energy which is roughly double that of the crystal and therefore about 100 meV. On the other hand, the central cell correction should scale with the wavefunction at the core and so should be about

Fig. 5.10. Low temperature equilibrium ESR spectra of heavily doped n-type and p-type a-Si:H. Dashed line indicates g-value of dangling bonds (Street and Biegelsen 1984).

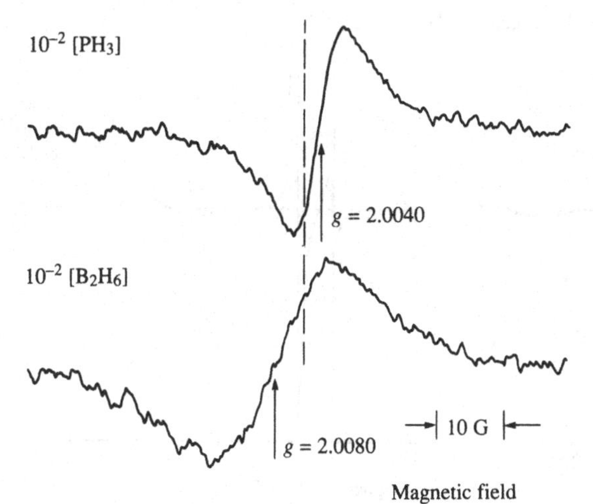

five times larger in amorphous silicon. This simple reasoning predicts that the arsenic donor in a-Si:H is roughly 25 meV deeper than phosphorus.

Tight binding theories of the donor states agree reasonably well with the effective mass estimates. Calculations based on an a-Si:H cluster model find that the phosphorus donor binding energy is about 0.2 eV and that for arsenic is about 0.05 eV larger (Nichols and Fong 1987). The boron acceptor level is calculated to be 0.35 eV above the valence band and both the donor and acceptor levels are broadened out by the local bond angle disorder, as would be expected. The tight binding calculations are likely to give larger binding energies than the effective

Fig. 5.11. ESR spectra of phosphorus and arsenic doped a-Si:H over a wide magnetic field range, showing the lines split by the dopant hyperfine interaction (Stutzmann and Street 1985).

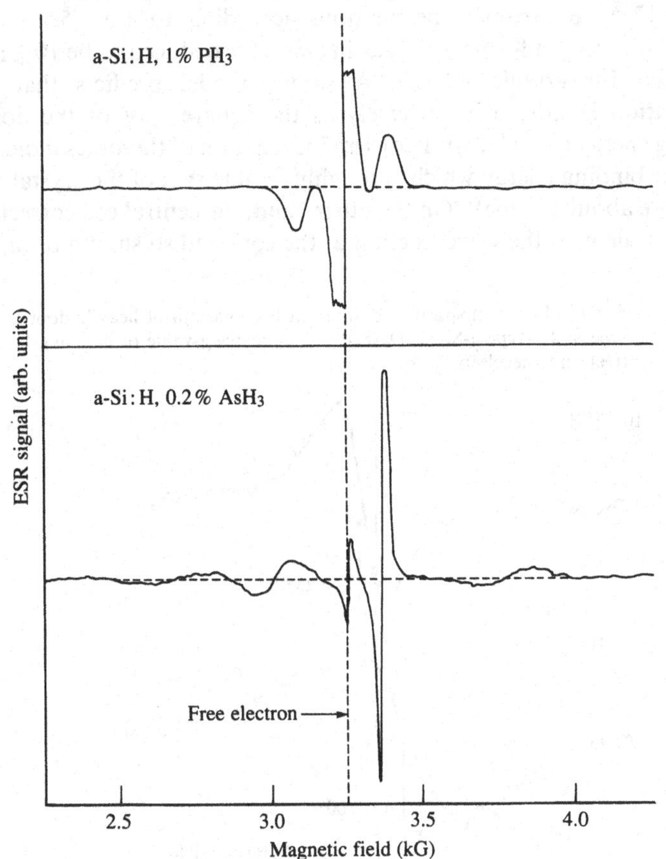

mass model, because the cluster is sufficiently small that the donor wavefunction can extend to the boundary.

The shallow donor levels are broadened out by disorder and form a band which overlaps with the intrinsic conduction band tail states. The overlap of the states is seen by the coexistence of the hyperfine and band tail resonances in Fig. 5.11. The concentrations of the two types of states are obtained from the intensities of the corresponding resonances and are shown in Fig. 5.12. The densities of neutral phosphorus donors and band tail electrons are approximately equal at low doping levels, but the donor states dominate at high doping. In contrast, the arsenic donors always have a much higher density than the band tail electrons. It is clear that the arsenic donor must be deeper than the phosphorus donor because the relative densities simply reflect the integrated densities of the two types of states below the Fermi energy. Assuming an exponential band tail with a slope of 25 meV, then a shift of the donor energy by 50 meV leads to a factor of about 10 larger contribution of the donor to the total spin signal. The measured difference between the two donors is about this amount, indicating that the energies differ by roughly 50 meV. This central cell correction is consistent with the estimates of the effective mass approximation.

The donor electrons are thermally excited into the larger density of conduction band states at elevated temperatures, reducing the neutral donor concentration. Donor ionization occurs in crystalline silicon near 20 K, but the corresponding effect in a-Si:H begins at about 200 K. The ESR spin density of the neutral phosphorus donor

Fig. 5.12. Doping dependence of the band tail and neutral donor densities in phosphorus and arsenic doped a-Si:H (Stutzmann, Biegelsen and Street 1987).

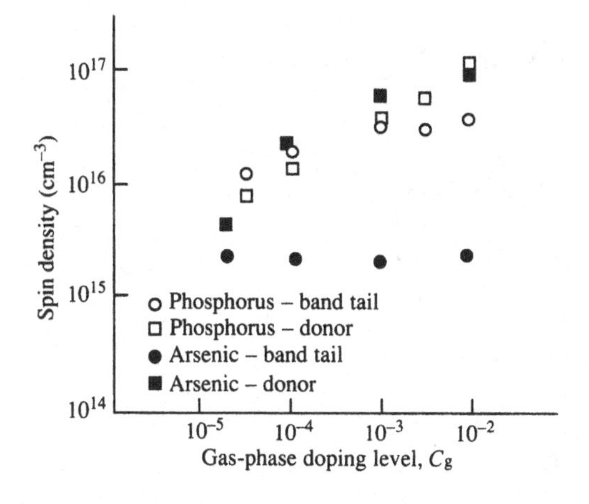

decreases with increasing temperature, vanishing near room temperature as is shown in Fig. 5.13 for phosphorus and arsenic. The density of band tail electrons increases, confirming the ionization of the donors. However, the sum of the two electron densities is not constant with temperature, so that evidently some of the electrons are not being detected by the ESR. It is not known whether the missing electrons occupy paired electronic states, or whether the ESR resonance of the high energy band tail states is rendered undetectable by a large lifetime broadening. Ionization of the arsenic donor occurs at a higher temperature because of its larger binding energy.

The doping dependence of the band tail and donor states, taken with the known position of the Fermi energy, allows an estimate of the density of states distribution of the donors, which is shown in Fig. 5.14 for a doping level of $C_g = 10^{-3}$. The average binding energies are about 100 meV for phosphorus donors and 150 meV for arsenic, measured from the mobility edge.

There is no sign of an ESR hyperfine interaction in boron-doped a-Si:H, so that there is little information about the acceptor states. It may be that boron acceptors have an unexpectedly small hyperfine interaction. A more likely explanation is that virtually all the acceptors are ionized. The valence band tail is much broader than the conduction band and the Fermi energy remains further from the band edge, so that the probability that a hole occupies an acceptor is much smaller. The

Fig. 5.13. The temperature dependence of the band tail (BT) and neutral donor (ND) density of phosphorus doped a-Si:H. The equivalent temperature dependence of the arsenic donors is shown by the dotted line (Stutzmann *et al.* 1987).

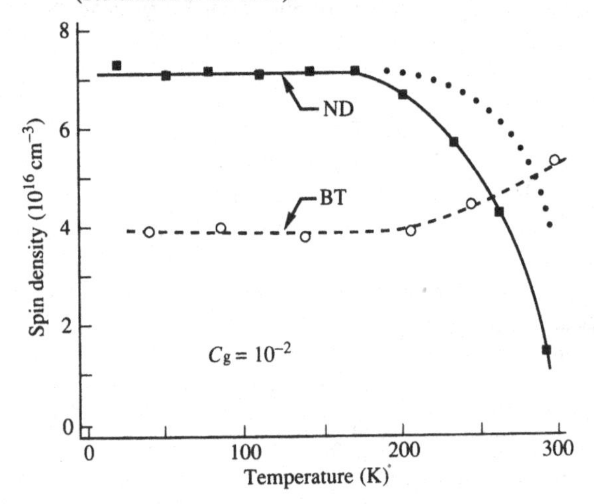

density of states model for the boron acceptors in Fig. 5.14 assumes an average binding energy of 100 meV and a broadening similar to that of the valence band tail. With these assumptions, the density of occupied acceptors is never greater than 1% of the occupied band tail states, which accounts for the absence of an observable acceptor hyperfine ESR signal.

A comparison of Fig. 5.12 and the defect density in Fig. 5.9 shows that the total density of the band tail electrons – neutral donors plus occupied intrinsic band tail states – is about ten times less than the density of deep defects induced by the doping. This is a remarkable result because it implies that almost all the donors are compensated by deep defects. However, before considering the consequences of this observation, it is helpful to discuss an alternate experimental technique for measuring the density of shallow electrons or holes, because of the possibility that ESR is missing some of the carriers due to electron pairing or broadening of the resonance.

Fig. 5.14. Density of states distribution estimated from the experimental information, showing the donor and acceptor states corresponding to a doping level of $C_g = 10^{-3}$. Dashed lines show the intrinsic band tail states (Street 1987b).

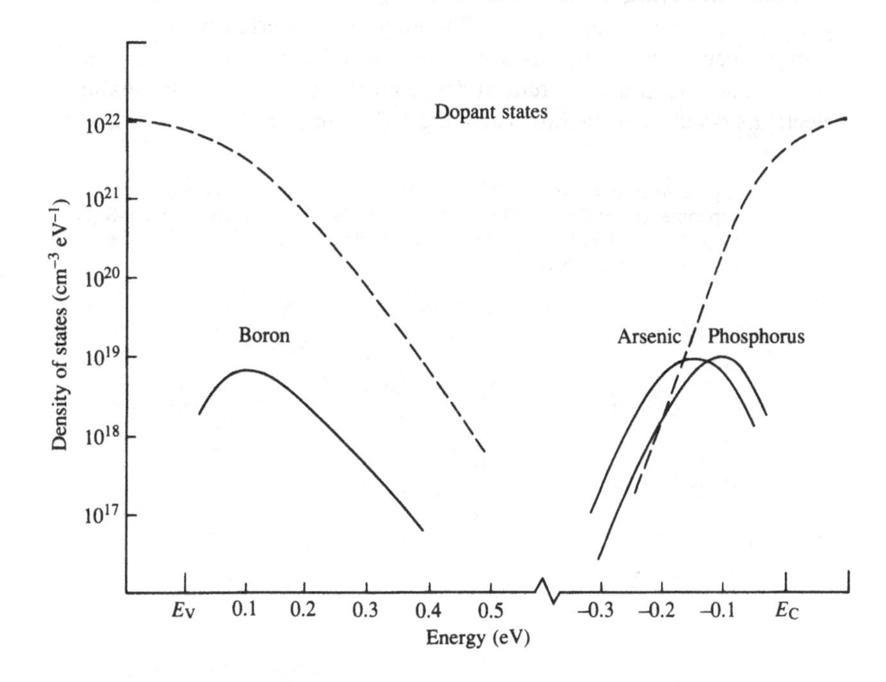

(2) *Electronic measurements of shallow states*

The sweep-out experiment provides an electrical measurement of the density of shallow occupied states (Street 1989). This technique uses a sample of a-Si:H in which a thin doped layer is embedded near one side of an otherwise undoped a-Si:H film – the sample structure is illustrated in Fig. 5.15(*a*). The measurement is performed by recording the transient current after the application of a voltage step. The result of a typical sweep-out experiment for a sample with a thin n-type layer is shown in Fig. 5.15(*b*). There is a current pulse due to the displacement current of the geometrical capacitance and also an additional current that typically lasts for about 10^{-6} s, before dropping to a very low value. The extra current is due to shallow electrons in the n-type layer which are swept across the thick undoped layer by the applied field. Only electrons which are shallow enough to be excited to the band edge move within a time, t, after the voltage is applied, and these occupy states of energy,

$$E < kT\ln(\omega_0 t) \tag{5.17}$$

Thus, within 1 μs, the room temperature current is entirely due to carriers occupying states shallower than about 0.3 eV, which are present only in the doped layer. The undoped material, with its Fermi energy deep in the gap, has a minimum carrier release time of order 10^{-2} s. The sweep-out current arises from the emission of the shallow electrons occupying the intrinsic band tail states, the donor states, and

Fig. 5.15(*a*) Illustration of the sample structure used in sweep-out experiments. (*b*) Example of the sweep-out current pulse for n-type a-Si:H. Q_D is the displacement charge of the sample electrodes, and Q_{SW} is the sweep-out charge (Street 1989).

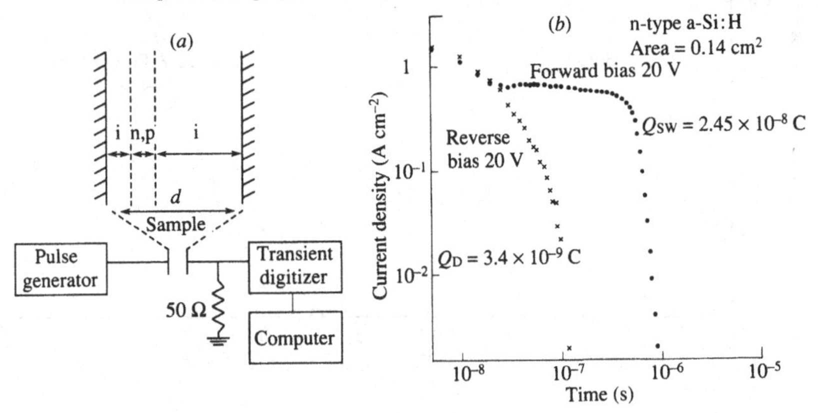

the small density of carriers above the mobility edge. These electrons are collectively denoted by the density n_{BT}. The precipitous drop in the current at the end of the pulse occurs because there is a very small density of states in the energy range between the conduction band tail and the deep states, so that there is a time range in which all the shallow carriers have been swept out, but few of the deep states can be excited. This measurement confirms the deep dip in the density of states which was first found by DLTS experiments and is shown in Fig. 4.18. In p-type material the distinction between the shallow and deep states is less clear, because the broader band edge merges more smoothly with the deep states. Nevertheless, the approximate density of shallow holes can also be obtained by sweep-out.

The value of n_{BT} is given by the time integral of the current pulse, Q_{SW}, after the displacement charge, Q_D, is subtracted. Since it is an electrical measurement, the technique does not distinguish between donor and band tail states, paired or unpaired states etc., but measures all the carriers within the allowed energy range. Experiments confirm that the sweep-out density is equal to the total ESR spin density in the different band edge resonances, within the experimental error (Stutzmann *et al.* 1987). Thus the low temperature ESR data is indeed measuring all the band edge carriers.

Fig. 5.16 shows the doping dependence of n_{BT} obtained from sweep out, for the phosphorus, arsenic, and boron dopants, compared to the dangling bond density. The results confirm that n_{BT} is about an order of magnitude less than the defect density and is even lower in the case of light boron doping. It should be noted that the value of n_{BT} actually depends quite sensitively on the thermal history of the sample, as is discussed in the next chapter, but it always remains less than the defect density. These data are for samples slowly cooled from the deposition temperature and stored at room temperature for an extended period and so are typical of samples that have not had any deliberate thermal treatment.

5.2.3 *The doping efficiency*

All the information needed to calculate the doping efficiency, η, of a-Si:H from Eq. (5.2) is provided by the experiments discussed above. η is obtained by equating the excess electron concentration with the density of donors and is defined in terms of either the gas-phase or solid-phase impurity concentration,

$$\eta_g = \frac{N_D + n_{BT}}{N_{Si}\, C_g}; \quad \eta_s = \frac{N_D + n_{BT}}{N_{Si}\, C_s} = \frac{\eta_g}{d_I} \tag{5.18}$$

The two values differ by the impurity distribution coefficient, d_I. A crucial assumption is that dangling bonds are the only deep defects which take up donor or acceptor electrons and holes. No other charged gap states of significant density have been found.

The electrons in the band edge states contribute only about 10% to the total (see Fig. 5.16), so that the simple approximate relation is

$$N_{donor} \approx N_D \tag{5.19}$$

Almost all the donors are compensated by deep defects, a result which is completely different from crystalline silicon in which there is virtually no defect compensation. The increased conductivity of doped a-Si:H results from the small excess band tail electron density which is present because N_{donor} is, in fact, slightly larger than N_D, so that

$$n_{BT} = N_{donor} - N_D \tag{5.20}$$

Fig. 5.16. Doping dependence of density of occupied band edge states, n_{BT}, for different dopants (Street 1987b).

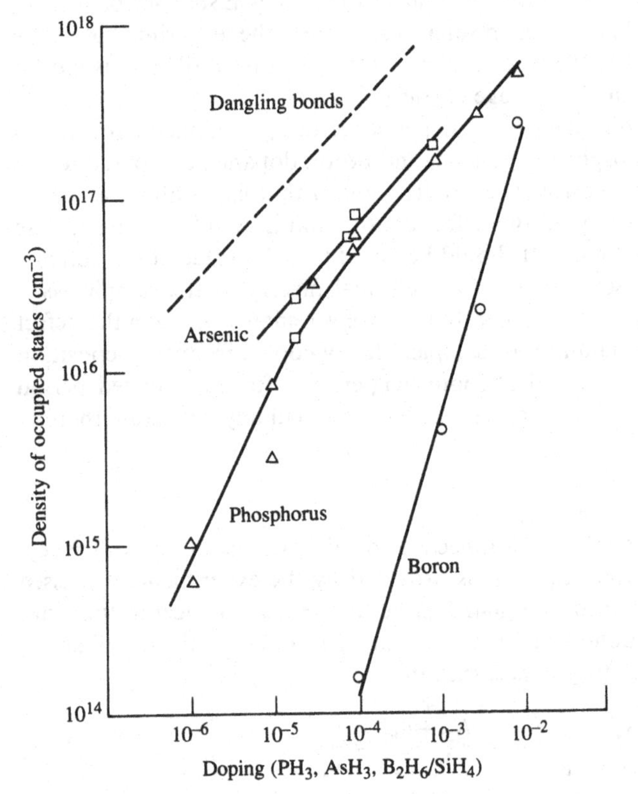

The defect density varies as the square root of the gas-phase phosphorus concentration, so that from Eq. (5.19),

$$N_{donor} \approx \text{const. } C_g^{\frac{1}{2}} \qquad (5.21)$$

and the gas phase doping efficiency is given by,

$$\eta_g = N_{donor}/C_g \approx \text{const. } C_g^{-\frac{1}{2}} \qquad (5.22)$$

Fig. 5.17 evaluates the doping efficiency of the different dopants in a-Si:H and a-Ge:H, based on measurements of the defect and band edge carrier densities.

The inverse square root dependence of the doping efficiency on C_g is evident in Fig. 5.17. An unexplained result is that all the dopants so far investigated have approximately the same doping efficiency. It is also apparent in Fig. 5.17(*a*) that when the doping efficiency is expressed in terms of the solid phase impurity concentration, the results do not follow a common doping dependence. This is particularly true for arsenic doping which has a large, doping-dependent, distribution coefficient. An explanation of the difference between the gas-phase and solid-phase doping efficiency is discussed in the Section 6.2.7.

The doping efficiency decreases continuously, dropping below 1 % at high doping levels. This result contrasts with the doping in crystalline silicon for which the doping efficiency is independent of the impurity concentration with a value of unity up to the solubility limit. The data also show how weak the doping effect is in a-Si:H. For example, at a gas phase doping level of 1 %, less than 1 % of the phosphorus is in the form of four-fold donor states, and of these donors, about 90 % are

Fig. 5.17. Evaluation of the doping efficiency of different dopants in a-Si:H and a-Ge:H, as a function of (*a*) the solid-phase concentration and (*b*) the gas-phase concentration (Stutzmann *et al.* 1987).

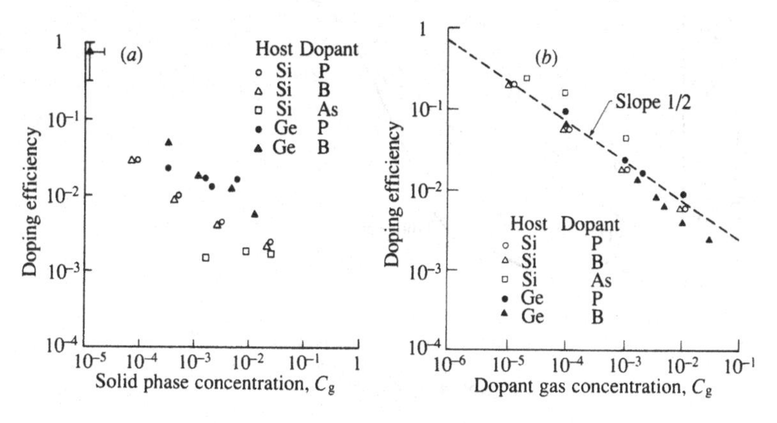

compensated by deep defects. Of the remaining 10% of the donor electrons at the band edge, most occupy localized states – at room temperature only about 10% are in conducting states above the mobility edge. Measured in terms of the free electron concentration, the doping efficiency is a mere 10^{-4}. Taking into account the much lower mobility of carriers in a-Si:H compared to crystalline silicon, it is not surprising that the maximum conductivity of n-type a-Si:H is more than five orders of magnitude below that of the crystal.

5.2.4 *Compensated a-Si:H*

Another fascinating aspect of doping in a-Si:H is that compensated material has strikingly different properties from the singly doped material. The attributes of doped a-Si:H are a low doping efficiency and the introduction of a large defect density, roughly equal to the active donor or acceptor concentration. The dopant states mostly lie within the existing band tail states and, because the doping efficiency is low, the band edge region is not greatly perturbed. The remaining impurities are inactive and contribute to the valence and conduction bands, although not significantly to the density of states at the band edge. Compensated a-Si:H in which there are roughly equal densities of boron and phosphorus is completely different. There are very few defects, but the band edge is greatly broadened out.

Fig. 5.18 compares the optical absorption spectra of phosphorus doped a-Si:H with the corresponding compensated material. The compensated a-Si:H has a substantially broader Urbach edge, but a

Fig. 5.18. Comparison of the optical absorption spectra of undoped, n-type and compensated a-Si:H (Jackson and Amer 1982).

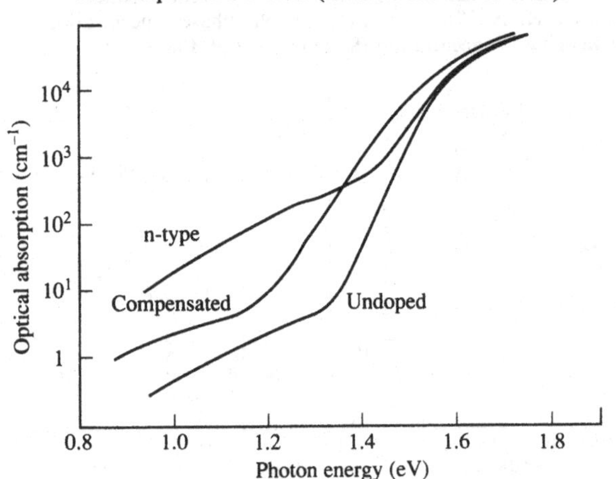

weaker defect absorption band. The shape of the Urbach tail is dominated by the valence band which is evidently broadened by compensation. A broadening of the conduction band is deduced from the time-of-flight data for the electron drift mobility shown in Fig. 5.19. There is a large decrease in the mobility as the doping level increases and a corresponding increase in the activation energy, as well as a large anomalous dispersion effect. The hole drift mobility is also reduced by 2–3 orders of magnitude at room temperature, with an increase in both the activation energy and the dispersion. Taken together these results show that both the conduction and valence band tails are broadened by compensation.

The low defect density in compensated material is apparent from the optical data in Fig. 5.18, which show a much reduced defect absorption band. The same result is deduced from time-of-flight and ESR data. Although the drift mobility is low, the mobility–lifetime product is comparable with the best undoped material, confirming the low defect density (see Fig. 8.24). The dangling bond density in the dark ESR experiment is about 4×10^{15} cm^{-3}, with little dependence on the doping

Fig. 5.19. Temperature dependence of the electron drift mobility of compensated a-Si:H. Data for a range of applied fields from 3×10^3 to 5×10^4 V cm^{-1} are shown (Marshall *et al.* 1984).

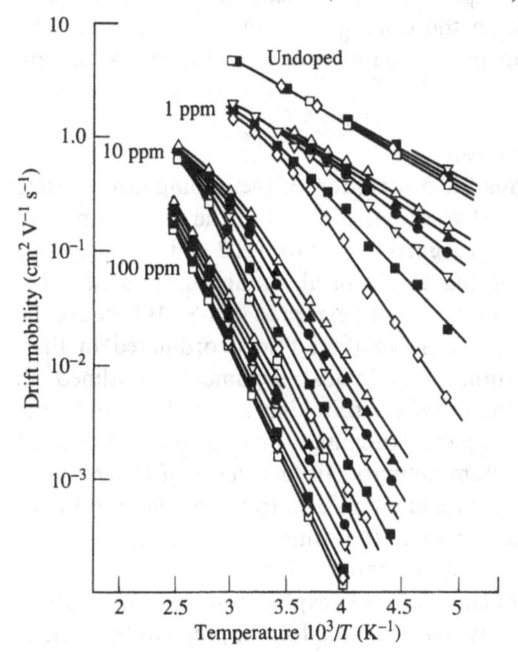

level and is close to the level in undoped a-Si:H. Few defects are observed in LESR and, at the higher doping levels, virtually all the electron density is in the hyperfine lines of neutral donors, with little spin density in the conduction band tail states. The hyperfine splitting is a little larger than in singly doped material, suggesting that the states are more localized, as would be expected from the broader band tail. Evidently, a large fraction of the states in the broadened conduction band tail originate from four-fold coordinated phosphorus atoms.

Boron–phosphorus complexes are one possible explanation of the broadening of the band tails in compensated a-Si:H. NMR data find that such close pairs exist and account for about half of the impurity atoms at high doping levels. It is not known whether these complexes form deep tail states and, in fact, it has been suggested that they do not, because boron phosphide has a wider band gap than a-Si:H, so that states well within the bands might be more likely. A second explanation of the broadening is that donors and acceptors are present in much larger concentrations than in the singly doped material and cause long range potential fluctuations. Initially there was no reason to suppose that the doping efficiency should increase in compensated material, and the boron–phosphorus complex model seemed the more plausible. However, as the mechanisms of doping in a-Si:H have become better understood, the second explanation has become more acceptable because such an increase in the doping is predicted by theory. The interpretation of the compensation data is discussed in the next section and in Chapter 6.

5.3 The doping mechanism

This section begins the description of the doping mechanisms in a-Si:H, first as a qualitative discussion and in the next chapter in more quantitative detail. It is argued in Section 1.2.2 and the beginning of this chapter, that according to chemical bonding arguments – the $8-N$ rule – substitutional doping is not expected in a-Si:H because the boron or phosphorus impurities are three-fold coordinated in their lowest energy configuration. Topological arguments, outlined in Section 2.2.2, also favor the lower coordination of the three-fold state rather than the four-fold dopants, because a-Si:H is overcoordinated compared to the ideal random network. In fact, most of the impurity atoms are in the three-fold state and conform to the predictions of the bonding and topology arguments. It is the small density of active donors and acceptors that requires explanation.

There are two approaches to the explanation of the doping properties. One possibility is that the four-fold doping configurations occur more or less accidentally at the surface during the growth. The excess energy from the plasma might allow the higher energy configurations, during the bonding reconstruction which takes place at the growing surface. It is difficult to evaluate this approach because little is known in detail about the surface of the growing film. However, the doping would certainly be sensitive to the details of the deposition conditions, which is not observed. The second possibility is that the four-fold doping configuration, although of higher energy than the three-fold states, is low enough in energy to be present in reasonable concentrations. This approach is successful in explaining both the doping and the associated defect formation.

Low energy dopant configurations may be explained by extending the $8-N$ rule approach to charged impurity states (Street 1982). For example, a positively charged phosphorus atom contains only four valence electrons instead of the usual five, and so its ideal bonding configuration is with a coordination of 4, P_4^+. The electronic configuration is the same as silicon, apart from the extra charge. Therefore, ionized donors obey the $8-N$ rule, even though neutral donors do not. The position of the Fermi energy determines whether or not a donor is ionized, so that when E_F is deep in the gap, the P_4^+ state is favored, but the donor density is suppressed when E_F is at the band edge. Since it is not possible to have ionized donors without an equal density of some other state to take up the excess charge, a low energy doping state should consist of ionized donors and compensating deep defects. The measurements find exactly this situation, with the dangling bonds being the compensating defects.

A schematic model of the donor and defect bonding states is shown in Fig. 5.20. The sp^3 hybridized states of the P_4 donor are split into bonding and the anti-bonding donor level, while the silicon defect has three bonding states and a dangling bond level in the middle of the gap. (The silicon anti-bonding states are not shown.) The transfer of the electron from the donor onto the dangling bond results in a gain of energy E_p–E_{d2}, which is 0.5–1 eV. This large transfer of energy makes the defect and dopant pair a low energy configuration. The main conclusion is that the donor and defect states are not independent consequences of the doping, but are inextricably linked. Substitutional doping in a-Si:H does not occur without the defects.

The dopants and defects form pairs in the sense of charge compensation, but are not necessarily physically paired on adjacent sites. The pairing at independent distant sites occurs because both states interact with the electron distribution, whose thermodynamic energy is defined by the Fermi energy. There is a close relation between the doping properties and the Fermi energy which is the basis of all the metastable phenomena described in the next chapter.

The role of the Fermi energy in the creation of defects and dopant states is shown as follows. The formation energy of a neutral donor is defined to be U_{po}. An ionized donor is created by transferring an electron from the donor level to the Fermi energy, so that the total formation energy is

$$U_p = U_{po} - (E_p - E_F) \tag{5.23}$$

where the energies are defined in Fig. 5.21. Similarly, the defect formation energy contains a term U_{do} for the neutral defect and the energy gained when an electron drops from the Fermi energy onto the two-electron defect state at E_{d2},

$$U_d = U_{do} - (E_F - E_{d2}) \tag{5.24}$$

Boron doping has equivalent properties, with the formation energies given by

$$U_b = U_{bo} - (E_F - E_b)$$

and

$$U_d = U_{do} - (E_{d1} - E_F) \tag{5.25}$$

Here the Fermi energy lies between the acceptor level and the defect state and electron transfer is from the defect to the acceptor. The appropriate energy level of the defect is therefore the one-electron state.

Returning to the n-type doping case, the total energy of the donor–defect pair is

$$U_p + U_d = U_{po} + U_{do} - (E_p - E_{d2}) \tag{5.26}$$

Fig. 5.20. Schematic energy diagram of the donors and dangling bond defects, showing the energy gain by the transfer of an electron from the donor to the defect.

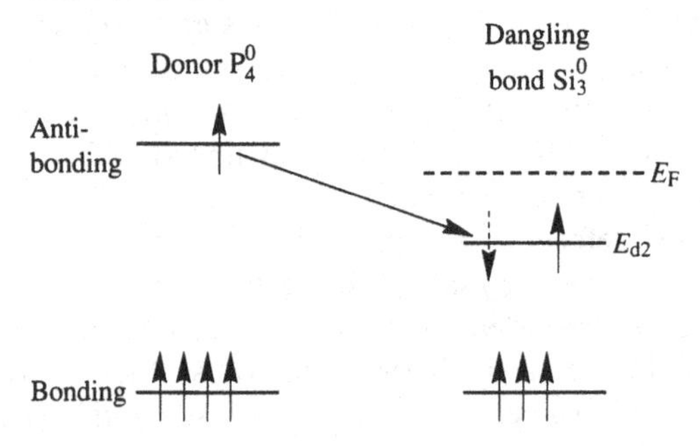

which shows that the formation energy is reduced by the transfer of the electron from the donor to the dangling bond.

Fig. 5.21 plots the formation energies of the donor and dangling bond as a function of the position of the Fermi energy, for assumed values of U_{po} and U_{do}. Increasing the position of E_F in the gap raises the donor formation energy, thus suppressing donor creation while simultaneously reducing the defect energy, causing a larger density. Chapter 6 discusses how this process is also responsible for the many metastable defect creation processes in a-Si:H.

The doping of a-Si:H can now be understood in terms of a hierarchy of possible impurity configurations shown in Fig. 5.22. Again taking the specific example of phosphorus doping, these are, in order of increasing formation energy; the non-doping state, P_3^0; the defect-compensated donor, $P_4^+ + D^-$; and the neutral donor, P_4^0. Accordingly

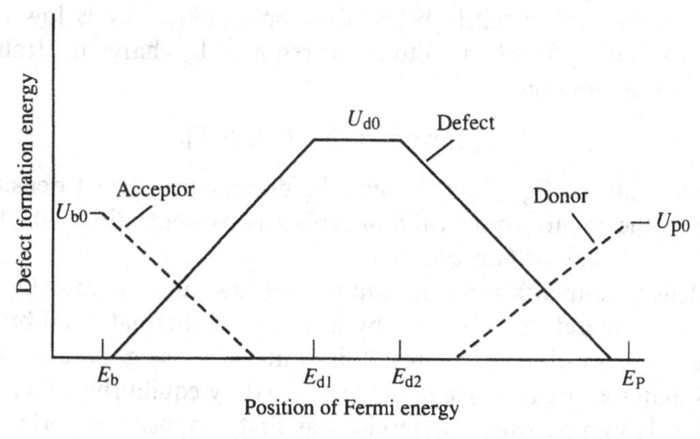

Fig. 5.21. Plot of the defect and dopant formation energies as a function of the Fermi energy position, assuming discrete gap states and formation energies.

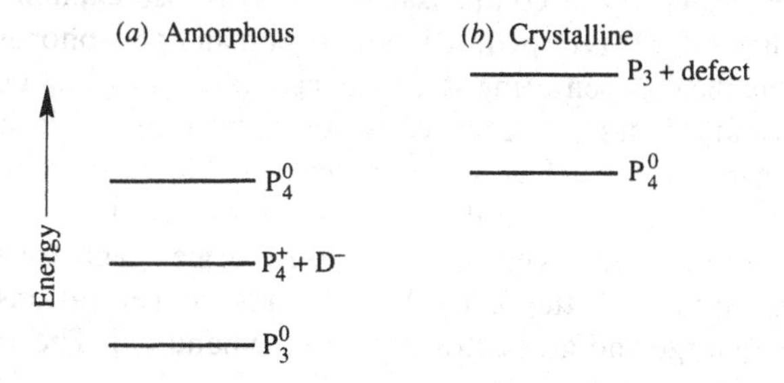

Fig. 5.22. Comparison of the formation energies of the various impurity bonding configurations in amorphous and crystalline silicon.

it is predicted that most of the phosphorus is three-fold coordinated, with most of the remainder in the form of ionized donors, but with very few neutral donors. This is exactly the situation found by experiment. Neglecting the high energy P_4^0 states, the balance between the three-fold and four-fold configurations can be written as the chemical reaction,

$$Si_4^0 + P_3^0 \leftrightarrows P_4^+ + D^-; \ (U_p + U_d) \tag{5.27}$$

where the term in the bracket is the energy of the reaction, and is given by Eq. (5.26).

In thermal equilibrium, the above reaction determines the concentrations of the different states through the law of mass action, with the result,

$$N_{P_4} N_D = K N_0 N_{P_3}; \ K = \exp[-(U_p + U_d)/kT] \tag{5.28}$$

where the different Ns denote the density of the different states, N_0 is the density of four-fold silicon sites, and K is the reaction constant. The derivation and application of the law of mass action is described in more detail in Section 6.2.1. When the doping efficiency is low (i.e. $N_{P_4} \ll N_{P_3}$), and N_{P_4} is set equal to N_D as required by charge neutrality, then Eq. (5.28) becomes,

$$N_{P_4} = N_D = (N_0 N_P)^{\frac{1}{2}} \exp[-(U_p + U_d)/2kT] \tag{5.29}$$

The main result of Eq. (5.29) is that the dopant and defect densities increase as the square root of the phosphorus concentration, which is the observed behavior (see Fig. 5.9).

The defect compensation, the square root law, and the low doping efficiency are therefore described by a model of thermal equilibrium bonding configurations. The result is quite surprising because disordered materials cannot usually be described by equilibrium thermodynamics. However, since the model was first proposed, experiments described in the next chapter have confirmed the equilibration of the dopants and defects.

The low defect density of compensated a-Si:H is also explained by the model. When a-Si:H is doped with both boron and phosphorus, the acceptors act as the compensating states instead of the dangling bonds. The defect density is suppressed because the Fermi energy is in the middle of the gap and the defect formation energy is large according to Eq. (5.24). The doping efficiency also increases; Fig. 5.21 and Eq. (5.23) show that the donor and acceptor formation energies decrease when E_F is in the middle of the gap. The dopant concentrations are correspondingly large and give extra states in the band tail. The energy gained by the transfer of an electron from the donor to the acceptor is approximately the band gap energy, 1.7 eV, and is larger than when the donors are compensated by defects, so that the total formation energy of the pairs is correspondingly reduced.

Although the doping model accounts for the general features of the experimental observations, several questions remain, which are addressed in the next chapter. Foremost is why and how thermal equilibrium concepts apply to a-Si:H which is apparently a non-equilibrium material. It is not obvious why the square root law for the defects applies to the gas-phase concentration, C_g and not always to the solid phase, C_s (see Fig. 5.17). The distinction is not easy to detect for phosphorus and boron doping, because C_s is almost proportional to C_g, but is more clear with arsenic doping, for which a very different concentration dependence would result. A further question is whether there is nearest neighbor pairing of the defect and donor states, as has been suggested by some experiments. The square root dependence of the defect density explicitly assumes that such pairing does not take place, because a paired defect would have a linear dependence on the phosphorus concentration (see Section 6.2.7). However the P_4^+ and D^- states do have a mutual Coulomb interaction, and so have some tendency to pair. Finally, do defects have similar equilibrium properties in undoped a-Si:H?

5.3.1 *Discussion of the doping model*

The main feature of the doping model is the thermal equilibrium distribution of the different possible bonding states. The low energy configurations are those in which the atomic coordination results in the pairing of electrons. There is a striking similarity between this doping model and the defect states in chalcogenide glasses. The P_4^+ and D^- states are analogous to a valence alternation pair (see Section 4.1.3), since the phosphorus is overcoordinated and the silicon undercoordinated. The change in coordination is accompanied by charge transfer, as in the chalcogenide defects, so that both atoms have full bonding and lone pair states, but empty anti-bonding states. This situation is anticipated by the $8-N$ rule for bonding (see Section 1.2.2). The common feature is that the lowest energy departure from the ideal network takes the form of a charged pair of defects. These are $(Se_3^+ + Se_1^-)$ in selenium, $(P_4^+ + D^-)$ in singly doped a-Si:H, and perhaps $(P_4^+ + B_4^-)$ in compensated a-Si:H. The doping of a-Si:H therefore falls within the expected properties of glasses.

Several consequences follow from the analogy with chalcogenide glasses. Valence alternation pairs pin the Fermi energy (see Section

4.1.2), so that E_F is also expected to be pinned between the donor and the dangling bond states. Conductivity experiments (Fig. 5.1) confirm that light phosphorus doping causes a large shift of E_F, but that higher doping causes only a small additional shift. The analogy with the chalcogenides also suggests that the defect and donor pair can be considered as states of a negative U defect. In one sense this is true, because the charge transfer reaction, Eq. (5.27), releases energy. However a negative U defect is one in which the addition of two electrons to the positive state immediately forms the negative defect. Here the positive and negative states are chemically different, so that conversion from one state to another does not normally take place. Furthermore, two neutral states are defined, P_4^0 and D^0, whereas a negative U defect has only one neutral state. The difference is, however, more quantitative than qualitative and depends on the time taken for a donor and a defect state to interconvert. A normal negative U defect changes coordination when charge state is altered in the time of a few lattice vibrations, with the process described by the configurational coordinate diagram of Fig. 4.2. The next chapter shows that the equivalent time for the interconversion of donors and defects in a-Si:H is about a year at room temperature, so that it is more useful to think of them as separate compensating defects.

The chemical bonding arguments expressed by the $8-N$ rule do not allow singly occupied electronic states. It follows that the density of states at the Fermi energy should be very low, because these states are singly occupied. Fig. 5.23 illustrates how one might expect the density of states to vary with doping under conditions in which the bonding interactions dominate, with the Fermi energy always in a region of low density of states. The DLTS experiments confirm that there are very few states between the defects and the band tail in n-type a-Si:H (Fig. 4.17) and the defect density is low in compensated a-Si:H when the Fermi energy is in the middle of the gap. Photoemission yield experiments indicate that E_F is always in a low density of states (10^{15} cm^{-3} eV^{-1}) for both boron and phosphorus doping. The predictions of the chemical bonding arguments therefore seem to work well in doped a-Si:H. These arguments introduce the idea, discussed further in Chapter 6, that the gap states in a-Si:H adjust their distribution depending on the position of the Fermi energy.

Finally it is interesting to contrast the doping mechanism in amorphous and crystalline silicon. The low doping efficiency and defect compensation of a-Si:H sets it apart from the normal doping of a crystalline semiconductor. The doping in a-Si:H is explained by the low energy bonding configurations, assuming thermal equilibrium.

This raises the question of why crystalline silicon should behave differently, despite being the equilibrium phase. The crystalline lattice constrains the impurity bonding, so that the three-fold state inevitably leaves an adjacent unsatisfied silicon bond. The additional energy of the broken bond ensures that the three-fold configuration has a higher energy than the four-fold state. The formation of compensating defects is similarly constrained by the lattice. The elementary defects in crystalline silicon are the vacancy and the interstitial, which have much larger formation energies than the dangling bond of a-Si:H. Fig. 5.22 compares the energies of the different impurity configurations in amorphous and crystalline silicon. The three-fold state has the lowest energy, but cannot be accessed in crystalline silicon due to the topological constraints. It is a curious observation that the thermal equilibrium crystalline structure prevents the formation of the lowest energy impurity configurations, whereas the non-equilibrium random network of a-Si:H allows the equilibrium impurity bonding states. At high enough concentrations the excess energy of the four-fold donors and acceptors in crystalline silicon is too large and precipitation occurs. Phosphorus-doped silicon precipitates as Si_3P_4, with three-fold coordinated phosphorus atoms, as our discussion leads us to expect. It

Fig. 5.23. Illustration of the doping dependence of the density of states distribution expected from chemical bonding arguments (Street 1985).

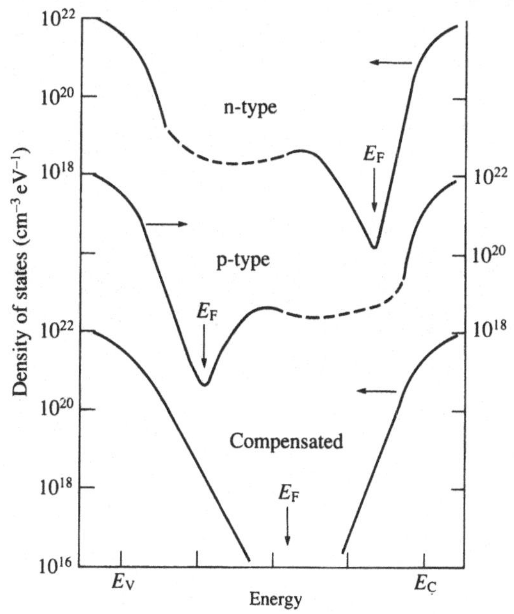

should also be clear from the defect compensation model of doping that no dopant precipitation is normally expected in an amorphous semiconductor and none is observed.

The doping model also helps explain the role of hydrogen in crystalline silicon. The formation energy of a three-fold impurity is greatly reduced by the presence of hydrogen, because the hydrogen can terminate any topologically induced broken bonds. The lower energy of the three-fold state is realized, with the effect that hydrogen passivates the doping, as observed. In boron-doped silicon, the local configuration appears to be a three-fold boron atom with a Si—H bond substituting for the fourth tetrahedral bond. Phosphorus-doped silicon may be different, with the hydrogen in an anti-bonding site, but dopant passivation occurs here too. The extra bonding options available to hydrogenated crystalline silicon suggest that the doping properties should be very similar to those of a-Si:H.

6 Defect reactions, thermal equilibrium and metastability

In Chapter 5, the doping properties of a-Si:H are described in terms of the alternative bonding structures of the dopants and the silicon network. The doping efficiency, η, is derived from the defect reaction, Eq. (5.27), by assuming thermal equilibrium between the different bonding configurations. This chapter applies the ideas of defect equilibrium to several phenomena, including a more detailed analysis of doping and compensation, the intrinsic defect density and the impurity distribution coefficients. The phenomenon of metastability, in which an external electronic excitation, such as light illumination, causes a reversible change in the density of electronic states, are also explained in the context of the defect reactions of the structure. The equilibration of the material represents a considerable simplification in our understanding of a-Si:H, because thermodynamic models can be used to predict the electronic properties.

It may seem surprising to apply thermal equilibrium concepts to amorphous silicon, because the amorphous phase of a solid is not the equilibrium phase. However, a subset of bonding states may be in equilibrium even if the structure as a whole is not in its lowest energy state. The attainment of equilibrium is prevented by bonding constraints on the atomic structure. The collective motion of many atoms is required to achieve long range crystalline order and the topological constraints are formidable. On the other hand the transformation of point defects requires the cooperation of only a few atoms. Therefore any partial thermal equilibrium may be expected at point defects or impurities.

Defect equilibrium is described by a reaction of the type

$$A + B \leftrightarrows C + D \tag{6.1}$$

where A–D are different configurations of point defects, dopants, electronic charges etc. The properties of interest are the equilibrium state and the kinetics of the reaction. Equilibrium is calculated from the formation energies of the various states, by minimizing the free energy. The kinetics are described by a relaxation time, τ_R, required for the structure to overcome the bonding constraints. τ_R is associated with an energy barrier, E_B, which arises from the bonding energies and is illustrated in Fig. 6.1 by a configurational coordinate diagram. The lower energy well may represent the fully coordinated network, and the higher energy well, a dangling bond defect. The energy difference is the defect formation energy, U_d, and determines the equilibrium defect density. The equilibration time, on the other hand, is given by the barrier height,

$$\tau_R = \omega_0^{-1} \exp(E_B/kT) \tag{6.2}$$

Fig. 6.1. Configurational coordinate diagram of the equilibration between two states separated by a potential energy barrier.

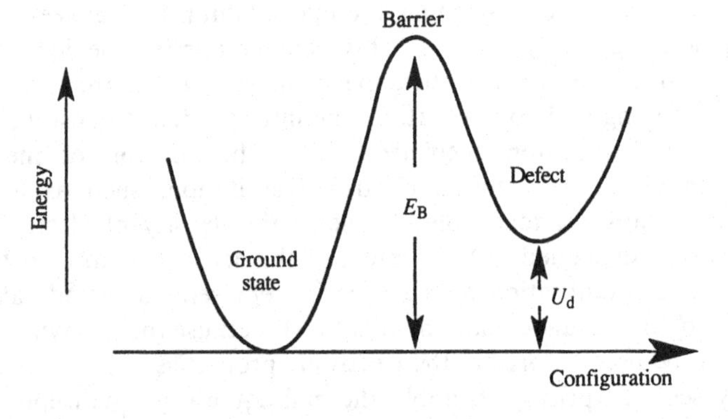

Fig. 6.2. Illustration of the thermal properties of a normal glass near the glass transition. The low temperature frozen state is kinetically determined and depends on the cooling rate.

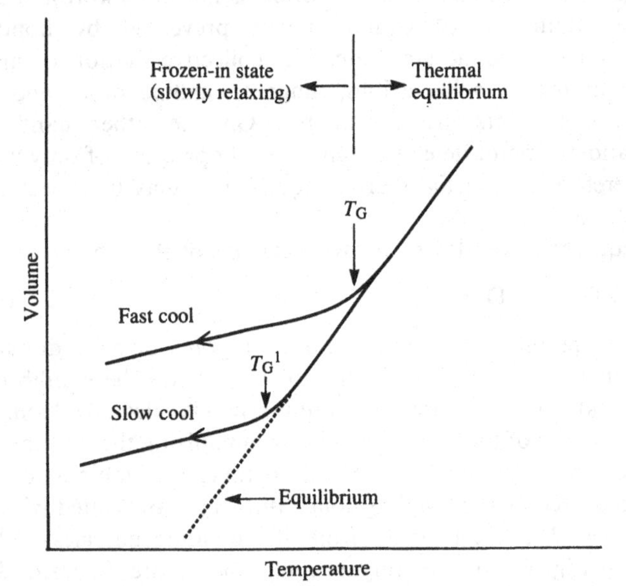

A larger energy barrier obviously requires a higher temperature to achieve equilibrium in a fixed time. The formation energy U and the barrier energy E_B are often of very different magnitudes. The equilibrium state and its kinetics are discussed separately in Sections 6.2 and 6.3.

Any defect reaction which has an energy barrier of the form shown in Fig. 6.1 exhibits a high temperature equilibrium and a low temperature frozen state. The temperature, T_E, at which freezing occurs is calculated from Eq. (6.2) by equating the cooling rate with $dT/d\tau_R$,

$$T_E \ln(\omega_0 k T_E^2 / R_C E_B) = E_B/k \tag{6.3}$$

where R_C is the cooling rate. An approximate solution to Eq. (6.3) for a freezing temperature in the vicinity of 500 K is

$$kT_E = E_B/(30 - \ln R_C) \tag{6.4}$$

from which it is easily calculated that an energy barrier of 1–1.5 eV is needed for $T_E = 500$ K, and that an order of magnitude increase in cooling rate raises the freezing temperature by about 40 K. Below T_E, the equilibration time is observed as a slow relaxation of the structure towards the equilibrium state.

There is a close similarity between the defect equilibration of a-Si:H and the behavior of glasses near the glass transition, which is useful to keep in mind when discussing the a-Si:H results. Fig. 6.2 illustrates the temperature dependence of the volume, V_0 (or other structural parameter), of a normal glass. There is a change in the slope of $V_0(T)$ as the glass cools from the liquid state, which denotes the glass transition temperature T_G. The glass is in a liquid-like equilibrium above T_G, but the structural equilibrium time increases rapidly as it is cooled. The transition occurs when the equilibration time becomes longer than the measurement time, so that the equilibrium can no longer be maintained and the structure is frozen. The transition temperature is higher when the glass is cooled faster and the properties of the frozen state depend on the thermal history. Slow structural relaxation is observed at temperatures just below T_G. The glass-like properties of a-Si:H are discussed further in Section 6.3.2.

6.1 Evidence of structural equilibration

The equilibration of the structure is manifested in reversible changes of the material properties as the temperature is changed. Most of the evidence for bulk thermal equilibration of defect and dopant states in doped a-Si:H comes from electronic transport measurements. Fig. 6.3 shows that the temperature dependence of the dc conductivity in n-type a-Si:H depends on the thermal treatment of the sample. For each of the curves in the figure, the sample is first annealed and then cooled to a low temperature and measured upon warming slowly. A high conductivity is obtained when the sample is annealed at 250 °C and rapidly quenched, but the conductivity is lower when the sample is annealed at 122 °C. Slow cooling from the higher temperature also results in a low conductivity. The conductivity has different properties above and below the temperature denoted by T_E, which for n-type samples is about 130 °C. Below T_E, the conductivity depends on the thermal history and is largest for the sample rapidly cooled from the highest temperature and smallest with slow cooling or a long relaxation

Fig. 6.3. The temperature dependence of the dc conductivity of n-type a-Si:H, after annealing and cooling from different temperatures and in a steady state equilibrium. The measurements are made during warming (Street, Kakalios and Hack 1988b).

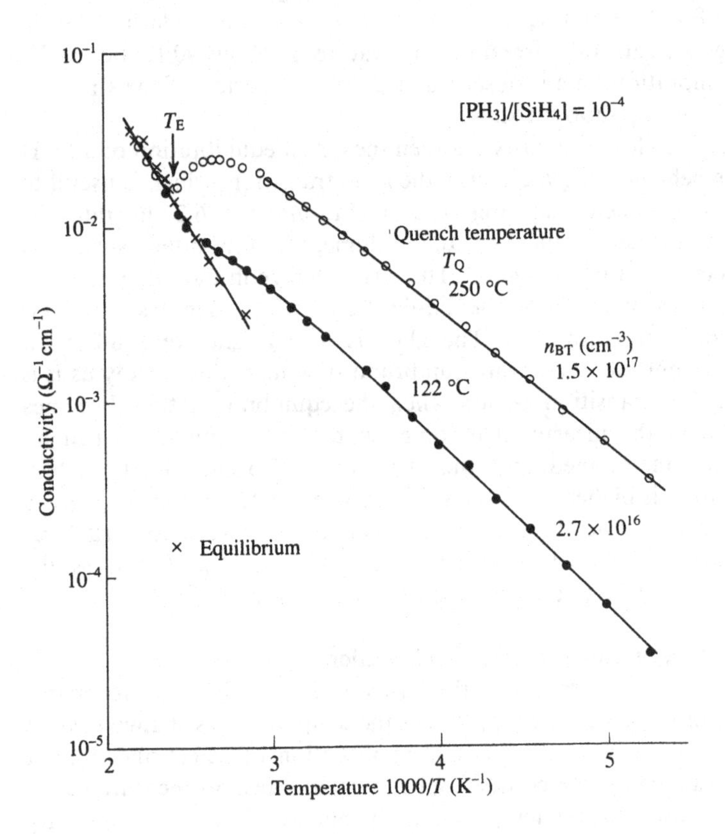

time. The conductivity has a larger activation energy above T_E and shows no sign of any dependence on the thermal history, as all the different curves merge into a single temperature dependence. There is a relaxation regime just below T_E, where the conductivity is time dependent and depends on the warming rate.

There are obvious qualitative similarities between the conductivity curves and the properties of a glass shown in Fig. 6.2, with a high temperature equilibrium state and a low temperature non-equilibrium state which depends on the cooling rate. The different conductivity is due to changes in the electron concentration, which is confirmed by direct measurements of the band tail carrier density, n_{BT}. Fast cooling from high temperature quenches in a larger n_{BT} than slow cooling and n_{BT} is proportional to the conductivity. The expected slow relaxation towards equilibrium at low temperature is also observed, as shown by time dependence of n_{BT} following a rapid quench from 210 °C in Fig. 6.4. n_{BT} slowly decays to a steady state, with the decay taking more than a year at room temperature, but only a few minutes at 125 °C. The temperature dependence of the relaxation time is plotted in Fig. 6.5 and

Fig. 6.4. Decay of the band tail carrier concentration n_{BT} in n-type a-Si:H at the indicated temperatures, following an anneal at 210 °C and rapid cooling (Street, Hack and Jackson 1988a).

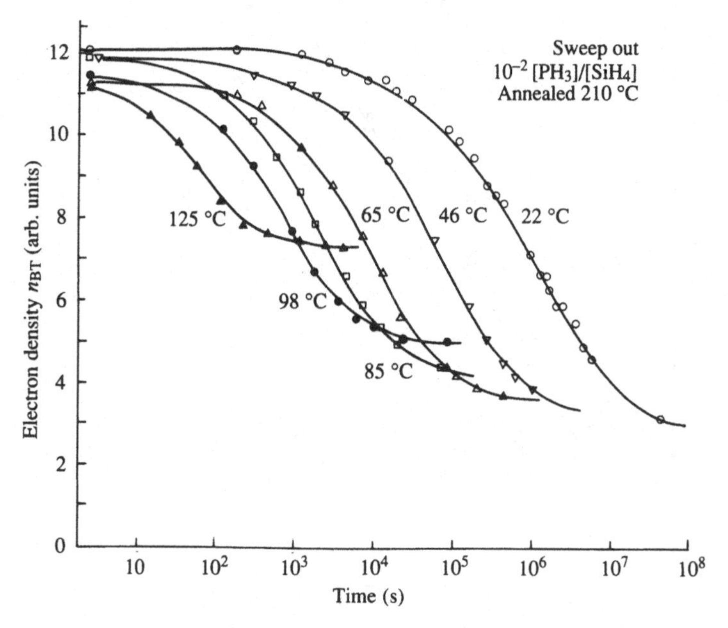

has an activation energy of about 1 eV, which is a measure of the barrier energy E_B. The relaxation is faster in p-type a-Si:H than in n-type. The equilibration temperature can be predicted from this activation energy using Eq. (6.4) and is 400 K for a heating or cooling rate of 0.1 K s^{-1}, in good agreement with the conductivity data.

The electronic transport therefore reflects an equilibrating structure, with an activated relaxation time. There is a high temperature equilibrium regime and a low temperature frozen state, separated by the temperature, T_E, which depends on the heating or cooling rate. The equilibrating states are the donors and dangling bonds, which are related to the band tail carrier density by charge neutrality, and to the conductivity, σ, according to

$$n_{BT} = N_{don} - N_D; \; \sigma = n_{BT} e \mu_D \qquad (6.5)$$

When no structural changes take place, both N_{don} and N_D are

Fig. 6.5. Temperature dependence of the relaxation time in n-type and p-type a-Si:H (Street *et al.* 1988a).

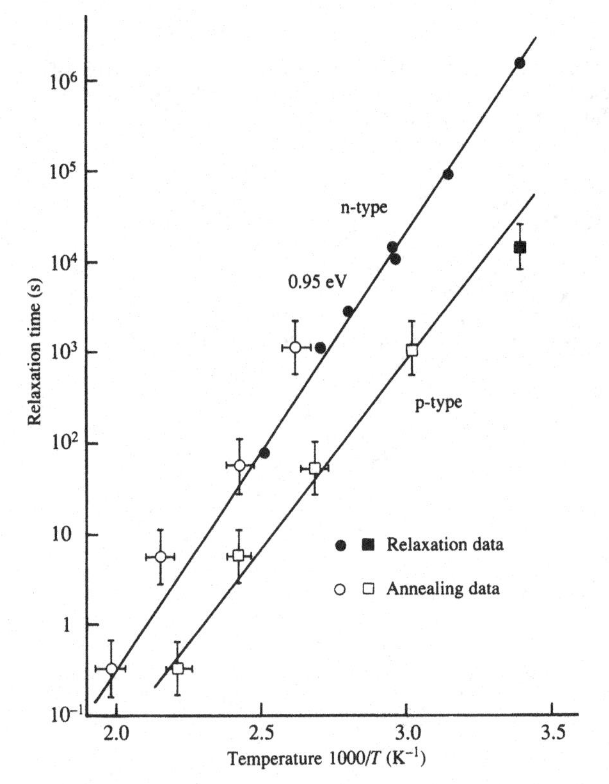

independent of temperature, and so n_{BT} is also constant. The variation of n_{BT} with thermal history implies that either or both of the donor or dangling bond states are the equilibrating species. Eq. (6.5) for n_{BT} relates the structural changes to the electronic properties and is the reason that the structural equilibrium is reflected in the transport data.

The temperature dependence of the equilibrium concentration of n_{BT} is shown in Fig. 6.6 at different doping levels. n_{BT} is thermally activated with an energy of 0.15–0.3 eV which decreases as the doping level increases. The overall change in n_{BT} is about one order of magnitude for the accessible range of equilibration temperatures. These measurements are performed at 300 K after annealing at the chosen temperature and cooling rapidly. The frozen-in value of n_{BT} equals that at the equilibration temperature, provided that the annealing time is long enough to reach a steady state and the cooling is fast enough to prevent

Fig. 6.6. The temperature dependence of the equilibrium conduction band tail electron concentration at different n-type doping levels (Street 1989).

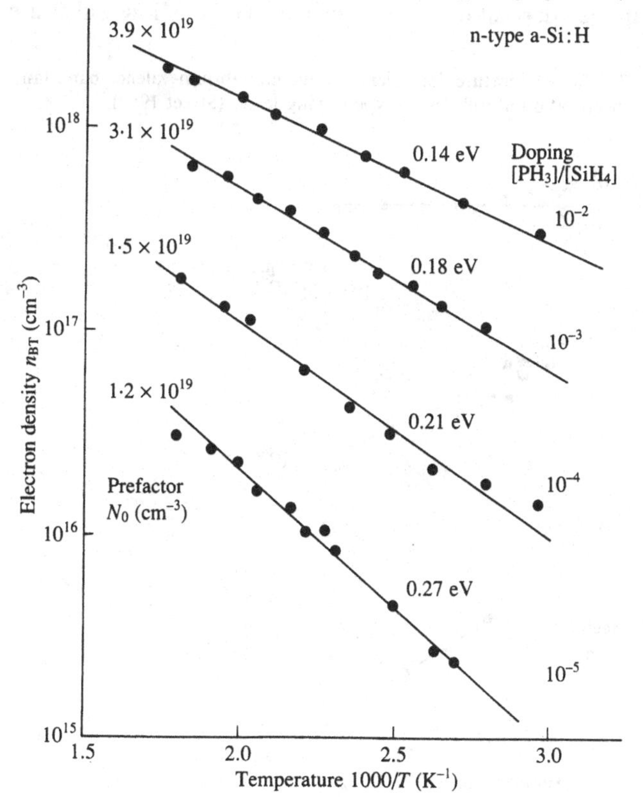

relaxation. A cooling rate of about 10^3 K s^{-1} is achieved by quenching with cold water, and raises the equilibration temperature to about 300 °C in n-type a-Si:H (Eq. (6.3)). The problem for the low temperature equilibrium data is not in fast cooling but in slow annealing, and times of up to a week are needed to ensure complete equilibration at 50–60 °C, as can be seen from Fig. 6.5.

The comparable equilibrium behavior of p-type a-Si:H films is shown in Fig. 6.7. The density of band tail holes, p_{BT} has a smaller activation energy than the n-type material and is constant at the highest doping levels. The increase in conductivity activation energy at the equilibration temperature is less obvious than in the n-type material (see Fig. 5.2), but the frozen state is identified by the dependence on thermal history. The equilibration temperature is lower than in n-type material by about 50 °C and the relaxation times in Fig. 6.5 are also correspondingly shorter.

The changes in n_{BT} and the conductivity are reversible and so are distinguished from other irreversible changes in the electrical properties. For example, deposition of n-type a-Si:H at a low substrate

Fig. 6.7. The temperature dependence of the equilibrium valence band tail hole concentration at different p-type doping levels (Street 1989).

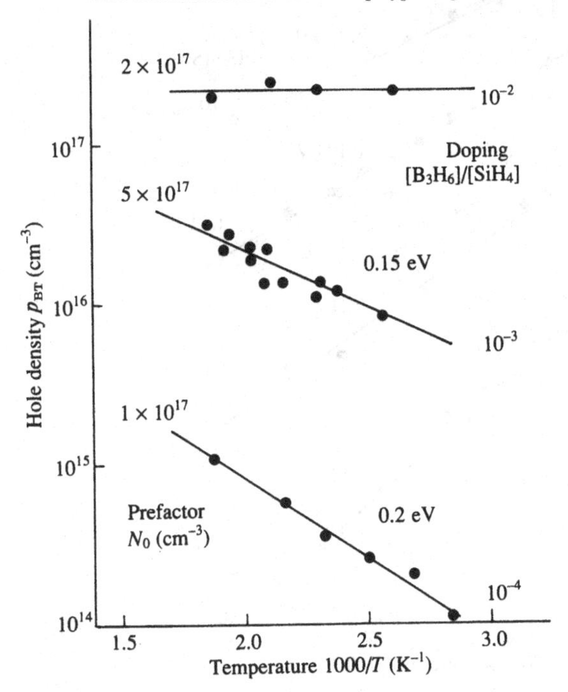

temperature results in a low conductivity which is increased by annealing to about 250 °C. The effect is irreversible and the conductivity state cannot be recovered by thermal cycling. Similarly, high temperature annealing leads to a loss of hydrogen from the films and an associated increase in the defect density which is irreversible unless hydrogen is diffused back into the material. Reversibility is an essential property of the thermal equilibrium state. However, the irreversible changes are closely related to the reversible ones and are discussed in Section 6.2.4.

The band tail carrier density n_{BT} is the difference between the donor and defect densities (Eq. (6.5)) and is generally smaller than either (see Fig. 5.16). n_{BT} is therefore a sensitive measure of the equilibration, and

Fig. 6.8. The temperature dependence of the equilibrium neutral defect density in undoped a-Si:H deposited with different deposition conditions (Street and Winer 1989).

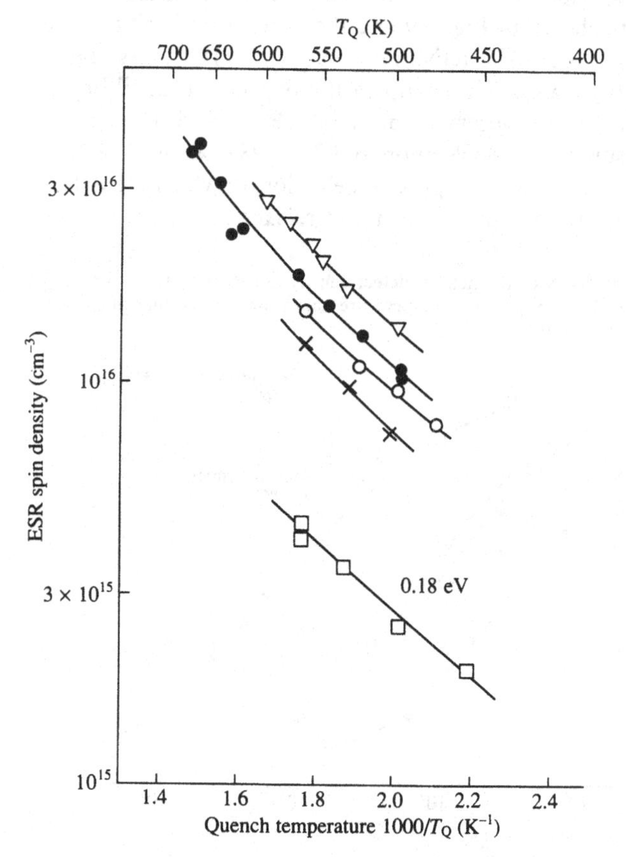

the separate measurements of the defects and dopant states are more difficult. PDS experiments find little temperature dependence of the defect density, from which it is inferred that the primary effect is a change in the density of the dopant states (Street *et al.* 1988a). The equilibration of donors is described further in Section 6.2.2.

Thermal equilibration effects are also present in undoped a-Si:H (Smith *et al.* 1986). The dangling bond density varies reversibly with temperature, as is shown in Fig. 6.8 for samples of different deposition conditions. The experiment is performed by rapidly cooling the material from different temperatures to freeze in the equilibrium configuration and measuring the ESR spin density of the $g = 2.0055$ resonance. The defect density increases four-fold between 200 °C and 400 °C, with an activation energy of 0.15–0.2 eV. The relaxation times are slower than in the doped material at the same temperature, and the equilibration temperature of about 200 °C, for a cooling rate of 0.1 K s^{-1}, is correspondingly higher than in the doped material. Some relaxation data are shown in Fig. 6.9 and the general similarity to the data in Fig. 6.3 is obvious. The activation energy is nearly 1.5 eV, again consistent with Eq. (6.4). The temperature-dependent equilibrium defect density has been observed by ESR, PDS and by photoconductivity measurements (McMahon and Tsu 1987, Xu *et al.* 1988). Some ESR data have found the opposite behavior, in which the defect density is reduced after a rapid quench and relaxes to a higher value

Fig. 6.9. The decay of the neutral defect density at the indicated temperatures following a high temperature anneal and rapid quenching (Street and Winer 1989).

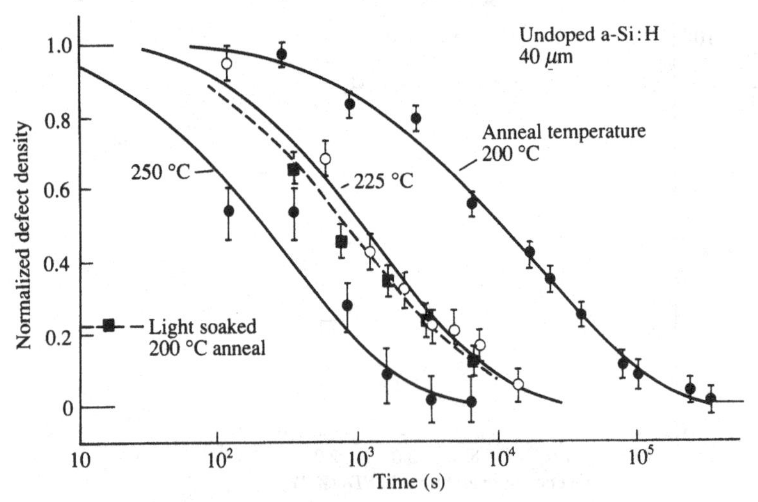

(Lee, Ohlsen and Taylor 1987). This phenomenon appears to be due to surface effects which are presently not understood.

6.2 Thermal equilibrium models

6.2.1 *Theory of chemical equilibrium*

This section briefly reviews some elementary aspects of the thermodynamics of chemical reactions, (e.g. Atkins (1978)) which are used to analyze a-Si:H. The thermodynamic equilibrium state of a system is described by a minimum of the Gibbs free energy function

$$G = H - TS \tag{6.6}$$

where $H = U - PV$ is the enthalpy, U is the formation energy, and S is the entropy. In a solid, the fact that most experiments are performed at constant pressure, P, rather than constant volume, V, leads to a negligible change in the energy, so that the enthalpy can be equated with the formation energy without significant error.

The entropy is defined from the statistical number of different ways in which the defects or dopants can be arranged. For n defects on a network of N_0 identical sites, the number of configurations is

$$W = N_0!/n!(N_0 - n)! \tag{6.7}$$

and using Stirling's approximation,

$$\ln n! \simeq n \ln n - n \tag{6.8}$$

the entropy, $S = k \ln W$, is evaluated as

$$S = k\left[N_0 \ln\left(\frac{N_0}{N_0 - n}\right) - n \ln\left(\frac{n}{N_0 - n}\right)\right] \tag{6.9}$$

Minimizing the free energy with respect to n, gives for the defect density,

$$n = \frac{N_0 \exp(-U/kT)}{1 + \exp(-U/kT)} \simeq N_0 \exp(-U/kT) \text{ when } U > kT \tag{6.10}$$

In a more complicated chemical reaction involving several species, it is useful to evaluate the separate contributions to the free energy of the individual species. The chemical potential, μ, is the extra free energy introduced by adding a defect (or dopant, molecule, electron, etc) at constant pressure and temperature,

$$\mu = \mathrm{d}G/\mathrm{d}n \tag{6.11}$$

where n is the concentration. In the example of defects in a network, the chemical potential is calculated by adding a small number of defects without reducing the number of non-defect sites. An evaluation of the change in free energy gives

$$\mu_{\text{defect}} = U + kT \ln (n/N_0) \tag{6.12}$$

and for the non-defect states,

$$\mu_{\text{nd}} = kT \ln [(N_0 - n)/N_0] \tag{6.13}$$

The formation energy of these states is defined to be zero.

The general expression for the chemical potential is,

$$\mu = \mu^{\circ} + kT \ln C \tag{6.14}$$

where C is the fractional concentration and μ° is the potential (i.e. the enthalpy) of the standard state – usually defined as unity concentration; the standard state of a gas is usually atmospheric pressure. As seen from the network defect example, μ° is the formation energy of the defect.

The law of mass action is perhaps the most useful expression to analyze chemical reactions and is derived as follows. Consider the reaction defined by

$$a\text{A} + b\text{B} \leftrightarrows c\text{C} + d\text{D} \tag{6.15}$$

where the capital letters denote molecular species, and the lower case letters the number of molecules in the reaction. The change in free energy when a small amount of the reactants on the left hand side change into the reaction products on the right hand side, is

$$\mathrm{d}G = \mu_{\text{C}}\,\mathrm{d}n_{\text{C}} + \mu_{\text{D}}\,\mathrm{d}n_{\text{D}} - \mu_{\text{A}}\,\mathrm{d}n_{\text{A}} - \mu_{\text{B}}\,\mathrm{d}n_{\text{B}} \tag{6.16}$$

which is written,

$$\mathrm{d}G = (c\mu_{\text{C}} + d\mu_{\text{D}} - a\mu_{\text{A}} - b\mu_{\text{B}})\mathrm{d}\zeta \tag{6.17}$$

where

$$\mathrm{d}\zeta = \mathrm{d}n_{\text{A}}/a = \mathrm{d}n_{\text{B}}/b = -\mathrm{d}n_{\text{C}}/c = -\mathrm{d}n_{\text{D}}/d$$

The free energy is a minimum at equilibrium, so that

$$\mathrm{d}G/\mathrm{d}\zeta = 0 = c\mu_{\text{C}} + d\mu_{\text{D}} - a\mu_{\text{A}} - b\mu_{\text{B}} \tag{6.18}$$

The total chemical potentials of the two sides of the reaction (weighted by the number of molecules in the reaction) are therefore equal at equilibrium. For the defect problem just discussed, it is easy to show that the defect density of Eq. (6.10) is obtained when Eqs. (6.12) and (6.13) are equated.

Substituting Eq. (6.14) for the chemical potentials into Eq. (6.18) results in the relations,

$$\frac{C_A{}^a C_B{}^b}{C_C{}^c C_D{}^d} = \exp\left(\frac{c\mu_C{}^\circ + d\mu_D{}^\circ - a\mu_A{}^\circ - b\mu_B{}^\circ}{kT}\right) = K = \exp\left(\frac{\Delta H}{kT}\right) \tag{6.19}$$

K is defined in terms of the chemical potentials of the standard states and so is a constant of the reaction, independent of the concentrations of the species. According to the definition of the chemical potential, ΔH is the difference in formation energies of the species on the two sides of Reaction (6.15). Eq. (6.19) is the law of mass action, which gives the equilibrium concentrations of the different species in terms of the reaction constant K.

There are many situations in which the details of the chemical reaction are not known. In the example of defects on the network, the density might be influenced by some unidentified secondary reaction. It is convenient to define an empirical chemical potential for each species as,

$$\mu = \mu^\circ + kT \ln(\gamma C) \tag{6.20}$$

where γ is a parameter known as the activity, which describes the departure from the ideal reaction. For example, consider the deposition of a species from gas phase to solid, described by the reaction,

$$A^{gas} \leftrightarrows A^{solid} \tag{6.21}$$

In equilibrium, $\mu_A{}^{gas} = \mu_A{}^{solid}$, and applying Eq. (6.20) gives,

$$C_s = (\gamma^{gas}/\gamma^{solid}) C_g \exp(-\Delta\mu/kT) \tag{6.22}$$

The factor $\gamma^{gas}/\gamma^{solid}$ represents the deviation from ideal behavior and may be a relative enhancement or reduction in the concentration of the solid. This approach is used to describe the doping distribution coefficient in Section 6.2.7.

6.2.2 *Defect and dopant equilibrium with discrete formation energies*

The random network of an amorphous material such as a-Si:H implies that the formation energy varies from site to site. A correct evaluation of the equilibrium state must include this distribution and also the width of the defect energy levels (Smith and Wagner 1987). It is instructive, however, first to solve the simpler problem in which the distributions are replaced by single formation energies and discrete gap states (Müller, Kalbitzer, and Mannsperger 1986, Street *et al.* 1988a).

The equilibrium is dominated by those states with the lowest formation energy, so that the approximation is made that there is a small subset of network sites at which a defect can be formed. Physically these are associated with weak bonds, at which the formation of a dangling bond or dopant is particularly easy. The actual distributions are considered in Section 6.2.3.

The formation of dangling bond defects in undoped a-Si:H is described by the reaction

$$\text{weak bond} \leftrightarrows \text{dangling bond} \tag{6.23}$$

The evaluation of the equilibrium density of neutral dangling bonds, N_D, is the same as the defect problem in the last section and is given by

$$N_D = \frac{N_{wb}\exp(-U/kT)}{1+\exp(-U/kT)} \approx N_{wb}\exp(-U/kT) \text{ when } U > kT \tag{6.24}$$

The measured activation energy of 0.15–0.2 eV for the defect density in Fig. 6.8 is the average defect formation energy U.

The thermodynamics of doped a-Si:H is a little more complicated, because both the defects and dopants are charged and so interact with the electron distribution whose chemical potential is the Fermi energy. The analysis is for the specific case of n-type doping with phosphorus and, following the model introduced in Chapter 5, it is assumed that both the phosphorus and silicon atoms may have either three-fold or four-fold coordination. The ground state configuration comprises the four-fold silicon and the three-fold phosphorus, and the defect reaction is,

$$Si_4^0 + P_3^0 \leftrightarrows P_4^+ + Si_3^- \tag{6.25}$$

The mass action solution is Eq. (5.28), but the analysis in Chapter 5 does not include the band tail carriers. This is done now by minimizing the free energy.

The formation energies of the defects and donors depend on E_F as,

$$U_d = U_{do} - (E_F - E_{d2}) \tag{6.26}$$

and

$$U_p = U_{po} - (E_p - E_F) \tag{6.27}$$

where U_{po} is the formation energy of the neutral donor, E_p and E_{d2} are the gap state energies, and the second term on the right hand side arises from the transfer of electrons to or from the Fermi energy when the states are charged.

The thermodynamic model is described by the energy level diagram in Fig. 6.10. Only the two-electron state of the dangling bond is included because the defects are negatively charged in n-type material. The intrinsic conduction band states of the silicon network enter the problem because they influence the position of the Fermi energy. The band tail states are the most significant because their occupancy is larger than the band states and so the simple model approximates these states as a single level of high density $N_T \gg N_{don}$, at the same energy as the donors. The total free energy of the donors and dangling bonds is,

$$G = N_{don}\{U_{po} - E_p + E_F + kT[\ln(N_{don}/N_p) - 1]\} + N_D\{U_{do} - E_F + E_{d2} + kT[\ln(N_D/N_0) - 1]\} \quad (6.28)$$

where N_p and N_0 are the number of available network sites of phosphorus and silicon atoms, with $N_D \ll N_0$ and $N_{don} \ll N_p$. The equilibrium condition is,

$$\partial G/\partial N_{don} = \partial G/\partial N_D = 0 \quad (6.29)$$

together with the condition for charge neutrality, Eq. (6.5), for which our density of states model gives for n_{BT}

$$n_{BT} = N_T \exp[-(E_p - E_F)/kT] \quad (6.30)$$

These equations are solved for the unknowns N_{don}, N_D and E_F. One form of the solution is,

$$N_D N_{don} = N_0 N_p \exp\{-[U_{po} + U_{do} - (E_p - E_{d2})]/kT\} \quad (6.31)$$

which is identical to the law of mass action solution, Eq. (5.28). The energy of the reaction (6.25) is the sum of the individual formation energies of the neutral state minus the energy gained by the electron transfer, as was deduced in the last chapter. The same results are obtained directly from the formation energies using the Boltzmann formula,

$$N_D = N_0 \exp[-(U_{do} - E_F + E_{d2})/kT] \quad (6.32)$$

Fig. 6.10. Energy level model used in the thermodynamic calculations described in the text, showing donor, defect and band tail states.

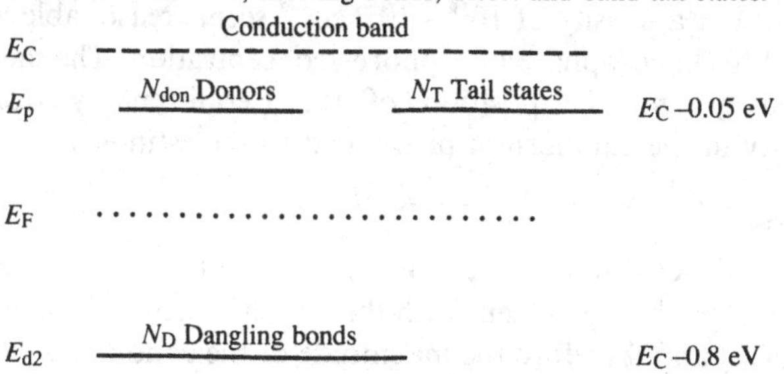

Conduction band

E_C

E_p — N_{don} Donors — N_T Tail states — E_C −0.05 eV

E_F

E_{d2} — N_D Dangling bonds — E_C −0.8 eV

and

$$N_{\mathrm{don}} = N_{\mathrm{p}} \exp[-(U_{\mathrm{po}} - E_{\mathrm{p}} + E_{\mathrm{F}})/kT] \tag{6.33}$$

The product of these expressions gives Eq. (6.31).

Measurements find that n_{BT} is quite small compared to N_{D} or N_{don}, so that the approximation can be made,

$$N_{\mathrm{don}} \approx N_{\mathrm{D}} \approx (N_0 N_{\mathrm{p}})^{\frac{1}{2}} \exp\{-[U_{\mathrm{p}} + U_{\mathrm{d}} - (E_{\mathrm{p}} - E_{\mathrm{d2}})]/2kT\} \tag{6.34}$$

which reproduces the square root dependence on the phosphorus concentration and shows that the individual formation energies at the equilibrium configuration are

$$U_{\mathrm{pe}} = U_{\mathrm{de}} = [U_{\mathrm{p}} + U_{\mathrm{d}} - (E_{\mathrm{p}} - E_{\mathrm{d2}})]/2 \tag{6.35}$$

This energy is the crossing point of the energy relations in Fig. 5.21. The position of the Fermi energy is

$$E_{\mathrm{F}} = \tfrac{1}{2}(E_{\mathrm{p}} + E_{\mathrm{d2}}) + \tfrac{1}{2}(U_{\mathrm{do}} - U_{\mathrm{po}}) - \tfrac{1}{2}kT \ln[(N_0/N_{\mathrm{p}})(1 - n_{\mathrm{BT}}/N_{\mathrm{D}})] \tag{6.36}$$

These expressions allow a physical interpretation of the experimental results of Figs. 6.3 and 6.6. In equilibrium, the Fermi energy is pinned between the donor and the dangling bond level. There is a close analogy between the expression (6.36) and the equivalent result for a negative U defect given by Eq. (4.10). The main difference is that here there is a donor and a defect, rather than a single defect, so that the Fermi energy contains an extra term $(U_{\mathrm{do}} - U_{\mathrm{po}})/2$ and a factor $\ln(N_0/N_{\mathrm{p}})$ in the kT term, but the physical reason for the pinning of the Fermi energy is the same. The band tail charge, n_{BT}, is predicted to be thermally activated, as can be seen from Eq. (6.30) and the pinning of the Fermi energy, and agrees with the observations in Fig. 6.6. The temperature dependence of the doping efficiency is the origin of the extra charge needed to allow n_{BT} to increase with temperature.

The parameters of the model can be estimated from the experimental data and some guesswork. The appropriate density of weak bonds is not known, but a density of $10^{19}-10^{20}$ cm^{-3} seems reasonable and N_{p} is set equal to the gas-phase phosphorus concentration. The measured defect density and the position of the Fermi energy from the conductivity in the equilibrium phase lead to the estimates,

$$U_{\mathrm{po}} = 0.3 \text{ eV and } U_{\mathrm{do}} = 0.6 \text{ eV} \tag{6.37}$$

The predicted activation energy of n_{BT} is about 0.15 eV with these values, in reasonable agreement with the measurements in Fig. 6.6. The position of E_{F} and therefore the magnitude of the conductivity depend in part on the relative values of U_{po} and U_{do} (see Eq. (6.36)). The relatively high conductivity of a-Si:H is apparently a consequence of the lower formation energy of donors compared with the defects. Note that the conductivity activation energy is predicted to be independent of the doping level. In practice, the activation energy changes with doping and is a feature of the data which is only explained when the distributions of the energies are included.

Equation (6.16), with a single formation energy of 0.6 eV and a weak bond density of 10^{20} cm^{-3}, predicts a defect density of 10^{15} cm^{-3} in undoped a-Si:H at a growth temperature of 250 °C, which is of the correct magnitude. Therefore the same parameters fit both the doped and undoped material. However, an activation energy of 0.6 eV should result in a much larger temperature dependence of N_{DB} than is actually observed (see Fig. 6.8). This discrepancy is also removed when the distribution of formation energies is included.

6.2.3 *Distributions of formation energies – the weak bond model*

The thermodynamic model in the previous section neglects the distribution of formation and gap state energies and the specific processes by which weak bonds are converted into defects. The analysis can be reformulated to take into account the distributions by generalizing expression (6.24),

$$N_D = \int \frac{N_0(U)\exp(-U/kT)}{1+\exp(-U/kT)}\,dU \tag{6.38}$$

where $N_0(U)$ is the distribution function of the formation energy and U contains the distribution of gap state energies according to Eqs. (6.26) and (6.27). A similar expression applies to the donor states. The evaluation of the equilibrium state is complicated because it depends on the distribution functions and also on whether there is a correlation between the formation energy and the corresponding gap state energy level (or donor level). The gap state distribution is reasonably well established by experiment, but less is known about the formation energies, particularly of the donors. A reasonable choice of the distributions has been shown to give a consistent fit to all the experimental data of the temperature and doping dependence of the defect density, conductivity and n_{BT} in n-type a-Si:H (Hack and Street 1988).

A physical model for the formation energies is more easily addressed for undoped a-Si:H because the dopants do not have to be included. The origin of the distribution is the disorder of the silicon network. Defect generation is expected to occur when a silicon–silicon bond breaks, and naturally the weakest bonds break most easily to give the lowest formation energy for defect creation. The model therefore focuses on the distribution of the weakest bonds.

There are no rigorous calculations of the distribution of formation energies, but the problem is addressed as follows (Stutzmann 1987a, Smith and Wagner 1987). Fig. 6.11 shows a molecular orbital model of a weak Si—Si bond, and a pair of dangling bonds. The conversion of the weak bond into the dangling bond is accompanied by the increase of the electron energy from a bonding state in the valence band to a non-bonding state in the gap. The formation energy of neutral defects is

$$U_{do} = E_{d1} - E_{WB} + \left[\sum_{VB \neq WB} (E_{VB}' - E_{VB}) + \Delta E_{ion} \right] \tag{6.39}$$

E_{d1} and E_{WB} are the gap state energies of the defect electron and valence band tail state associated with the weak bond. E_{VB} and E_{VB}' are energies of the states of the valence band electrons, before and after the bond is broken. The sum is the change in energy of all the valence band states other than from the broken weak bond. ΔE_{ion} is the change in energy of the ion core interaction for the structure with and without the defect.

The weak bond model assumes that the terms in the square bracket in Eq. (6.39) are negligible, so that

$$U_{do} = E_{d1} - E_{WB} \tag{6.40}$$

The distribution of formation energies therefore follows the density of valence band tail states. The dangling bond energy level is estimated to be about 0.8 eV above the valence band mobility edge and the band tail extends at least 0.5 eV into the gap, so that the energies are reasonably consistent with the known defect densities. The only justification for the model is that the creation of the defect occurs without a large

Fig. 6.11. An illustration of the change in the electronic states when a weak Si-Si bond is converted into a dangling bond.

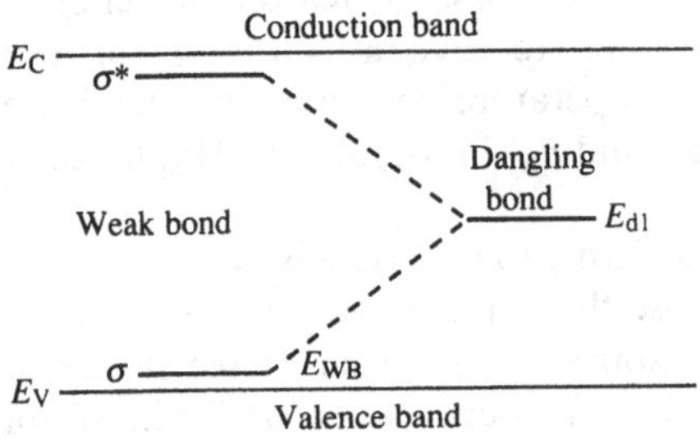

perturbation of any valence band states other than of the weak bond which is broken. The validity of the model therefore depends on the local bonding around a defect being the same as the bonding of the defect-free network, because any change in the bond energies contributes to the defect formation energy. This may be a good approximation for the dangling bond if the defect wavefunction maintains the sp^3 hybridization of the network and there is not too much relaxation of the bonds.

The weak bond model is useful because the distribution of formation energies can be evaluated from the known valence band and defect density of states distributions. Fig. 6.12 illustrates the distribution of formation energies, $N_0(U)$. The shape is that of the valence band edge given in Fig. 3.16 and the position of the chemical potential of the defects coincides with the energy of the neutral defect gap state. Fig. 6.12 also shows that in equilibrium virtually all the band tail states which are deeper than E_D convert into defects, while a temperature-dependent fraction of the states above E_D convert.

The defect creation process has to be considered in more detail to complete the calculation of the defect density (Street and Winer 1989).

Fig. 6.12. The distribution of formation energies according to the weak bond model. The shape is proportional to the valence band density of states.

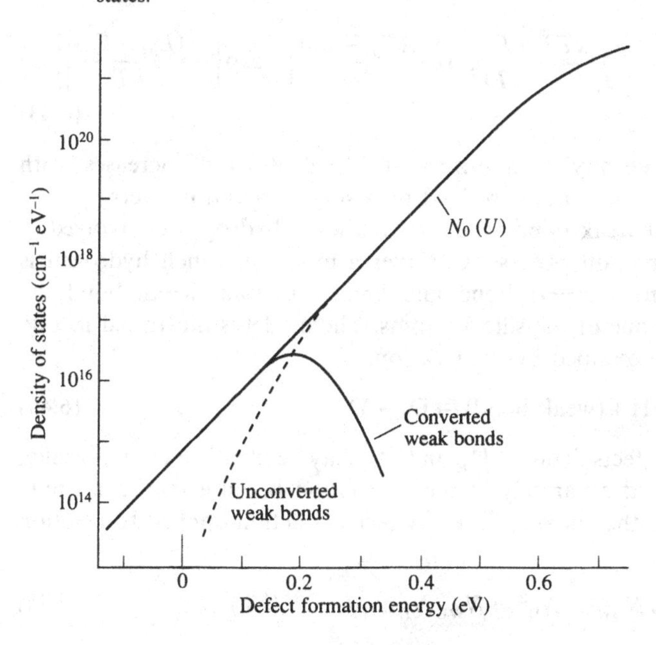

Reaction (6.23) makes no assumptions as to the specific mechanism by which weak bonds are converted into defects. The mechanism determines the entropy of the reaction by defining the sites where the defects can reside. For example, the breaking of one weak Si—Si bond creates two defects, but the equilibrium defect density is different if the two defects are allowed to remain close together as a pair, than it is if they are able to diffuse apart. Alternatively, the defect creation can be mediated by hydrogen diffusion which allows the defects to occupy Si—H sites from which the hydrogen is removed. The different results are illustrated by calculations for two specific models.

The first model is of defects which are created at weak bond sites and are allowed to diffuse apart. The defect density is given from Eqs. (6.38) and (6.40) by

$$N_D = \int \frac{N_V(E_{WB}) \exp[-(E_{d1} - E_{WB})/kT]}{1 + \exp[-(E_{d1} - E_{WB})/kT]} \mathrm{d}U \tag{6.41}$$

For simplicity, the energy level distribution of the defect is neglected, but is considered in Section 6.2.5. Integration of Eq. (6.41) with an exponential valence band of slope,

$$N_V(E_{WB}) = N_{VO} \exp[-(E_{WB} - E_V)/kT_V] \tag{6.42}$$

yields approximately,

$$N_D = \frac{N_{VO} kTT_V}{T_V - T} \left\{ \frac{T_V}{T} \exp\left[-\frac{(E_{d1} - E_V)}{kT_V} \right] - \exp\left[-\frac{(E_{d1} - E_V)}{kT} \right] \right\} \tag{6.43}$$

The defect density is non-zero at $T = 0$ K, and increases with temperature, but is not described by a single activation energy.

A different weak bond model results when hydrogen is involved in the defect creation process. Consider a model in which hydrogen is released from a Si—H bond and breaks a weak Si—Si bond, by attaching to one of the silicon atoms. The model is illustrated in Fig. 6.13, and is described by the reaction,

$$\text{Si—H} + (\text{weak bond}) \leftrightharpoons D_H + D_W \tag{6.44}$$

The two defects denoted D_H and D_W may be identical electronically, but are treated separately in the calculation because they contribute differently to the entropy. The law of mass action applied to reaction (6.44) gives,

$$N_{DH} N_{DW} = N_D{}^2 = N_{WB} N_H \exp(-2U/kT) \tag{6.45}$$

where the Ns are the densities of the different species and by the construction of the model, $N_{DH} = N_{DW} = N_D$. (Note that the total defect density is therefore $2N_D$.) The distribution of defect formation energies again associates a different density of valence band tail states with each formation energy. However, the model assumes that only the weak bonds have a distribution of energies and that the defects formed by the removal of hydrogen from Si—H are equivalent (although a different distribution of the Si—H sites could be added). Consider a small portion of the valence band distribution $\delta N_V(U)$, for which the defect formation energy is U. The density of defects $\delta N_{DW}(U)$ created from these states is given by

$$N_{DH}\,\delta N_{DW}(U) = [\delta N_V(U) - \delta N_{DW}(U)]N_H \exp(-2U/kT) \tag{6.46}$$

The total number of defects on the weak bonds equals the density of those on Si—H sites, so that

$$\int dN_{DW}(U) = N_D \tag{6.47}$$

It follows, therefore, that N_D is the solution to the integral equation,

$$N_D = \int \frac{N_H N_V(E_{WB})\,dE_{WB} \exp(-2U/kT)}{N_D + N_H \exp(-2U/kT)} \tag{6.48}$$

where the integral is over the distribution of tail states. Substituting for Eq. (6.42) gives

$$N_D = N_H N_{VO} \exp\left[-\frac{(E_{d1} - E_V)}{kT_V}\right] \int \frac{\exp(U/kT_V)\,dU}{N_H + N_D \exp(2U/kT)} \tag{6.49}$$

The two weak bond models predict different defect densities, the

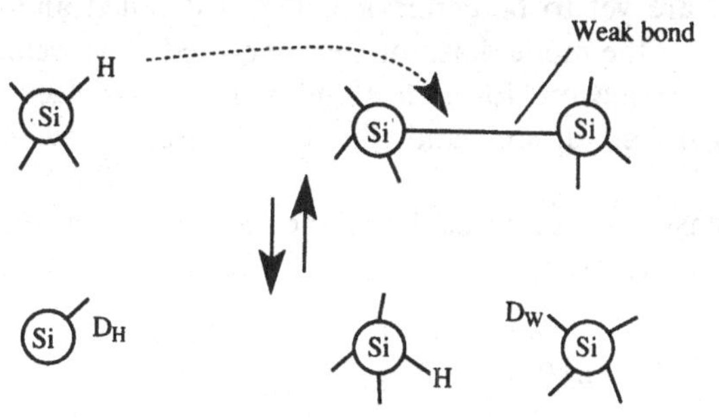

Fig. 6.13. Illustration of the hydrogen-mediated weak bond model in which a hydrogen atom moves from a Si—H bond and breaks a weak bond, leaving two defects (D_H and D_W) (Street and Winer 1989).

origin of which is the entropy contribution to the free energy. The second model allows the defects to move amongst a large density of Si—H sites, whereas in the first model the defects are constrained to be on the weak bond sites only.

Fig. 6.14 compares the calculated temperature dependence of the defect density for four different weak bond models, including the two just described (*A* and *C* in the figure). Of the two others, model *B* has only silicon sites and does not allow the defects to diffuse apart after the weak bond breaks. Model *D* involves hydrogen motion and assumes that hydrogen is released from a second Si—H bond so that the weak bond is saturated with two hydrogen atoms. Other models of the same general type can be constructed. The details of the model influence the defect density and its temperature dependence. The two models involving hydrogen have larger entropy than the others, which is reflected in the increasingly large difference in the defect density as the temperature is raised. The defect densities converge to the same values at low temperature because the entropy is a decreasing contribution to the free energy.

Evaluation of the defect density requires a knowledge of the valence band tail slope and the defect energy level. The optimum growth conditions give a slope of about 500 K and the defect energy level is about 0.8 eV above the valence band edge. With these values and $N_{VO} = 2\times10^{20}$ cm^{-3} eV^{-1}, the predicted defect density for the hydrogen-mediated model *C* is close to 10^{16} cm^{-3} at 500 K. Both the magnitude and the temperature dependence are in good agreement with the experimental data (see Fig. 6.14). The weak temperature dependence of N_D can now be understood as arising from the distribution of energies.

6.2.4 *The role of the band tails and deposition conditions*

The useful feature of the weak bond model is that it provides a first-principles calculation relating the defect density to measurable quantities, T_V and E_d. Further development of the model is needed, as the assumptions are yet to be confirmed, and a detailed analysis of doped a-Si:H using the model has not been reported. The weak bond model provides a framework for understanding how the defect density varies with growth conditions and alloying, aspects of which are discussed next.

The defect density in the weak bond model (e.g. Eqns. (6.43) or (6.49)) is conveniently expressed as

$$N_D = \xi(T, T_V, N_H)\, N_{VO}\, kT \exp\left[-\frac{(E_{d1}-E_V)}{kT_V}\right] \tag{6.50}$$

where $\xi(T, T_V, N_H)$ represents the entropy factor which differs for each specific model, but which is a slowly varying function. The equilibrium defect density is primarily sensitive to T_V and E_{d1}, through the exponential factor. Raising T_V from 500 K to 1000 K increases the defect density by a factor of about 100.

The sensitivity to the band tail slope accounts for the change in defect density with deposition conditions and annealing. Low deposition temperatures, below the usual 200–250 °C, result in an increased defect density, reaching above 10^{18} cm^{-3} with room temperature deposition. The band tails are much broader at the low deposition temperatures, so that Eq. (6.50) predicts the higher defect density. When the low deposition temperature material is annealed, the defect density is reduced. In Section 2.3.4, this behavior was taken as evidence that the material was not initially in equilibrium. The model

Fig. 6.14. Predicted temperature dependence of the equilibrium defect density for four different weak bond models, the details of which are described in the text (Street and Winer 1989).

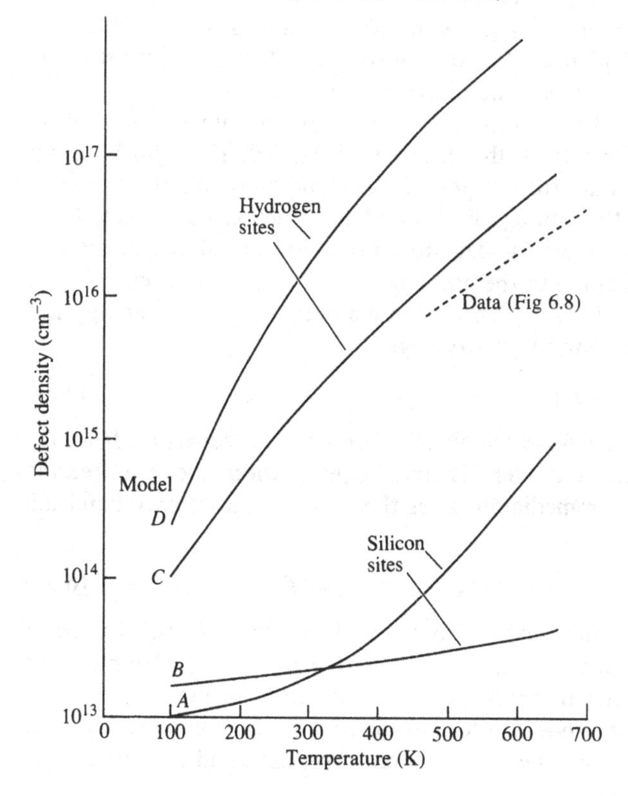

just described suggests a different interpretation, that the reduction in defect density is because the band tails become steeper, with defect equilibrium being maintained all along.

The slope of the Urbach tail increases at deposition temperatures well above 250 °C and is associated with a lower hydrogen concentration in the film (see Fig. 3.20). It seems probable that the equilibrium temperature also increases and both effects lead to an expectation of a higher defect density, in agreement with observations (see Fig. 2.2). The correlation between T_V and N_D predicted in Eq. (6.50), is observed over the complete range of deposition conditions (see Fig. 3.22). Thus, according to the weak bond model, the low defect densities in the best a-Si:H films are a direct consequence of the steep band tails.

Both E_d and T_V may change in alloys of a-Si:H. There is a larger defect density in low band gap a-SiGe:H alloys, which is predicted from the reduced value of E_d accompanying the shrinking of the band gap. In the larger gap alloys, such as a-SiC:H, the expected reduction in N_D due to the larger E_d seems to be more than offset by a larger T_V, so that the defect density is again greater than in a-Si:H. There is, however, no complete explanation of why a-Si:H has the lowest defect density of all the alloys which have been studied.

The weak bond model assumes a non-equilibrium distribution of weak bonds arising from the disorder of the a-Si:H network. It has been proposed that the shapes of the band tails are themselves a consequence of thermal equilibrium of the structure (Bar-Yam, Adler and Joannopoulos 1986). The formation energy of a tail state is assumed proportional to the difference in the one-electron energies, so that the energy, U_{vb}, required to create a band tail state of energy E_{VB} from the valence band mobility edge is

$$U_{vb} = (1+s_V)E_{VB} \tag{6.51}$$

where $s_V < 1$ is a parameter which represents the departure from the simple one-electron model. Thermal equilibration above a freezing temperature, T_E, immediately gives the shape of the valence band tail as,

$$N(E_{VB}) = N_0 \exp(-E_{VB}/kT_V); \; T_V = T_E/(1+s_V) \tag{6.52}$$

The attraction of this model is that it explains the exponential slope of the band tail, which has proved to be so difficult to derive from models of non-equilibrium disorder. Also, the similarity of the valence band tail slope and the deposition temperature is no longer coincidental, but is a predicted result. The steeper conduction band tail is explained by

Table 6.1. *Comparison of the equilibrium effects in a-Si:H and chalcogenide glasses, in terms of the length scales of the structural units involved*

	Short range order (*defects*)	Intermediate range order (*band tails*)	Long range order (*crystal*)
a-Si:H	Equilibrium (T-dependent defect density)	Partial equilibrium (annealing)	No equilibrium (disorder)
Glasses	Equilibrium (T-dependent defect density)	Equilibrium (T-dependent drift mobility)	No equilibrium (disorder)

a larger value of s, which is justified by a comparison of the deformation potentials of the valence and conduction bands in crystalline silicon (Bar-Yam *et al.* 1986).

There is evidence of bulk equilibration of the band tails in chalcogenide glasses, but apparently not in a-Si:H. The drift mobility

Fig. 6.15. Measurements of the drift mobility showing that the band tail slope is unaffected by annealing and quenching in a-Si:H, but changes reversibly in chalcogenide glasses.

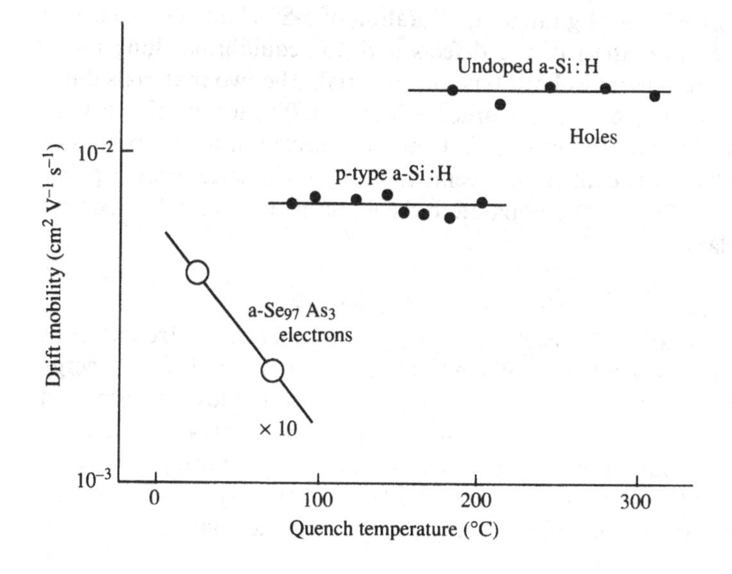

of selenium decreases as the glass transition temperature is raised by rapid quenching, indicating that the band tails broaden reversibly and are in thermal equilibrium (Abkowitz 1985; Kasap, Polischuk, Aiyah and Yannacopoulos 1989). Similar experiments in a-Si:H do not find the equivalent change of slope for thermal quenching temperatures up to 300 °C; a comparison of the selenium and a-Si:H results is shown in Fig. 6.15. Furthermore, it should be noted that this model of equilibrating band tails is inconsistent with the preceding model of defect formation from the band tail states. Both cannot be simultaneously correct, since if the band tail density is in equilibrium, Eq. (6.38) for N_D reduces to a simple exponential with formation energy $E_{d1} - E_V$.

However, some relaxation of the band tail disorder in a-Si:H does occur. Deposition at low temperature gives a large band tail slope which is decreased by annealing. Evidently the bonding structures which define the band tail states can move towards equilibrium, even if equilibrium is not attained.

The interplay of equilibrium and non-equilibrium states can be explained in terms of atomic motion of different sized atomic clusters. Defects occur at single atom sites and their equilibration does not require the correlated motion of many atoms. In contrast, the strain in a weak Si—Si bond is determined by the configuration of a small cluster containing about 5–15 atoms around the bond, corresponding to the intermediate range order (see Section 2.2.2) and crystallization involves the ordering of all the atoms. Table 6.1 compares the short, intermediate, and long range equilibration of a-Si:H and chalcogenide glasses. Equilibration of the defects and non-equilibrium long range disorder are present in both types of material. The two materials differ in the intermediate range order which equilibrates in the glasses (Abkowitz 1985) but not in a-Si:H and is reflected in the properties of the band tails. The difference seems to be a consequence of the rigidity of the a-Si:H network compared to the liquid-like structural relaxation in the glasses.

6.2.5 *Doping dependence of gap state energies*

So far it has been assumed that the defect electronic states form a narrow band at E_d. When there is a broad distribution of energy levels, the defect equilibration favors those with the lowest formation energy, which are the states lowest in the gap (see Eq. (6.40)). The effect of the distribution is particularly significant in comparing the defect states in differently doped materials, because the contribution of the gap state to the formation energy depends on the charge state (Bar-

Yam *et al.* 1986). Combining Eq. (6.40) with Eqs (6.26) and (6.27) gives for the defect formation energy

$$U_d(D^+;\text{ p-type}) = E_F - E_{WB} \quad (6.53)$$
$$U_d(D^\circ;\text{ undoped}) = E_{d1} - E_{WB} \quad (6.54)$$
$$U_d(D^-;\text{ n-type}) = E_{d1} + E_{d2} - E_{WB} - E_F \quad (6.55)$$

The one- and two-electron gap state energies, E_{d1} and E_{d2}, enter once for each electron on the defect, and the different sign of E_F is because an electron is added to the defect in n-type a-Si:H making it negative, but removed in p-type material. The defect reaction in p-type material selects defects equally from the available distribution because E_d does not enter into Eq. (6.53). However, n-type doping favors defect states that are particularly low in the gap because these have a reduced formation energy. The magnitude of the shift in the energy level depends on the width of the energy level distribution. If the unoccupied defects have a gaussian distribution of width σ_D, then the charged states are shifted in energy by approximately

$$\Delta E = -\sigma_D{}^2/kT_V \quad (6.56)$$

for each electron added (Street and Winer 1989). Fig. 6.16 illustrates the expected energies of the one-electron and two-electron states for the different doping types. The measurements of the defect band in Figs. 4.17 and 4.18 indicate that $\sigma_D \sim 0.1$ eV, so that the predicted shift of the peak is about 0.2 eV. Some of the experiments discussed in chapter 4 indicate that such a shift is present, although there is sufficient disagreement between the different measurements that the effect is not yet confirmed.

6.2.6 *Compensated a-Si:H*

The defects which are induced in doped a-Si:H reduce the dopant formation energy by providing electronic compensation. Defects are no longer needed to lower the energy when a-Si:H is doped with both boron and phosphorus. Instead charged dopant states result from the reaction

$$P_3{}^0 + B_3{}^0 \leftrightarrows P_4{}^+ + B_4{}^- \quad (6.57)$$

This defect reaction explains the different properties of compensated a-Si:H. The law of mass action gives

$$N_{P_4} N_{B_4} = N_{P_3} N_{B_3} \exp\{-[U_{po} + U_{bo} - (E_p - E_b)]/kT\} \quad (6.58)$$

where for simplicity the calculation assumes discrete formation energies. The energy $E_p - E_b$ is roughly 1.5 eV because the donor and acceptor levels are close to their respective band edges. The energy gain by the transfer of the electron is much larger than in singly doped a-Si:H, and consequently the total dopant formation energy is smaller and may even be negative. The equilibrium doping efficiency is predicted to be larger than for singly doped material and the defect density smaller. The Fermi energy is pinned in the middle of the gap and is relatively insensitive to deviations from compensation. When there is partial compensation, for example, with more phosphorus than boron, the tendency of E_F to move to the conduction band suppresses the phosphorus doping efficiency and equalizes the active dopant concentrations.

The electrical data described in Section 5.2.4 are consistent with this expected behavior. The low defect density and the pinning of E_F are both observed. The higher doping efficiency explains the wider band tails and the associated reduction in the carrier mobility, because the many extra dopant states directly add localized states to the band tails. In addition, the large Coulomb potential fluctuations due to the high concentration of charged dopants lead to a further broadening of the band tails.

6.2.7 *Defect and dopant pairing*

Some defect spectroscopy measurements in n-type a-Si:H have been interpreted in terms of close $P_4^+D^-$ pairs (Kocka 1987) and there is also evidence of boron–phosphorus complexes in compensated a-Si:H (Boyce *et al.* 1987). The driving force for pairing is the Coulomb interaction between oppositely charged states. Such a paired state must be treated as a distinct species in the defect reaction, which takes the form,

$$P_3 + \mathrm{Si}_4 \leftrightarrows (\mathrm{P}_4^+\mathrm{D}^-)_{\mathrm{pair}} \qquad N_{\mathrm{pair}} = K_p N_{\mathrm{P}_3} N_{\mathrm{Si}_4} \tag{6.59}$$

Fig. 6.16. Diagram of the doping dependence of the defect gap state distribution predicted by the defect equilibrium model.

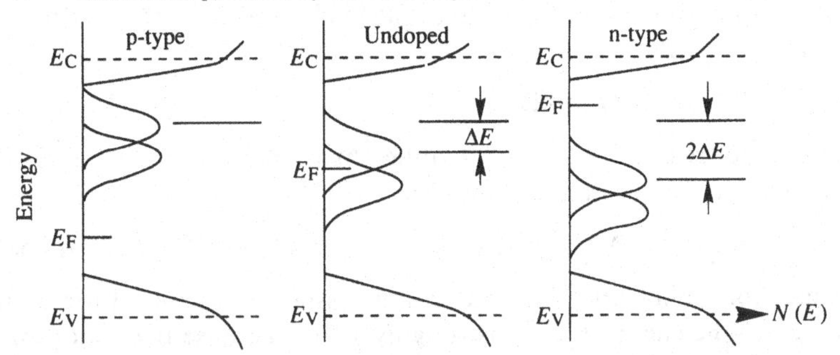

The second equation applies the law of mass action with rate constant K_p. The equilibrium density of pairs is proportional to the phosphorus concentration. Therefore, the square root law for the defects in Fig. 5.9 is inconsistent with the formation of pairs, if the equilibrium model is valid. The different concentration dependences for paired and isolated states imply that pairing is most likely at the highest phosphorus or boron concentrations.

The pairing is described by the distribution function, $G(R)$, for the nearest neighbor separation between a P_4^+ and a D^- state. With a random separation of N_D defects and donors (Reiss 1956)

$$G(R) = 4\pi R^2 N_D \exp(-4\pi R^3 N_D/3) \tag{6.60}$$

which is shown in Fig. 6.17 by the solid line. The function $G(R)$ has a peak at

$$R_{av} = (2\pi N_D)^{-\frac{1}{3}} \tag{6.61}$$

The Coulomb attraction between the donor and the defect leads to a non-random distribution, with an enhanced concentration of close pairs. The modified distribution is

$$G'(R) = \beta G(R) \exp(e^2/4\pi\varepsilon\varepsilon_0 RkT) \tag{6.62}$$

provided the exponential term is not too large. The exponential factor represents the enhanced probability of the nearer states due to the Coulomb interaction and β is a normalization factor. The modified

Fig. 6.17. The nearest neighbor distribution function for randomly dispersed sites (solid line). The associated pair distribution (dashed line) applies when there is a Coulomb attraction between oppositely charged states.

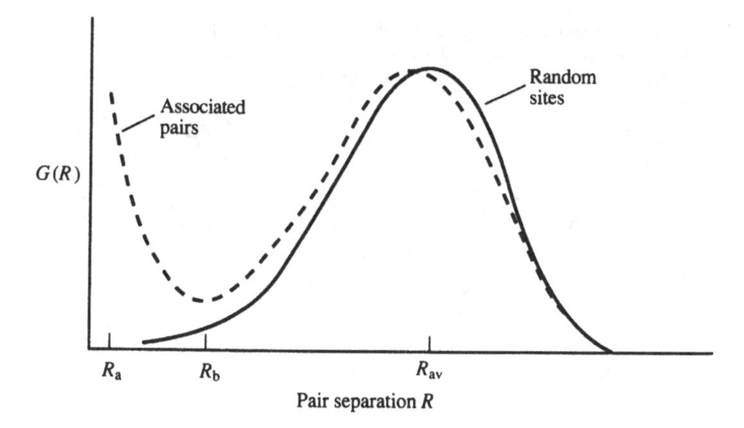

distribution is indicated in Fig. 6.17. A low temperature obviously favors pairing because the Coulomb energy is large relative to kT. The distribution is separated into two regions by the minimum at R_b. The distant pairs at $R > R_b$ are only slightly perturbed from the random distribution. The close pairs for which $R < R_b$ are strongly modified by the Coulomb interaction and their density is given by

$$N_{pair} = \int_{R_a}^{R_b} N_D G'(R)\, dR \simeq 4\pi\beta N_D^2 \int_{R_a}^{R_b} R^2 \exp\left(\frac{e^2}{4\pi\varepsilon\varepsilon_0 RkT}\right) dR \tag{6.63}$$

where R_a is the minimum separation, illustrated in Fig. 6.17. The pair density varies as N_D^2, so that pairing is enhanced at high defect concentrations. The N_D^2 term also implies that if the random defect density increases as the square root of the phosphorus concentration, then the pairs are linear in concentration, as was deduced from the mass action equations. The temperature dependence of N_{pair} suggests that thermal equilibrium at a low temperature leads to the highest density of pairs. So far, neither the expected concentration nor temperature dependence has been reported, indicating that there is little defect–dopant pairing.

The integral in Equation (6.63) is prevented from diverging by the minimum separation, R_a, of the pair. Pairing is suppressed when the nearest neighbor separation is unstable and it seems likely that the $(P_4^+D^-)$ pair is such an example. Fig. 6.18 illustrates that a rearrangement of a single bond converts the pair into a neutral three-fold phosphorus atom surrounded by a fully coordinated silicon network. The P_3^0 state has a lower energy than the pair, so that the bonding change should occur whenever possible. The distribution

Fig. 6.18. Model showing the annihilation of a nearest neighbor donor–defect pair by a bond switching mechanism. The double line indicates the bond which changes position.

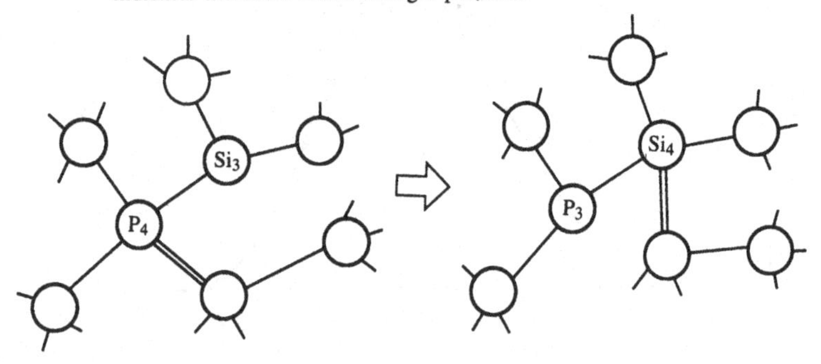

function therefore has a cut-off at least at the second neighbor separation, which reduces the expected density of pairs.

Boron–phosphorus pairs, in the form of $(B_4^-P_4^+)$, are more probable than the defect–dopant pairs. The thermal equilibrium model in Section 6.2.3 predicts that the concentrations of ionized dopants is much higher in compensated a-Si:H than in singly doped material, which favors pairing according to Eq. (6.63). In addition, the nearest neighbor pairs are probably stable. Although it is easy for the four-fold coordinated pair to break a bond and change to a $B_3^0 P_3^0$ pair, it seems plausible that the four-fold state has the lower energy (and even if the bond does break, the chemical pairing remains). Assuming that the close pair is four-fold coordinated and charged, the density is

$$N_{bp} \approx 4(N_{don}^2/N_{Si})\exp(E_{Cou}/kT) \tag{6.64}$$

where N_{don} is the concentration of phosphorus donors (and boron acceptors) and E_{Cou} is the Coulomb energy. The factor of 4 comes from the four possible nearest neighbor bonds. NMR experiments confirm boron–phosphorus pairs, with relative concentration, N_{bp}/N_{don}, of about 0.4 when the impurity concentration is 1% (Boyce *et al.* 1987). The Coulomb energy needed to give this concentration is about $2.5kT$, which is about 100 meV for the deposition temperature of 250 °C. It is difficult to estimate the Coulomb energy for a close pair because it is very sensitive to the shape of the wavefunction. However, since 100 meV is about the estimated binding energy of a donor electron, the Coulomb energy for the pair should be at least this large.

6.2.8 *The dopant distribution coefficient and equilibrium growth*

A different application of chemical equilibrium leads to an explanation of how the incorporation of defects and dopants depends on the growth conditions (Winer and Street 1989). Section 5.1 describes the unexpected rf power and gas concentration dependence of the dopant distribution coefficient, particularly for arsenic doping. A schematic diagram of the growth process is shown in Fig. 6.19, in which three-fold and four-fold silicon and dopants are deposited from the gas phase. The deposition reactions proposed for arsenic doping are

$$\mathrm{Si^{gas}} \leftrightarrows \mathrm{Si_4} \qquad N_{Si_4} = K_1 C_g^{Si} \simeq K_1 \tag{6.65}$$

$$\mathrm{Si^{gas}} \leftrightarrows \mathrm{D^-} - \mathrm{e} \qquad n_e^{-1} N_{Si_3} = K_2 C_g^{Si} \approx K_2 \tag{6.66}$$

$$\mathrm{As^{gas}} \leftrightarrows \mathrm{As_3} \qquad N_{As_3} = K_3 C_g^{As} \tag{6.67}$$

$$\mathrm{As^{gas}} \leftrightarrows \mathrm{As_4^+} + \mathrm{e} \qquad n_3 N_{As_4} = K_4 C_g^{As} \tag{6.68}$$

The subscripts denote the coordination in the a-Si:H film, e is an electron, C is the gas-phase mole fraction of the species, and the equations are the mass action expressions for the reactions with reaction constants k_1–k_4. The defect and donor concentrations can be solved in terms of either solid or gas-phase concentrations,

$$\text{gas: } N_D^- \simeq N_{As_4}^+ = (K_2 K_4 C_g^{Si} C_g^{As})^{\frac{1}{2}} \approx (K_2 K_4 C_g^{As})^{\frac{1}{2}} \tag{6.69}$$

$$\text{solid: } N_D^- \simeq N_{As_4}^+ = \left(\frac{K_2 K_4}{K_1 K_3} N_{Si_4} N_{As_3}\right)^{\frac{1}{2}} \simeq \left(\frac{K_2 K_4 N_{As}}{d_{As} N_{Si}}\right)^{\frac{1}{2}} \tag{6.70}$$

where d_{As} is the arsenic distribution coefficient, defined by

$$d_{As} \equiv \frac{N_{As}}{N_{Si}} \frac{C_g^{Si}}{C_g^{As}} \simeq \frac{N_{As_3}}{N_{Si_4}} \frac{1}{C_g^{As}} = \frac{K_3}{K_1} \tag{6.71}$$

The approximations are valid because the concentration of band tail electrons is small and most of the arsenic and silicon atoms are three-fold and four-fold coordinated, respectively.

Eqs. (6.69) and (6.70) are the doping reactions related to the gas-phase and solid-phase concentrations of silicon and phosphorus. The square root law applies to the gas-phase arsenic concentration provided that k_2 and k_4 are regular (i.e. independent of arsenic concentration). However, the law only applies to the solid-phase concentrations (Eq. (6.70)) if, in addition, d_{As} is also constant. The variation of the arsenic distribution coefficient with rf power and gas concentration (Fig. 5.4) explains the deviations from the doping law for the solid phase which are shown in Fig. 5.17. The phosphorus distribution coefficient has a much weaker dependence on the rf power and gas concentration, so

Fig. 6.19. Illustration of the deposition processes in which silicon and dopants are incorporated into the material in their different possible bonding configurations.

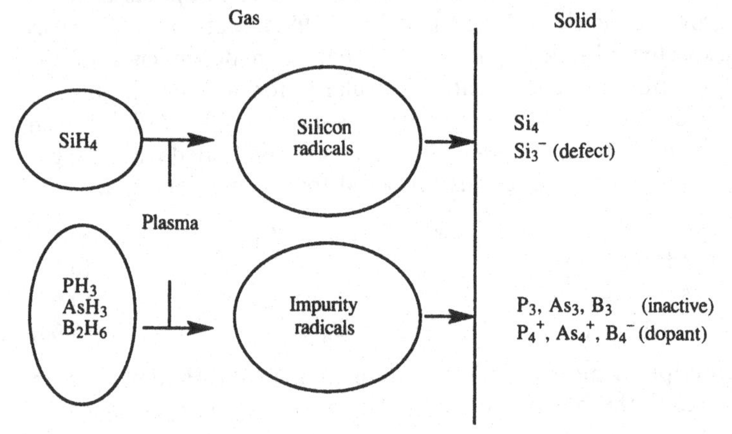

that the square root law applies well to both gas and solid concentrations. The irregular properties of the reaction constant k_3, indicate that the deposition of three-fold coordinated arsenic is complicated (Winer and Street 1989).

The distribution coefficient can be expressed in terms of the relative deposition rates for silicon and arsenic, r_{Si} and r_{As}

$$d_{As} = r_{As}/(r_{Si}\, C_g^{As}) \tag{6.72}$$

The film growth rate is dominated by the silicon deposition which is proportional to the rf power, $r_{Si} = \alpha W_{rf}$, so that from Eq. (5.6),

$$r_{As} = d_{reg}\, C_g^{As}\, \alpha W_{rf} + d_{irr}\, C_g^{As\frac{1}{2}} \tag{6.73}$$

The arsenic deposition rate contains two terms. First is a normal plasma-enhanced growth process which is proportional to the rf power and to the arsenic concentration. The second term is the origin of the irregular distribution coefficient and is independent of rf power and proportional to the square root of the gas concentration. It is unclear what surface chemical reaction leads to the second term, but whatever the mechanism, the result is arsenic incorporated into inactive sites which cannot form donors through the bulk equilibration process. This type of analysis does not give the specific chemical reaction during growth, but is helpful in establishing the general properties of the reaction. At present the identity of the arsenic configurations is unclear.

Hydrogen provides another example of an irregular distribution coefficient which indicates complex surface chemical reactions. The ability of hydrogen to move in and out of the growth surface is noted in Section 2.3.4. Its incorporation into the growing film may be described by the chemical reaction,

$$\mathrm{Si}^{gas} + \mathrm{H}^{gas} \leftrightarrows (\mathrm{SiH})^{solid} \tag{6.74}$$

Equilibrium growth occurs when the chemical potential of the hydrogen is equal in the gas and in the film. If the hydrogen behaves as an ideal gas and solute, then the concentration in the film should increase with the partial pressure of atomic hydrogen in the gas phase, but experiments find that this is generally not the case (Johnson *et al.* 1989). Furthermore, the result of adding hydrogen to the plasma during growth of a-Si:H is different from the effects of post-hydrogenation. In the first case, the hydrogen concentration of the film decreases, whereas in the second it increases. The complex behavior of hydrogen during growth occurs because there is extensive reconstruction of the bonding at the growth surface, so that changes in the growth conditions alter the silicon network. The crystallization which occurs at high hydrogen dilution of the plasma is a clear indication that hydrogen promotes structural change. On the other hand, post-hydrogenation does not produce such large changes in the silicon network because the bonding constraints are much greater in the bulk than at the growth surface. The effects of hydrogen on weak Si—Si bonds can perhaps be described in the two cases as

$$\text{growth} \qquad (\text{Si—Si})_{\text{weak}} + \text{H}^{\text{gas}} \leftrightarrows (\text{Si—Si})_{\text{strong}} + \text{H}^{\text{gas}} \qquad (6.75)$$

$$\text{post-hyd.} \qquad (\text{Si—Si})_{\text{weak}} + \text{H}^{\text{gas}} \leftrightarrows 2\text{Si—H} \qquad (6.76)$$

which illustrate the different behavior.

6.3 Kinetics of structural relaxation and equilibrium

The description of the equilibrium state in terms of formation energies does not consider the rate at which the structure equilibrates. The configurational coordinate diagram of Fig. 6.1 illustrates the energy barrier which must be overcome in order to reach equilibrium. The excitation over the barrier corresponds to the motion of atoms which, for example, allow a four-fold silicon atom to transform into a three-fold dangling bond. The energy barrier is manifested in the kinetics of equilibration and structural relaxation.

Measurements of the equilibrium state make use of the observation that rapid cooling freezes the structure, allowing the characteristics of a high temperature state to be measured at a lower temperature. The freezing is incomplete, however, and the structure slowly relaxes to the equilibrium appropriate to the lower temperature. Figs. 6.4 and 6.9 give examples of the relaxation in doped and undoped a-Si:H, and the data of Fig. 6.4 are replotted in a normalized form in Fig. 6.20. The relaxation time is about a year at room temperature, so that most experiments performed at or below this temperature can be considered to be completely in the frozen state.

The relaxation of the electronic properties corresponds to a change in the bonding coordination of silicon or dopant atoms within the network. The process therefore involves some atomic motion of either silicon, hydrogen or the dopants. Hydrogen is known to diffuse easily in a-Si:H, as is obvious from the fact that it is evolved from a-Si:H at a fairly low temperature. On the other hand, the silicon network is rigid and both silicon and the dopants have a much lower diffusion rate (Persans and Ruppert 1987, Scholch, Kalbitzer, Fink and Behar 1988). Hydrogen motion therefore seems the obvious candidate for the mechanism by which the thermal equilibration takes place. Two other possibilities have also been considered, which do not involve the real diffusion of atoms. The coordination of a state can be changed simply by breaking a bond, with a local distortion, but no diffusion of atoms. The difficulty with this model is that coordination defects must occur in pairs, and Section 6.2.7 shows that the thermodynamic properties of paired defects are different from separate random defects; the experiments indicate that the dopants and defects are separated. Defects may diffuse apart by a bond switching mechanism which does not involve atomic motion and the floating bond defects discussed in Section 4.3 were proposed to have this property. However, measurements find that the defect diffusion is negligible near the defect equilibration temperature (Jackson 1989).

6.3.1 *Stretched exponential relaxation*

The relaxation data of Fig. 6.20 are not described by a single time constant, but instead extend over more than four orders of magnitude in time. The time dependence follows a stretched exponential relation,

$$\Delta n_{BT} = n_0 \exp[-(t/\tau)^{\beta}] \tag{6.77}$$

The solid lines in the data of Fig. 6.20 are fitted to this expression, with values of β ranging from 0.45–0.70 as indicated.

Stretched exponential relaxation is a fascinating phenomenon, because it describes the equilibration of a very wide class of disordered materials. The form was first observed by Kohlrausch in 1847, in the time-dependent decay of the electric charge stored on a glass surface, which is caused by the dielectric relaxation of the glass. The same decay is observed below the glass transition temperature of many oxide and polymeric glasses, as well as spin glasses and other disordered systems.

Fig. 6.20. The time-dependent relaxation of the band tail carrier denisty (Fig. 6.4) plotted in normalized form. The solid lines are fits to the stretched exponential with parameter β as indicated (Kakalios, Street and Jackson 1987).

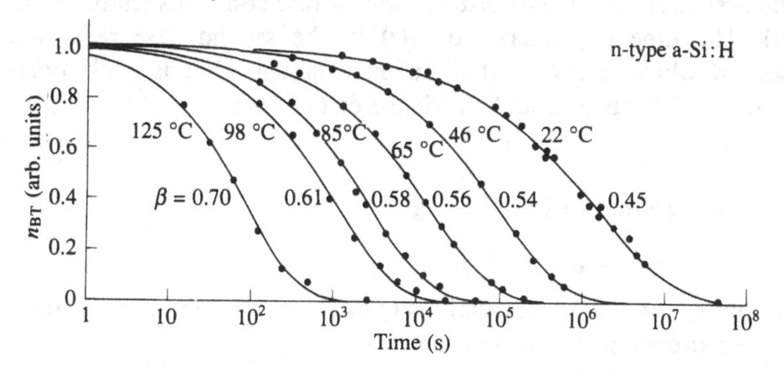

It is apparently a general characteristic of glassy disorder, although there has been considerable debate over the relation between the stretched exponential decay and the microscopic relaxation mechanisms.

The bonding disorder of a glass suggests that a decay with a single time constant is not expected, but instead an average over the structural configurations. One possibility is a local variation in decay rates described by a distribution of time constants $F(\tau)$, so that

$$n(t) = \int F(\tau) \exp(-t/\tau)\, \mathrm{d}\tau \tag{6.78}$$

An appropriate choice of the distribution function can account for any reasonable shape of the relaxation data. The unexplained feature of this model is that there is no apparent reason why the resulting decay should be a stretched exponential, as this does not result from any obvious distribution of lifetimes, such as a gaussian. The integral in Eq. (6.78) contains the implicit assumption that the relaxation events occur independently at each site. An alternative model describes the relaxation in terms of a heirarchy of processes (Palmer, Stein, Abrahams and Anderson 1984), in which increasingly large configurations of atoms are involved in the relaxation as time progresses. The relaxation processes are then no longer independent of each other, but evolve with time. Palmer *et al.* (1984) were able to obtain the stretched exponential from some estimates of the way such a structure would evolve.

These alternative descriptions of the relaxation are simply expressed by rate equations of the form,

$$\mathrm{d}N/\mathrm{d}t = -k(x)N \tag{6.79}$$

and

$$\mathrm{d}N/\mathrm{d}t = -k(t)N \tag{6.80}$$

In the first case, the spatial distribution of rate constants leads to Eq. (6.78). The time dependence of $k(t)$ in the second case reflects a relaxation which occurs within an environment that is itself time-dependent. If k has a time dependence of the form,

$$k(t) = k_0 t^{\beta-1} \tag{6.81}$$

then the integration of Eq. (6.80) gives

$$N(t) = N_0 \exp[-(t/\tau)^{\beta}] \tag{6.82}$$

Kohlrausch (1847) used this analysis to explain his original observation of stretched exponential relaxation.

Eq. (6.81) implies that the atomic motion which causes the structural relaxation has a power law time dependence. The temperature dependence of the parameter β, measured from the relaxation data of Fig. 6.20, is given in Fig. 6.21 and follows the relation,

$$\beta = T/T_R \text{ with } T_R \approx 600 \text{ K} \tag{6.83}$$

It has already been suggested that the motion of the bonded hydrogen may be the origin of the structural relaxation. Hydrogen is, indeed, observed to have a diffusion coefficient, D_H, which decreases as a power law in time; the data for boron-doped a-Si:H are shown in Fig. 2.22. The time dependence of D_H has the right form and magnitude to account for the stretched exponential relaxation. Fig. 6.21 shows that the experimental values of β obtained from the decay of n_{BT} agree well with the dispersion parameter of the hydrogen diffusion. This comparison establishes a quantitative link between the structural relaxation and the diffusion of bonded hydrogen.

It is natural to look for a connection between the time-dependent hydrogen diffusion and the similar dispersive motion of electrons and holes in a band tail. Section 3.2.1 shows that the trap-limited carrier mobility has as a power law time dependence with a dispersion

Fig. 6.21. Temperature dependence of the stretched exponential parameter, β, from the data of Fig. 6.20 and from the ESR data in Fig. 6.29. Also shown is the dispersion parameter for hydrogen diffusion (Kakalios *et al.* 1987).

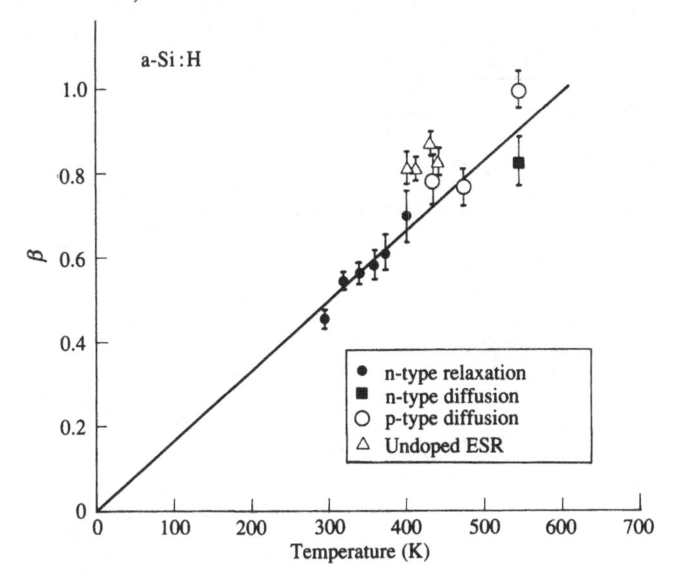

parameter given by $T/T_{C,V}$. Dispersive electronic transport occurs because the carriers are progressively trapped in deeper and deeper states of an exponential band tail. This immediately suggests that the stretched exponential relaxation occurs because of hydrogen diffusion in an equivalent exponential distribution of trapping sites.

The hydrogen diffusion is, indeed, described in Section 2.3.3 in terms of the trapping model shown in Fig. 6.22. The Si—H bonds and the tail of weak Si—Si bonds provide the hydrogen traps, while hydrogen is mobile in the higher energy interstitial states. Thus the diffusion is explained by a multiple trapping model in which hydrogen is thermally activated from a distribution of traps to a mobile state. The dispersive diffusion follows, provided the distribution of states is roughly exponential. The distribution of weak bonds is associated with the shape of the valence band tail in Section 6.2.3, and the values of T_R for hydrogen diffusion and hole transport are not very different.

This analogy between the motion of electronic carriers and of hydrogen cannot be taken very far because the hydrogen diffusion has some added complications. Dispersive diffusion of electrons or holes occurs with a low density of carriers in a large density of traps and the trapping events are largely independent. Hydrogen diffusion takes

Fig. 6.22. Schematic model showing the distribution of hydrogen binding energies. The more distorted Si—Si weak bonds trap hydrogen in deeper states. The dispersive hydrogen diffusion corresponds to the trapping and release from the weak bonds and Si—H bond sites (Street *et al.* 1988a).

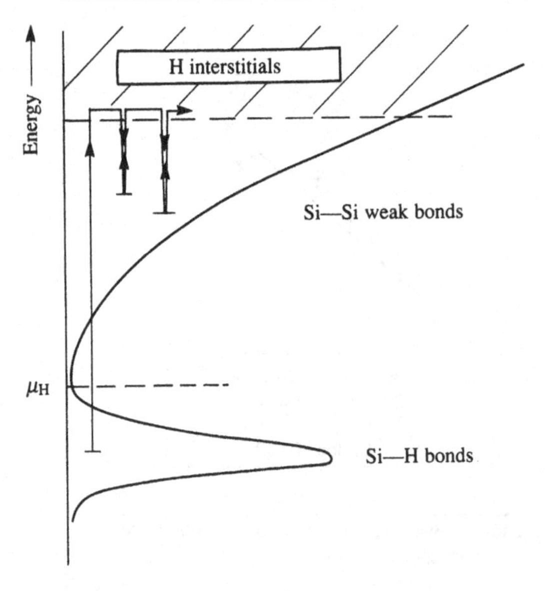

place with a high hydrogen concentration compared to the density of traps and there are probably cooperative effects, in which a particular atom cannot move unless a neighboring atom also moves. It may be that the exponential distribution of traps should be viewed as configurations of groups of atoms rather than as traps of a single hydrogen atom.

The analysis of dispersive electronic transport in Section 3.2.1 assumes that all the carriers move with the same time-dependent mobility, described by a demarcation energy which changes with time. This approximation is equivalent to the assumption of a single time-dependent rate constant in Eq. (6.81). The more complete theory of dispersive transport recognizes that the dispersion actually originates from a very broad distribution of individual hopping times. Shlesinger and Montroll (1984) have applied the original concept of a continuous time random walk (Scher and Montroll 1975), which so successfully explained the dispersive transport, to the case of atomic relaxation. Their analysis confirms the prediction of a stretched exponential relaxation of the form given by Eqs. (6.82) and (6.83).

In dispersive electronic transport, the time dependence of the drift mobility leads to some unusual experimental observations, for example, an apparent thickness and field dependence of the mobility. These same effects enter in the relation between the dispersive hydrogen diffusion and the electronic relaxation time. Following Jackson, Marshall and Moyer (1989), the time-dependent diffusion coefficient is described by

$$D_{\mathrm{H}}(t) = D_{00}(\omega_0 t)^{\beta-1} \tag{6.84}$$

The evaluation of D_{H} from a measurement of the diffusion profile gives

$$D_{\mathrm{H}} = L^2/4t \tag{6.85}$$

where L is the diffusion length. The temperature dependence of D_{H} depends on whether the measurement is performed with a constant diffusion time or diffusion length. In the time-of-flight experiment of the electron mobility, the transport length is constrained by the thickness of the sample. There is no such absolute constraint in the diffusion experiments, but it is often convenient to hold the diffusion length approximately constant, and this is the case for most of the published data. Eliminating the time from Eqs. (6.84) and (6.85) to express the diffusion coefficient as a function of diffusion length gives

$$D_{\mathrm{H}}(L) = (4D_{00}/\omega_0 L^2)^{1/\beta}\,\omega_0 L^2/4 \tag{6.86}$$

The rate constant, $k(t)$, for the relaxation is $D_{\mathrm{H}}(t)/a^2$, where a is a hopping length of atomic dimensions. The solution to Eq. (6.80) then gives the relaxation time, $(k_0/\beta)^{1/\beta}$, in terms of the hydrogen diffusion coefficient. When the thermally activated relaxation time is expressed as

$$\tau = \omega_0^{-1} \exp(E_B/kT) \tag{6.87}$$

it follows from Eq. (6.84) that

$$E_B = kT_R \ln(\beta a^2 \omega_0 / D_{00}) \tag{6.88}$$

The relation between τ and D_H is,

$$\tau = \frac{1}{D_H(L)} \left[\frac{4^{1-\beta} \beta a^2}{L^{2(1-\beta)}} \right]^{\frac{1}{\beta}} \tag{6.89}$$

The various measurements of the relaxation time and D_H in Fig. 6.23 show that Eq. (6.89) is obeyed quite well. The data include n-type, p-type and undoped a-Si:H and cover a range of temperatures.

The hydrogen diffusion activation energy increases with time because of the dispersive hydrogen motion. The relaxation of defects and dopants typically involves a much shorter hydrogen diffusion distance than in measurements of the diffusion coefficient, so that the activation energy of τ is smaller than of D_H and is given by

$$E_B = E_{DH} - 2kT_R \ln[L/2\beta a] \tag{6.90}$$

Fig. 6.23. A test of Eq. (6.89) relating the relaxation time to the hydrogen diffusion coefficient. Many different manifestations of the equilibration process are included in the data (Jackson and Moyer 1988).

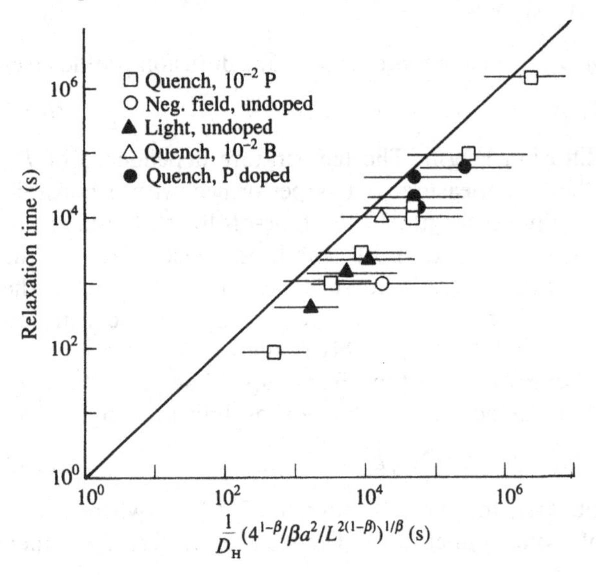

The observed difference of 0.2–0.5 eV is consistent with $L \sim 100$–1000 Å and $a \sim 10$–100 Å.

The quantitative agreement between the relaxation and the diffusion data is strong evidence that the hydrogen is the mechanism of structural relaxation to equilibrium. The hydrogen mechanism also leads to a structural interpretation of the stretched exponential decay in terms of an exponential band tail of trapping site. Although the hydrogen mechanism is clearly specific to a-Si:H, the general mechanism of dispersive diffusion can apply to other amorphous materials. Dispersive transport of either electronic carriers or atoms is a consequence of the disorder of the energy levels and is an effect of wide generality.

6.3.2 *The hydrogen glass model*

There is considerable evidence that hydrogen diffusion is the mechanism of thermal equilibration. The evidence is necessarily indirect because the hydrogen apparently does not participate directly in the defect or dopant states, but acts as a catalyst to the motion. For example, the movement of a hydrogen atom from one Si—H bond at site A to a neighboring dangling bond at site B is described by

$$(\mathrm{SiH})_A + \mathrm{D}_B \leftrightarrows \mathrm{D}_A + (\mathrm{SiH})_B \tag{6.91}$$

The result of the reaction is to move a defect from A to B, but the energy of the Si—H bond does not enter into the reaction energy, because it is preserved on both sides of the reaction. On the other hand a large energy is needed to release the hydrogen into a mobile interstitial site so that it can move from A to B and this energy determines the rate of the reaction. Hydrogen motion can create or annihilate defects or dopant states by similar reactions. For example, hydrogen might break a weak bond by the process described in Eq. (6.44). The energy of the reaction again does not contain any contribution from the Si—H bond energy, because the bond is preserved. Furthermore, the two defects can diffuse apart via reaction (6.91). Equivalent reactions can be written for the creation, annihilation and migration of dopant states.

At the beginning of this chapter, the behavior of glasses near the glass transition temperature was compared with the changes in conductivity of a-Si:H at the equilibrium temperature. Stretched exponential relaxation is also a characteristic of glasses. These similarities lead to the proposal that the bonded hydrogen in a-Si:H can be considered to be a glass (Street *et al.* 1987a). This model divides

Table 6.2. *Comparison of the measured and calculated equilibration temperatures for different doping types*

Material	T_E (measured) (°C)	T_E ($D_H = 10^{20}$ cm^2 s^{-1}) (°C)
p-type	80	70
n-type	130	110
Undoped	200	160

the a-Si:H network into two components: a rigid silicon network and a glass-like network of bonded hydrogen. The assumed properties have some similarities to those of a super-ionic conductor in which there is a rigid crystal containing a liquid-like sublattice of highly mobile atoms.

Whether or not the hydrogen-glass is a valid description depends in part on the definition of a glass, which is a rather loose term. Certainly a-Si:H has similar kinetic properties to those of a glass, with a high temperature equilibrium and a low temperature frozen state. The glass analogy has proved to be useful and was the reason that the stretched exponential form for the relaxation was explored. Two other connections can be made with the properties of normal glasses, through the viscosity and the specific heat.

The viscosity of a glass is about 10^{13} poise at the glass transition

Fig. 6.24. Differential scanning calorimetry measurements of the heat released during equilibration of n-type a-Si:H after (*a*) fast and (*b*) slow cooling (Matsuo *et al.* 1988).

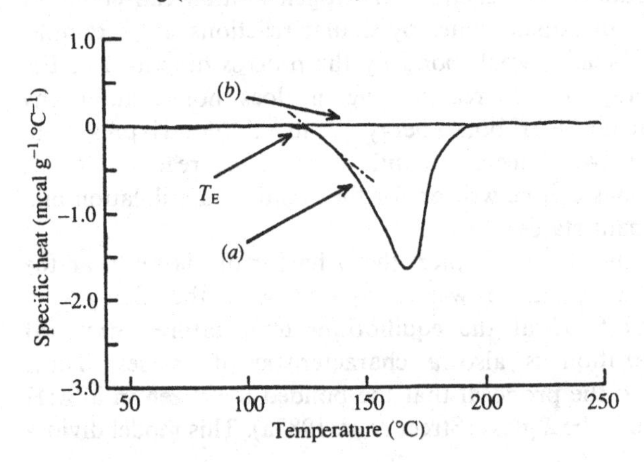

temperature. The viscosity, η, of a liquid is related to the diffusion coefficient by the Stokes–Einstein relation,

$$\eta D = kT/6\pi a_D \tag{6.92}$$

where a_D is the hopping distance of the diffusing species. The glass transition temperature is therefore expected at a diffusion coefficient of about 10^{-20} cm^2 s^{-1}, for a hopping length of a few angströms. Table 6.2 compares the equilibration temperature measured from the conductivity experiments (e.g. Fig. 6.3) with the predicted value obtained by extrapolating the hydrogen diffusion coefficient data of Fig. 2.20. The two temperatures agree quite well and, in particular, account for the changes in the equilibration temperature for different doping types.

The hydrogen diffusion coefficient is given by,

$$D_H = \tfrac{1}{6} a_D^2 p_H \tag{6.93}$$

where p_H is the hopping rate of the hydrogen atoms. The relaxation time is approximately 1000 s at the equilibrium temperature of a typical conductivity measurement. When D_H is set equal to 10^{-20} cm^2 s^{-1} and $a_D = 5$ Å, then $p_H \approx 2 \times 10^{-5}$ s^{-1}, and the motion of about 10^{20} cm^3 hydrogen atoms is needed to equilibrate the structure. The changes in the defect and dopant densities are typically in the range 10^{16}–10^{17} cm^{-3}, suggesting that about 1000 hydrogen hops are needed to equilibrate each state.

The glass transition of a conventional glass is characterized by a change in the specific heat, which is observed over an extended temperature range in a calorimetry measurement. The hydrogen glass model suggests that a similar observation may be expected in a-Si:H. An endothermic feature of this type has been reported in n-type a-Si:H and is shown in Fig. 6.24 (Matsuo *et al.* 1988). There is a broad peak in the temperature range 120–180 °C. The temperature of the feature and its dependence on the cooling rate identify it with the equilibration effects. The onset temperature of the endotherm is observed to have the same dependence on substrate temperature during growth as the equilibration temperature measured from the conductivity. The change in specific heat is small, because only hydrogen atoms participate in the structural equilibration.

6.4 Metastability

The phenomenon of metastability is closely related to the defect equilibrium properties. Some external excitation – illumination, charge, current flow, energetic particles, etc. – induces defects (or dopants) which are subsequently removed by annealing. The metastable defects anneal at 150–200 °C which is the temperature at which equilibration occurs in a few minutes, immediately suggesting a connection between metastability and defect equilibration. Annealing is simply the restoration of thermal equilibrium which is disturbed by the external excitation. Similar effects occur in chalcogenide glasses (Tanaka 1976) and also in crystalline semiconductors (Kimerling 1978).

The external excitation may change both the reaction rate and the equilibrium state of a defect reaction. Fig. 6.25 illustrates the expected temperature dependence of the reaction rate in different situations. The activation energy of a purely thermal process is E_B, which is about 1 eV in a-Si:H and is associated with hydrogen motion. An external excitation provides a non-thermal source of energy to overcome the barrier – the non-radiative recombination of excess cariers is often identified as the energy source. The recombination-enhanced reaction rate has a lower activation energy of $E_B - E_X$, where E_X is the energy provided by recombination (Kimerling 1978). The reaction rate is athermal when E_X is equal to or larger than E_B, but is thermally activated when E_X is smaller.

Defects or dopants are created because the external excitation drives the defect reaction away from the initial equilibrium. The formation energy of charged defects, and therefore their equilibrium density, depends on the position of the Fermi energy according to Eqs. (6.26) and (6.27). Thus, the equilibrium defect density is

$$N_D = N_{D0} \exp(\Delta E_F / kT) \tag{6.94}$$

where ΔE_F is the shift of the Fermi energy, and N_{D0} is the neutral defect density corresponding to an unshifted Fermi energy. Any process which causes the Fermi energy to move, disturbs the equilibrium of the states and tends to change their density. The examples of metastability discussed below cover most of the possible ways in which the Fermi energy (or quasi-Fermi energy) can be moved by an external influence.

Most electronic devices – solar cells, transistors, sensors, etc. – depend for their operation on a shift of E_F by an external source. The metastable creation of defects, which is associated with the operation of these devices, causes a degradation of the device characteristics and is a significant technological problem. An example is the reduction of the solar cell efficiency after prolonged illumination. The degradation is not too severe, because the defect creation rate is very slow, but there is a large research endeavour aimed at studying and mitigating the effects.

6.4.1 *Light-induced defects*

The first metastable effect found in a-Si:H and the most widely studied is the creation of defects by prolonged illumination of absorbing light. This was discovered in 1977 and is the cause of the slow degradation of solar cells as they are exposed to sunlight (Staebler and Wronski 1977). The first measurements found a decrease in the photoconductivity during the illumination and a drop in the dark conductivity after the illumination. Staebler and Wronski (1977) suggested that both observations were due to the creation of additional localized gap states. The effect has been studied by most of the techniques used to measure gap states, but there is considerable debate about its phenomenology and interpretation (e.g. see Fritzsche (1987)).

Fig. 6.25. The expected temperature dependence of the reaction rate for a recombination-enhanced defect creation process.

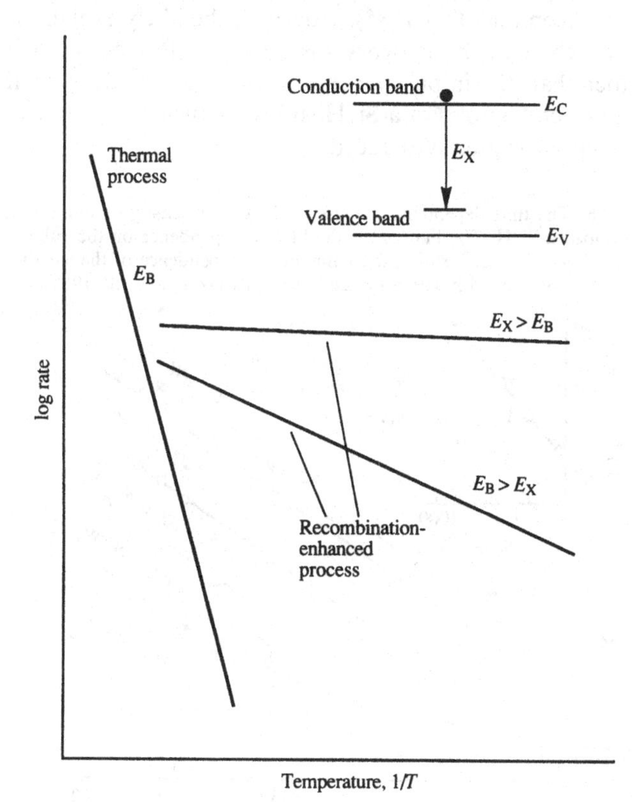

One issue is whether one or more type of defect is involved. Most reports conclude that the created defects are dangling bonds and the results of PDS, ESR and time-of-flight after light soaking are indistinguishable from unilluminated a-Si:H which has a high defect density due to non-optimum growth conditions. Some other experiments, notably the annealing studies of Han and Fritzsche (1983), conclude that there are at least two distinct effects, because the defect-induced optical absorption and photoconductivity anneal differently. One proposed explanation is based on the expectation that light creates defects with a distribution of gap states energies (Shephard, Smith, Aljishi and Wagner 1988). The difference between the annealing of absorption and photoconductivity may be explained by recombination properties and annealing rates which depend on the gap state energy.

The role of impurities in the defect creation is also controversial. There is no doubt that the density of metastable defects increases when the concentration of oxygen or nitrogen is above about 1 at% (Stutzmann, Jackson and Tsai 1985). However, the likely explanation is that alloying changes the network disorder to allow easier defect creation, rather than the impurity being associated directly with the light-induced defect. Samples of a-Si:H still show the effect even when the impurity density is greatly reduced.

Fig. 6.26. The time-dependent increase of the defect density by illumination of undoped a-Si:H. The lines show the fit to a dependence on the cube root of time. The insert shows the temperature dependence of the steady state defect density with activation energy, E_a (Stutzmann *et al.* 1985).

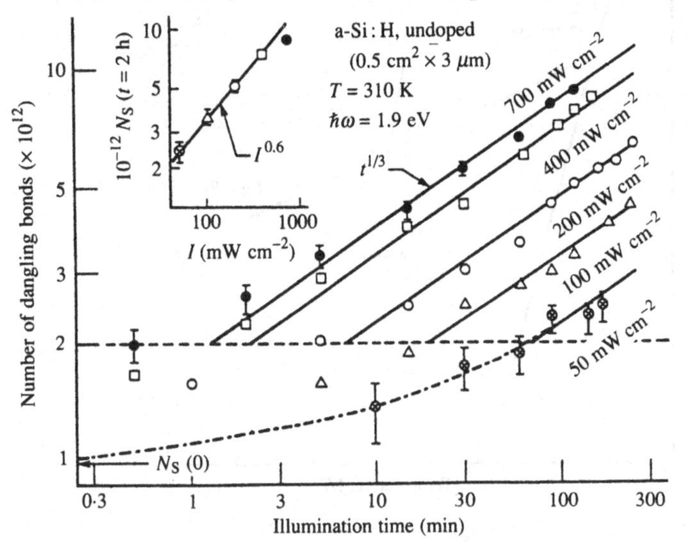

Another question is whether the observed effects represent the creation of new defects or the change in occupancy of existing states (Adler 1981). The light-induced effects are characterized by very slow creation and annihilation rates – illumination with sunlight results in the creation of about 10^{17} cm^{-3} defects after many hours. The effective capture cross-section for the defect creation is estimated to be no more than 10^{-25} cm^2, assuming an electron trapping mechanism, which is ten orders of magnitude lower than normal capture cross-sections for an electronic state. Similarly, since the induced defects in undoped a-Si:H are stable almost indefinitely at room temperature and are only annealed above 150 °C, the rate of annealing must be much slower than any normal excitation rate of electrons or holes to the band edge. The low creation and annealing rates strongly suggest that a structural change is taking place rather than a change in occupancy.

The following discussion concentrates on the creation of bulk dangling bond defects, which may not be the only process, but is almost certainly the dominant one. Dersch, Stuke and Beichler (1980) were the first to show that illumination causes an increase in the $g = 2.0055$ paramagnetic defect and concluded that the Staebler–Wronski effect was the creation of dangling bonds. The metastable defect creation and annealing is described by the potential well model shown in Fig. 6.1, except that the barrier is overcome by the recombination energy from

Fig. 6.27. The time dependence of light-induced defect density at different temperatures, showing that a steady state is reached. The inset shows the steady state defect density (Stutzmann *et al.* 1985).

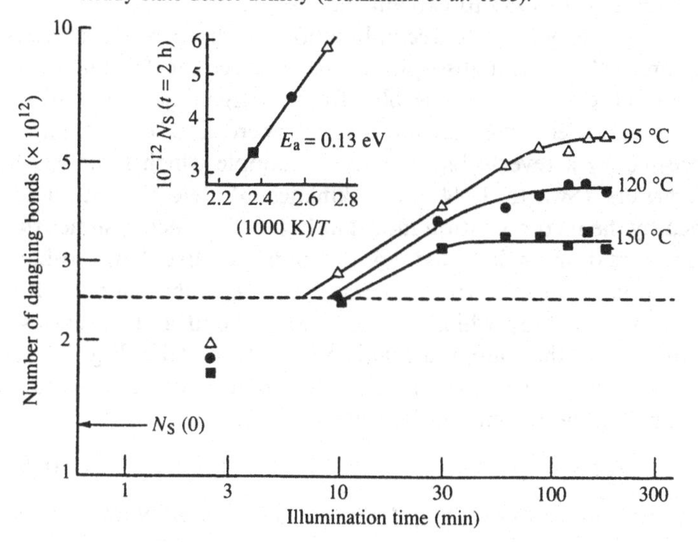

the excited carriers, rather than thermally. The kinetics of defect creation therefore give information about how the light excitation causes the defects and the thermal annealing kinetics describes the return to the equilibrium state.

Fig. 6.26 plots the illumination time dependence of the number of dangling bonds. When the defect density, N_D, is more than twice the initial value, N_D increases as the cube root of time, and so is strongly non-linear. The figure shows no sign of saturation after 2–3 hours of strong illumination at room temperature, but a steady state is obtained at higher light intensities or longer illumination times (Park, Liu and Wagner 1990). The steady state is reached more quickly when the experiment is performed at elevated temperatures. Fig. 6.27 shows that the limiting value of N_D decreases as the temperature is raised. The same figure shows that the rate of defect creation is insensitive to the temperature. The almost athermal creation process indicates that all of the energy needed to overcome the barrier is supplied by the light excitation. The dependence on the illumination intensity G is also non-linear, having a power close to 0.6, as shown in Fig. 6.26. The defect creation kinetics are therefore described by

$$N_D(t) = \text{const. } G^{0.6} t^{0.33} \tag{6.95}$$

It is immediately clear that the reciprocity law ($N_D(t) \propto Gt$) is not obeyed and that the defect creation is not proportional to the total light flux. The implication is that the new defects modify the defect creation process, which is the essence of the following model proposed by Stutzmann *et al.* 1985 to explain the results.

Defects are created by the recombination of photoexcited carriers, rather than by the optical absorption. The evidence for this conclusion is that defect creation also results from charge injection without illumination (see Section 6.5.2) and that defect creation by illumination is suppressed by a reverse bias across the sample which removes the excess carriers (Swartz 1984). The kinetics of defect creation are explained by the recombination model in Fig. 6.28, which assumes that the defect creation is initiated by the non-radiative band-to-band recombination of an electron and hole. The recombination releases about 1.5 eV of energy which breaks a weak bond and generates a defect. In terms of the configurational coordinate model of Fig. 6.1, the energy overcomes the barrier E_B. The defect creation rate is proportional to the recombination rate

$$dN_D/dt = c_d np \tag{6.96}$$

where c_d is a constant describing the creation probability and n and p

are the electron and hole concentrations. The defect creation process represents only a small fraction of the recombination, most of which is by trapping at the dangling bond defects. The recombination is quite complicated because there are four possible transitions to the two energy levels of the defect and in steady state the rates balance to keep the defect occupancy constant; the recombination is discussed in more detail in Chapter 8. Provided that the illumination intensity G is high enough, the carrier densities n and p at the conduction and valence band edges are given to a good approximation by

$$n = p = G/AN_D \tag{6.97}$$

where A is an average recombination constant. Photoconductivity experiments, which measure the carrier concentration n, confirm the expected dependence on G and N_D, although at low defect density there is some deviation from the simple monomolecular decay.

The time dependence of the defect density is obtained by combining Eqs. (6.96) and (6.97) and integrating to give

$$N_D{}^3(t) - N_D{}^3(0) = 3c_d G^2 t/A^2 \tag{6.98}$$

where $N_D(0)$ is the initial equilibrium defect density. At sufficiently long illumination times such that $N_D(t) > 2N_D(0)$, Eq. (6.98) approximates to

$$N_D(t) = (3c_d/A^2)^{1/3} G^{2/3} t^{1/3} \tag{6.99}$$

which closely describes the intensity and time dependence of the experimental results.

The defect creation is a self-limiting process. As the defect density increases, the additional recombination through the new states reduces the concentration of band edge carriers and suppresses the creation

Fig. 6.28. An illustration of the recombination model used to explain the creation of defects by illumination.

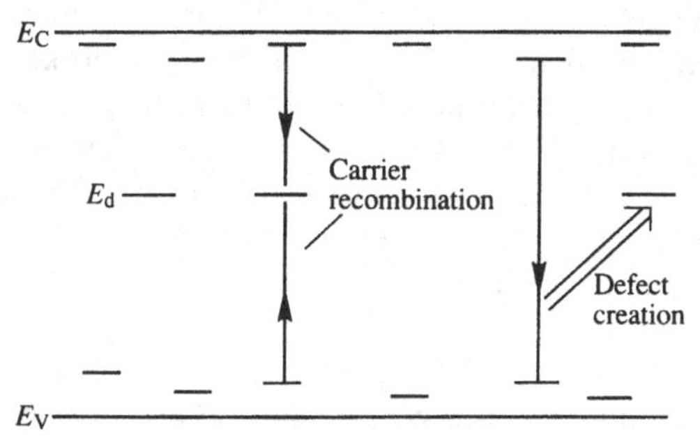

mechanism. The most rapid rate of defect creation is in samples with the lowest defect density and the slow down of the rate explains the strongly sublinear time dependence. The model predicts a similar time dependence for the photoconductivity, the observation of which confirms that the creation mechanism involve the band edge carriers (Stutzmann *et al.* 1985).

The parameter c_d is the susceptibility of the material to defect creation and is roughly 10^{-15} cm^3 s^{-1}. It is the product of the rate of band-to-band transitions and the probability that such a transition creates a defect, but no experiment has yet been able to measure these two terms separately. The magnitude of c_d is sensitive to the material properties and, for example, is much larger in a-Si:H films deposited below 100 °C compared to the usual 200–250 °C. However, little is known about the details of how the electron–hole recombination gives up its energy to the lattice to cause defect creation.

The annealing kinetics of the light-induced defects are shown in Fig. 6.29. Several hours at 130 °C are needed to anneal the defects completely, but only a few minutes at 200 °C. The relaxation is non-exponential, and in the initial measurements of the decay the results were analyzed in terms of a distribution of time constants, Eq. (6.78) (Stutzmann, Jackson and Tsai 1986). The distribution is centered close to 1 eV with a width of about 0.2 eV. Subsequently it was found that the decay fits a stretched exponential, as is shown in Fig. 6.29. The parameters of the decay – the dispersion, β, and the temperature dependence of the decay time, τ – are similar to those found for the thermal relaxation data and so are consistent with the same mechanism of hydrogen diffusion. The data are included in Fig. 6.23 which describes the general relation between τ and D_H. The annealing is therefore the process of relaxation to the equilibrium state with a low defect density.

The association of the relaxation kinetics with the hydrogen diffusion rate is evidence that the defect creation process is the result of hydrogen motion. Fig. 6.30 illustrates one proposal for the microscopic mechanism (Stutzmann *et al.* 1985). A Si–Si bond is broken and a hydrogen atom moves from an adjacent site to stabilize the broken bond. The deformation of the bond which initiates the process is assumed to result from the bond weakening when an electron and hole are localized close together. Some variations on this basic model have been suggested. For example, Carlson (1986) suggests that the defects occur at the surface of microvoids where there is perhaps more flexibility to allow a broken bond and more hydrogen to stabilize it. So far there is little definite evidence which relates the defect to specific sites in the material. In both of these models the hydrogen moves, but remains bonded to silicon before and after the reaction and so acts as a catalyst.

The light-induced and thermal equilibrium defect reactions are aspects of the same general process. Indeed, the structural models proposed are virtually identical (compare Figs. 6.13 and 6.30). In the two-well description of Fig. 6.1, excitation over the barrier in either direction can, in principle, be thermal or by an external excitation. The

Fig. 6.29. The decay of the normalized light-induced defect density at different annealing temperatures, showing the stretched exponential behavior (Jackson and Kakalios 1988).

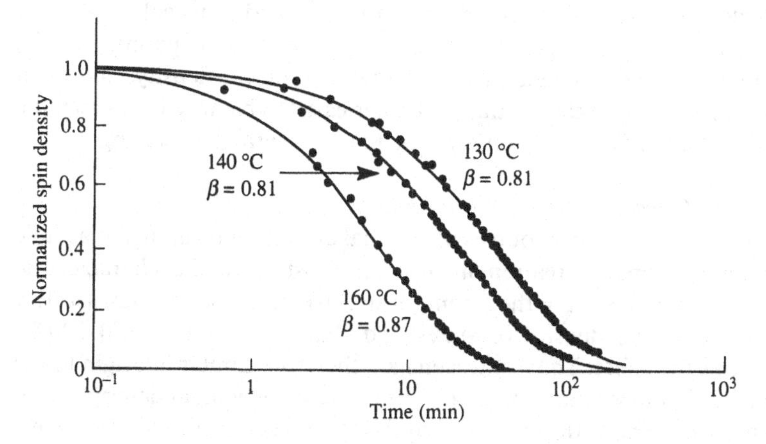

Fig. 6.30. One of the models proposed to explain the light-induced defect creation, in which the broken Si—Si bond is stabilized by hydrogen motion (Stutzmann *et al.* 1985).

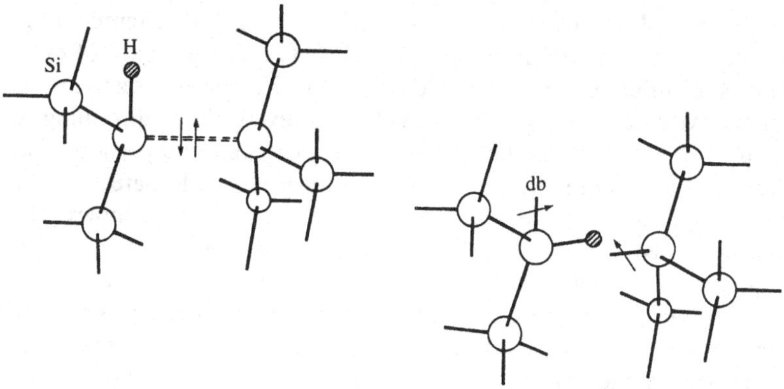

Staebler–Wronski effect corresponds to an external excitation over the barrier from the ground to excited state induced by carrier recombination, and its annealing is thermal excitation in the reverse direction. The thermal equilibrium state occurs when there is a balance between the thermal excitation rates over the barrier in both directions. Redfield (1987) points out that illumination can cause the reverse process of defect annihilation and some experiments have reported the effect (Delahoy and Tonon 1987). When the illumination is kept on for a long enough time, the forward and reverse rates balance and a steady state is reached. At low temperature, the steady state density is determined by the external excitations, whereas at high temperatures the thermal transitions dominate. At intermediate temperatures, the effects are a combination of both thermal and external excitation, depending on the particular conditions of the experiment. The temperature dependence of the steady state defect density, shown in Fig. 6.27, suggests that in this particular case the balance is between the forward external excitation and the reverse thermal annealing.

6.4.2 *Other metastable phenomena*

Several other sources of external excitation result in metastable defect or dopant creation in a-Si:H. Most have the characteristic property that a shift in the Fermi (or quasi-Fermi) energy causes a slow increase in the density of states and that annealing to 150–200 °C reverses the effect. The phenomena are therefore similar in origin to the optically-induced states and fall within the same general description of departures from the thermal equilibrium state induced by excess carriers.

(1) *Current injection*

A space charge limited current flow in a p^+–p–p^+ structure results in a slow increase in the defect density at room temperature (Kruhler, Pfleiderer, Plattner and Stetter 1984). The presence of extra defects is inferred from the reduction in the current seen in the conductivity data in Fig. 6.31. The effect is reversible by annealing to about 150 °C. The results are interesting because they show that electron–hole recombination is not necessary to create defects, since there are virtually no band edge electrons in the p^+–p–p^+ samples. The energy needed to form the defects must instead come from the hole trapping. The energy from electron trapping in n–i–n samples is evidently insufficient to create defects at room temperature since the effect is not observed. Possibly this difference is related to the lower hydrogen diffusion coefficient in n-type a-Si:H.

(2) *Charge accumulation at interfaces*

The Fermi energy near a dielectric interface is controlled by the field across the dielectric. The transfer characteristics (source-drain current versus gate voltage) of a field effect transistor changes slowly with time when the transistor is held on with a large accumulation charge. For some time it was believed that the effect was entirely due to charge trapping in the dielectric (usually silicon nitride). However, as the properties of the nitride improved, it became clear that defect creation in the a-Si:H film near the interface was also a degradation mechanism (van Berkel and Powell 1987). An interface defect density change of ΔN cm^{-2} causes a shift of the threshold voltage, ΔV, in the device characteristics, given by

$$\Delta V = e\Delta N/C_g \tag{6.100}$$

where C_g is the capacitance of the gate dielectric. Fig. 6.32 shows the time dependence of the normalized threshold voltage shift at different measurement temperatures. The defect creation rate is thermally activated and the rate increases by a factor 100 between 360 K and 420 K. The time dependence of the defect creation follows the stretched exponential form and the parameters are again similar to those found in the thermal equilibration and optical creation experiments. The diffusion of hydrogen is therefore once more implicated in the defect creation.

Fig. 6.31. Conductivity data for a p$^+$–p–p$^+$ device. A is the annealed state, B is after a current has flowed for 24 hours, and P is after prolonged illumination. The reduction in space charge limited current is attributed to defect formation (Kruhler *et al.* 1984).

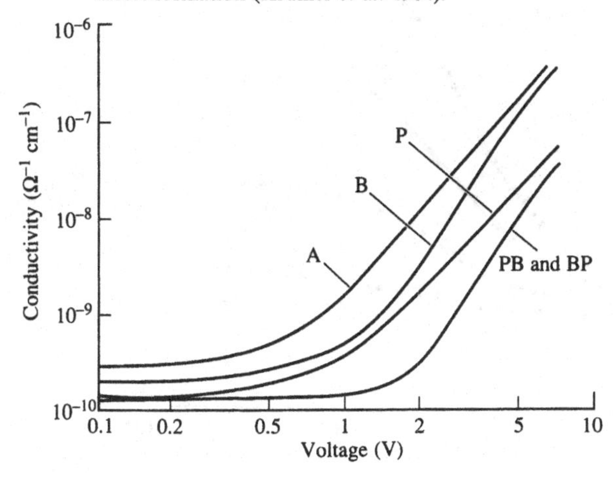

A point to note is that the defect creation at the interface is quite strongly thermally activated, unlike the Staebler–Wronski effect which is almost athermal. The likely explanation is that the light excitation provides all the energy needed to overcome the barrier, whereas the interface bias reduces the barrier by the shift of E_F, but a thermal activation energy is still needed. These different situations are illustrated in Fig. 6.25.

(3) *Particle bombardment*

In the early work on a-Si:H there were many studies of defect creation by charged particle bombardment, such as silicon or helium ions or electrons. In this case the defects are not induced by a shift in the Fermi energy position, but by atomic impacts which break bonds and displace atoms. The bombardment induces dangling bonds and is again reversed by annealing to 150–200 °C (Stuke 1987). The annealing has an activation energy of about 1 eV and a non-exponential form, but there have been no reports showing whether the decay fits a stretched exponential. Hydrogen diffusion was the suggested interpretation of these results and the comparison with the other effects seems to confirm this.

Fig. 6.32. Time dependence of the induced defect density near the dielectric interface of a field effect transistor after the application of a gate bias. The measurement is of the threshold shift, ΔV, which is proportional to the defect density (Jackson and Moyer 1988).

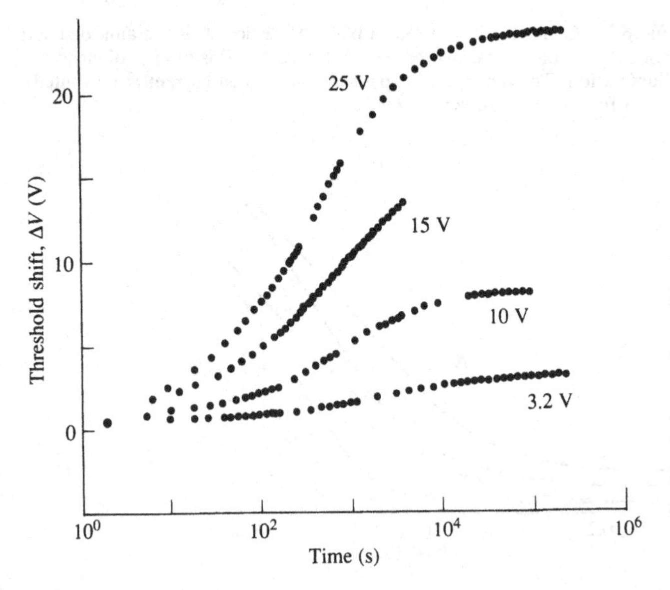

(4) *Metastable changes in the doping efficiency*

Thermal equilibrium of doped a-Si:H involves both the defects and dopant states and metastable changes in the dopant density by external excitation are therefore expected. One manifestation of the equilibrium is a large increase of the junction capacitance when an n-type Schottky barrier is annealed under reverse bias, indicative of an increase in the doping efficiency (Lang *et al.* 1982b). The result was difficult to understand when the measurements were first reported, but is now explained as the expected consequence of the equilibration. A reverse bias lowers the position of the Fermi energy in the gap of doped a-Si:H. The thermodynamic model predicts that the defect density is suppressed and the dopant concentration raised. Both effects increase the density of band tail carriers and so increase the apparent doping efficiency. The changes occur only in the depletion layer, where E_F is different from its unbiassed position. The increase in doping efficiency is confirmed by sweep-out measurements, which find that a large increase in n_{BT} is induced by a reverse bias anneal (Street 1989). A consequence of the bias annealing effect is that the doping efficiency should be larger near a Schottky contact even in the absence of an applied bias, because of the shift of E_F due to the band bending. No direct confirmation of the effect has been reported, although doping does change the barrier properties (see Section 9.1.4).

An increase in the dopant concentration with prolonged illumination is also deduced from ESR experiments (Stutzmann 1987b). The illumination causes a substantial increase in the density of deep defects as measured by PDS. If no other changes took place, the Fermi energy would move down in the gap and the ESR signals from the band edge state would decrease rapidly. In practice, there is only a small reduction and after illumination the donor states have a higher relative concentration than the intrinsic band tail states. The explanation offered is that both the donor density and the dangling bond density are increased. Illumination is reported to increase the conductivity of heavily doped p-type a-Si:H, also indicating a light-induced increase in the doping efficiency (Jang *et al.* 1988).

7 Electronic transport

The previous chapters are concerned with the structure and the density of states distribution of a-Si:H, describing the effects of growth, doping and defect reactions. The remainder of the book addresses the various electronic phenomena which result from the electronic structure. Foremost amongst these properties is the electrical conductivity, σ. The distinction between localized and extended electronic states is one of the fundamental concepts in the study of amorphous semiconductors. At zero temperature, carriers in extended states are conducting, but in localized states are not. Most of the experimental measurements of the localized state distribution rely on this property. Although both the concept of electrical conduction and its measurement seem simple, it is a complex process. The conductivity is a macroscopic quantity which represents an average property of the carriers as they move from site to site. The calculation of the conductivity therefore involves the transfer rate, scattering and trapping processes, as well as the appropriate average over the distribution of states. The averaging due to the disorder is the most difficult and leads to many of the interesting effects. The theory of conductivity near a mobility edge in disordered systems has been debated for many years and is still not completely agreed. The theory has applications beyond the boundaries of amorphous semiconductors, for example, in doped crystals, amorphous metals and low-dimensional materials. The discussion of the theory in this book is necessarily abbreviated and describes only the main ideas and how they apply to a-Si:H.

The conductivity is the product of the carrier density and the carrier mobility,

$$\sigma = ne\mu \tag{7.1}$$

The contributions to σ are summed over the density of states,

$$\sigma = \int N(E)e\mu(E)f(E,T)\,\mathrm{d}E \tag{7.2}$$

where $f(E,T)$ is the Fermi function. The integral contains contributions from electron transport above E_F and hole transport below E_F. When conductivity takes place far from E_F by a single type of carrier, non-degenerate statistics can be applied

$$\sigma = \int N(E) e\mu(E) \exp\left[-\frac{(E-E_F)}{kT}\right] dE \tag{7.3}$$

This equation is more conveniently written as,

$$\sigma = \frac{1}{kT}\int \sigma(E) \exp\left[-\frac{(E-E_F)}{kT}\right] dE \tag{7.4}$$

where

$$\sigma(E) = N(E) e\mu(E) kT \tag{7.5}$$

$\sigma(E)$ is the conductivity when $E_F = E$ and its evaluation is one main topic of this chapter.

The dominant conduction path is determined by the density of states, the carrier mobility and the Boltzmann factor and these are illustrated in Fig. 7.1 for states near the band edge. When there is a very high defect density in the middle of the band gap, as in unhydrogenated amorphous silicon, conduction takes place by hopping at the Fermi energy. The much lower defect density in a-Si:H prevents this mechanism from contributing significantly and instead conduction takes place by electrons or holes at the band edges where both the density of states and the mobility increase with energy. Thus $\sigma(E)$ is of the form shown in Fig. 7.1, increasing monotonically with energy. For the particular case in which $\sigma(\mathrm{E})$ increases abruptly from zero to a finite value σ_{min} at the mobility edge energy, E_C, then evaluation of Eq. (7.4) gives

$$\sigma = \sigma_{min} \exp[-(E_C - E_F)/kT] \tag{7.6}$$

σ_{min} is referred to as the minimum metallic conductivity and is given by

$$\sigma_{min} = N(E_C) e\mu_0 kT \tag{7.7}$$

where μ_0 is the free carrier mobility at E_C. The early theories predicted that the conductivity does increase abruptly at E_C (Mott and Davis, 1979, Chapter 2), but there is now considerable doubt about the sharpness or even the existence of a mobility edge. Nevertheless, virtually all the conductivity experiments are analyzed in terms of Eq. (7.6). This remains a reasonable approximation even if $\sigma(E)$ does not change abruptly, provided that it increases rapidly over a limited energy range. The experiments inevitably measure an average of $\sigma(E)$ over an energy range of at least kT. Thus the sharpness of the change in $\sigma(E)$ is best observed at very low temperatures. Unfortunately the

Fermi energy in a-Si:H cannot be brought closer than about 0.1 eV to E_C, so that the conductivity can hardly be measured below about 100 K and there is only limited information about the sharpness of the mobility edge. Most of the detailed tests of the mobility edge theories are made on disordered crystals and in metals in which the Fermi energy can be made to cross the mobility edge, giving measurable conductivity at low temperatures (Thomas 1985).

The central problem in studying the conductivity is to find the energy and temperature dependence of $\sigma(E)$ and the related $\mu(E)$ and to understand the physical processes involved in the transport. The motion of the carriers at non-zero temperatures may be either in extended states or by hopping in localized states and the magnitude of the conductivity is determined by the elastic and inelastic scattering mechanisms. In addition, when there is any local inhomogeneity of the

Fig. 7.1. Illustration of the density of states at the band edge, together with the electron distribution $n_{BT}(E)$, the conductivity $\sigma(E)$ and mobility $\mu(E)$: $\sigma(E)$ and μ(E) may change abruptly (*a*), or gradually (*b*) at the mobility edge E_C.

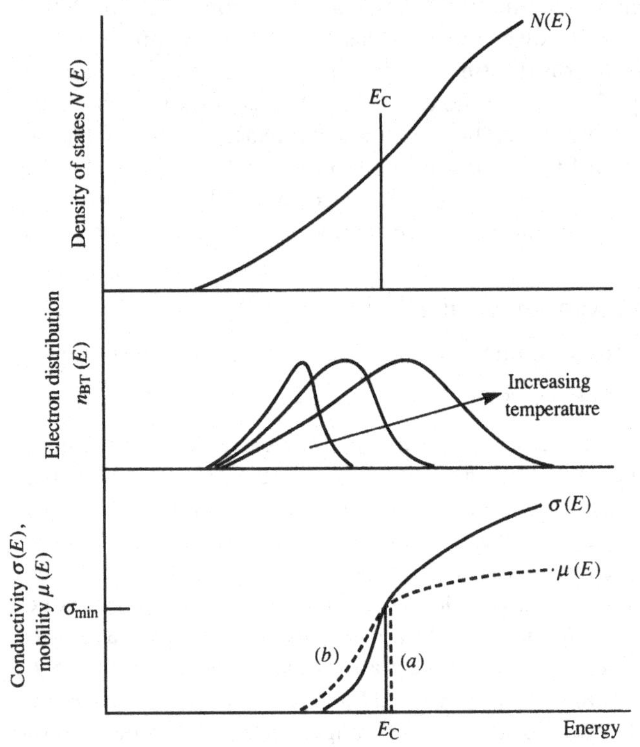

material, the measured conductivity is an average value over the distribution of the local values of σ.

There are four experimental techniques that are commonly used to gain information about the electronic transport in a-Si:H; dc conductivity, the drift mobility, thermopower, and the Hall effect. Sections 7.1 and 7.2 describe these measurements and how the information about $\sigma(E)$ and $\mu(E)$ can be extracted, and are followed by a discussion of the theory of electronic conduction in Sections 7.3 and 7.4.

7.1 Measurements of dc conductivity

The conductivity of a-Si:H is usually thermally activated, at least over a limited temperature range and is described by

$$\sigma(T) = \sigma_0 \exp(-E_\sigma/kT) = \sigma_0 \exp[-(E_{TR}-E_F)/kT] \qquad (7.8)$$

where E_{TR} is defined as the average energy of the conducting electrons and σ_0 is the conductivity prefactor. Some examples of the conductivity are given in Figs. 5.1, 5.2 and 7.4. E_{TR} coincides with E_C for the model of conductivity above an abrupt mobility edge, but other transport mechanisms are possible for which $E_{TR} \neq E_C$, such as band tail hopping. The energies E_σ and E_{TR} are illustrated in Fig. 7.2 in relation to the density of states and the electron distribution. Comparing Eqs. (7.6) and (7.8), one might imagine that a measurement of $\sigma(T)$ immediately gives the location of the mobility edge $(E_C - E_F)$, and that the prefactor gives the conductivity at the mobility edge. However, the measurements are

Fig. 7.2. Schematic diagram of the density of states distribution showing the conductivity activation energy, E_σ, the average conduction energy, E_{TR}, with respect to the mobility edges, and the Fermi energy. The temperature dependence parameters, γ_F, γ_G and γ_T are indicated.

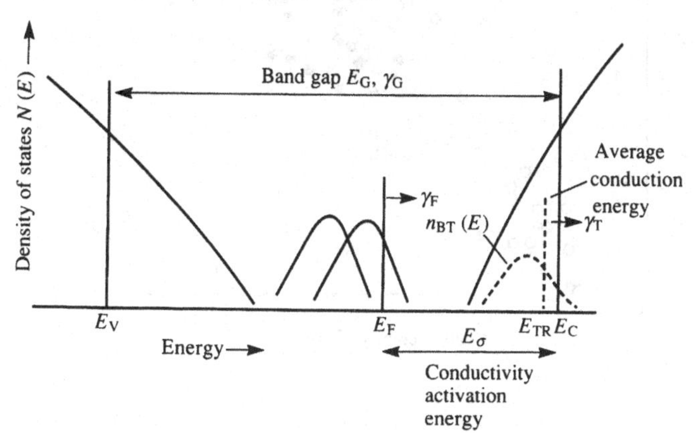

singularly unhelpful in this regard as there is a huge variation in the values of σ_0. This is shown by the results in Fig. 7.3 which includes data for undoped and doped a-Si:H. The correlation between σ_0 and the activation energy, E_σ, is referred to as the Meyer–Neldel relation (after its first observation in polycrystalline materials by Meyer and Neldel (1937)) and is described by,

$$\ln \sigma_0 = \ln \sigma_{00} + E_\sigma/kT_m \tag{7.9}$$

where σ_{00} is a constant with a value of about 0.1 Ω^{-1} cm^{-1} and kT_m is ~ 50 meV. The Meyer–Neldel relation is only approximately obeyed, as is evident from the scatter in the data.

Combining Eqs. (7.8) and (7.9) gives

$$\ln \sigma(T) = \ln \sigma_{00} + E_\sigma/kT_m - E_\sigma/kT \tag{7.10}$$

Fig. 7.3. Measured values of the conductivity prefactor σ_0 versus the conductivity activation energy, showing the Meyer–Neldel relation. Data from several different laboratories are shown (Tanielian 1982).

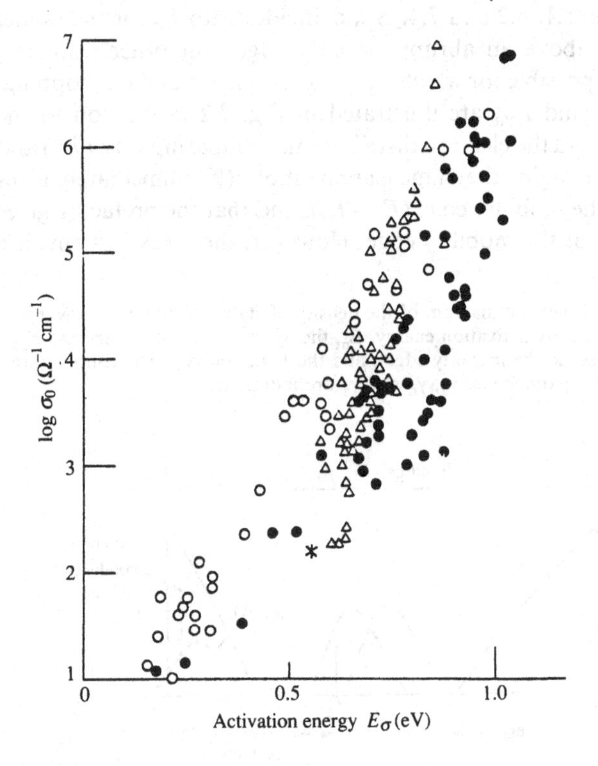

so that

$$\sigma(T_m) = \sigma_{00} \tag{7.11}$$

The Meyer–Neldel relation is a statement that the temperature dependence data for different samples all intersect at the same value of conductivity, σ_{00}, at the temperature $T_m \approx 600$ K. There is still no completely satisfactory explanation of why this occurs.

The most obvious mechanism by which the prefactor is different from the expected value is when the activation energy is temperature-dependent

$$E_{TR} - E_F = (E_{TR} - E_F)_0 - \gamma T \tag{7.12}$$

where a constant value of γ represents the first order approximation to the temperature dependence, and $(E_{TR} - E_F)_0$ is the value extrapolated to zero temperature. Substituting Eq. (7.12) into Eq. (7.8) gives

$$\sigma = \sigma_0 \exp(\gamma/k) \exp[-(E_{TR} - E_F)_0/kT] \tag{7.13}$$

Therefore, within the linear approximation, the conductivity activation energy E_σ measures the value of $(E_{TR} - E_F)_0$ and the prefactor contains the additional factor $\exp(\gamma/k)$. The temperature dependence of the energies accounts for the Meyer–Neldel relation provided that

$$\gamma/k = \ln(\sigma_{00}/\sigma_0) + E_\sigma/kT_m \tag{7.14}$$

There are three easily identifiable contributions to the temperature dependence factor γ. The locations of E_{TR} and E_F may change with respect to the density of states distribution. In addition, the band gap is temperature-dependent, so that the different gap states move with respect to each other as the temperature is varied. The three contributions to the temperature dependence are illustrated in Fig. 7.2 and combine to give

$$(E_{TR} - E_F) = (E_{TR} - E_F)_0 - (\gamma_F + \gamma_G + \gamma_T)T \tag{7.15}$$

The temperature dependence of E_F is referred to as the statistical shift and is present whenever the Fermi energy lies within an asymmetrical density of states distribution. The position of E_F is determined from the electron concentration, n_e, by

$$n_e = \int N(E) f(E, E_F T)\, dE \tag{7.16}$$

At zero temperature all the states are filled up to E_F, but at an elevated temperature there are empty states below E_F and filled ones above. As indicated by Fig. 7.3, $E_F(T)$ moves in the direction of the lower $N(E)$ as the temperature is raised, when E_F lies in a monotonically varying density of states. The density of states in a-Si:H changes very rapidly near the band tails, so that the statistical shift, γ_F, can be large.

The second component, γ_G, is from the temperature dependence of the energy levels with respect to each other. Optical absorption data find that the band gap, E_G, decreases with increasing temperature with a coefficient of $\gamma_{G0} = 5k = 4.4 \times 10^{-4}$ eV K^{-1} (Tsang and Street 1979). The Fermi energy is never further than half the band gap energy from the main conduction path and consequently γ_G is at most $\frac{1}{2}\gamma_{G0}$. There is no detailed information about the temperature dependence of the defect levels and it is usually assumed that γ_G is proportional to $(E_{TR} - E_F)$,

$$\gamma_G = \gamma_{G0}(E_{TR} - E_F)/E_G \tag{7.17}$$

E_F is in the middle of the gap in undoped samples, and $\gamma_G \sim \frac{1}{2}\gamma_{G0}$. In doped samples, when E_F is in the band tail, this scaling does not apply, as the band tails are broader at high temperature rather than steeper, but in any event γ_G is small.

The temperature dependence of E_{TR} is more difficult to evaluate. The Boltzmann factor in Eq. (7.4) increases the contribution to the conduction from higher energy states as the temperature increases. This gives a negative value to γ_T, with a magnitude which depends on how rapidly $\sigma(E)$ increases with energy. There is another contribution γ_T if the shape of $\sigma(E)$ is temperature-dependent. Section 7.4.3 describes a model in which γ_T is positive because of phonon-induced delocalization of states below the mobility edge.

A more detailed analysis of the conductivity data is needed to explain the prefactor and to understand the Meyer–Neldel relation. The magnitudes of γ_F and γ_G depend on the location of the Fermi energy, the density of states distribution and the thermal state of the material. There are three situations which can be analyzed easily.

7.1.1 *Doped a-Si:H above the equilibration temperature*

Conductivity data for n-type and p-type samples above the equilibration temperature are shown in Fig. 7.4 and similar data are in Fig. 5.2. The conductivity is activated with a prefactor of 100–200 Ω^{-1} cm^{-1} for both doping types and the activation energy is 0.3–0.4 eV in n-type material and 0.4–0.6 eV in p-type. In the thermal equilibrium regime, the Fermi energy is pinned by the defect and dopant states and consequently the statistical shift is small, as is discussed in Section 6.2.2. γ_F may be calculated from a numerical integration of Eq. (7.16), knowing the density of states distribution in the band tail and the temperature dependence of the band tail electron density n_{BT}, which is given in Fig. 6.6. The results of such a calculation for different n-type doping levels find that γ_F is indeed small (Street *et al.* 1988a). The explanation of the small statistical shift is as follows. The temperature dependence of E_F is

$$\gamma_F = \frac{dE_F}{dT} = \left(\frac{dE_F}{dT}\right)_{n_{BT}} + \frac{dE_F}{dn_{BT}}\frac{dn_{BT}}{dT} \tag{7.18}$$

The first term on the right hand side of Eq. (7.18) is the usual statistical shift determined at a constant electron density. However, the electron density is itself temperature dependent above the equilibration temperature because of the structural changes. The second term on the

Fig. 7.4. Temperature dependence of the conductivity above the equilibration temperature. The top left corner shows the high temperature extrapolation of the data.

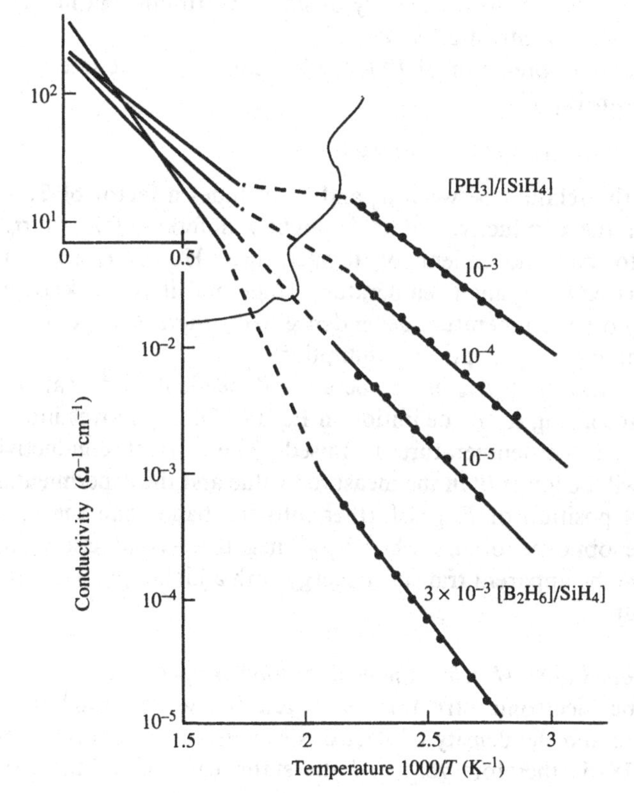

right hand side in Eq. (7.18) is non-zero and almost exactly cancels out the first term. The cancellation can be derived from Eq. (6.30) for the equilibrium value of n_{BT}.

The Fermi energy in doped samples is within the band tail, so there should be no significant contribution to the conductivity from the temperature dependence of the band gap. Thus, in the absence of any temperature dependence of E_{TR}, σ_0 is equal to the measured conductivity prefactor,

$$\sigma_0 = 100\text{–}200\,\Omega^{-1}\,\text{cm}^{-1} \tag{7.19}$$

Having calculated the Fermi energy position, the energy of the conduction path within the density of states distribution may be deduced from a comparison of the calculated conductivity with the experimental activation energies. When this comparison is made, the density of states at the conduction path, $N(E_{TR})$, is found to be $(2\text{–}4) \times 10^{21}\,\text{cm}^{-3}\,\text{eV}^{-1}$ and the conductivity energy lies just above the exponential tail. There is an indication that the conduction energy increases with respect to the density of states distribution at higher n-type doping levels (Street *et al.* 1988b).

A conductivity prefactor of $100\text{–}200\,\Omega^{-1}\,\text{cm}^{-1}$ gives the free carrier mobility from Eq. (7.5),

$$\mu_0 = \sigma_0/N(E_T)ekT = 10\text{–}15\,\text{cm}^2\,\text{V}^{-1}\,\text{s}^{-1} \tag{7.20}$$

Note that the relation between μ_0 and σ_0 includes a factor of T. The analysis of the conductivity data is slightly different if μ_0 or σ_0 is assumed to be independent of temperature. However, since the experiments cover a small temperature range and it is not known if there is any other temperature dependence to σ_0, there is no guide as to which is the more appropriate assumption.

Suppose next that E_{TR} is temperature-dependent and that γ_T is positive. According to the definition in Eq. (7.15), E_{TR} moves into the band gap as the temperature is raised. The correct conductivity prefactor will be lower than the measured value and the experimentally determined position of E_{TR} is further into the band than the actual value. The opposite occurs when γ_T is negative – conduction takes place above the apparent transport energy with a higher prefactor than that measured.

7.1.2 *Doped a-Si:H below the equilibration temperature*

The electronic structure is frozen below the equilibration temperature and the density of electrons is constant. The second term in Eq. (7.18) is therefore zero and the statistical shift of the Fermi energy now makes a large contribution to the conductivity. Fig. 6.3 shows data for n-type a-Si:H in this regime, in which the different results correspond to quenching the sample from different temperatures. The conductivity prefactor is in the range 1–10 Ω^{-1} cm^{-1} which with the small activation energies of 0.1–0.2 eV is consistent with the Meyer–Neldel relation. The Fermi energy remains in the band tail so that there should be no correction to the conductivity prefactor from the temperature dependence of the gap.

The effect of the statistical shift can be calculated from Eq. (7.18) and the density of states distribution, just as for the equilibrium state except that the density of electrons is now constant. The resulting values of γ_F are in the range $3k$–$6k$, so that the correction to the prefactor of $\exp(\gamma_F/k)$ from the statistical shift is a factor of 10–100. The calculated conductivity is shown in Fig. 7.5 and a good fit to the data is found with a conductivity prefactor of 100 Ω^{-1} cm^{-1} and the same transport energy as was obtained from the analysis of the equilibrium regime. The modeling of the conductivity therefore shows that the only difference

Fig. 7.5. Comparison of calculated and measured conductivity of n-type a-Si:H in the frozen state. The calculation assumes a conductivity prefactor of 100 Ω^{-1} cm^{-1} (data from Fig. 6.3).

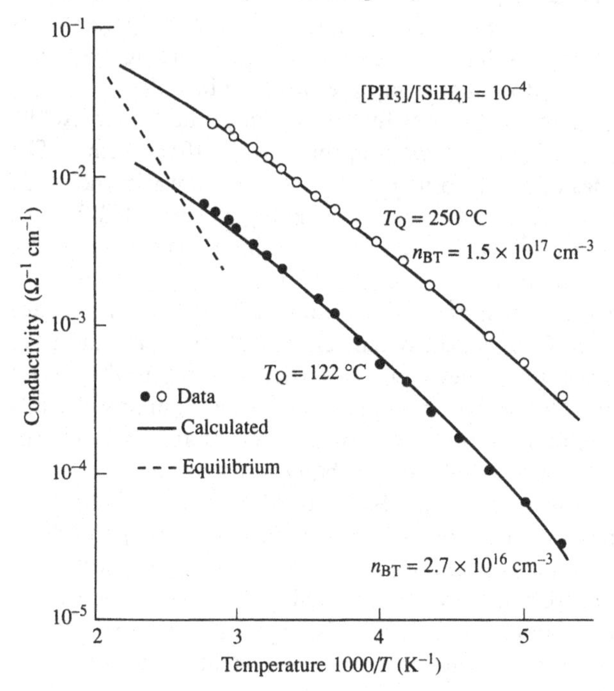

between the equilibrium and frozen state comes from the statistical shift of the Fermi energy.

7.1.3 *Undoped a-Si:H*

Undoped a-Si:H contributes the largest range of values in the Meyer–Neldel plot in Fig. 7.2. The Fermi energy is far from the band tails and so there may be a significant contribution from all three γ factors. However, the statistical shift does not contribute if the Fermi energy is pinned by a sufficiently large density of states. Taking this approach, Dersch measured the conductivity in samples which have been bombarded by electrons to raise the defect density and then progressively annealed to remove the defects (see Stuke (1987)). Fig. 7.6 shows that, at the highest defect densities, the activation energy is 0.85 eV and the prefactor is 2300 Ω^{-1} cm^{-1}. These values remain constant in the initial stages of annealing, but subsequently the activation energy decreases and the prefactor increases, indicating that the statistical shift becomes significant. It is concluded that there is no statistical shift in the initial stages of annealing. For a conductivity activation energy of 0.85 eV, Eq. (7.17) gives $\gamma_G \approx \gamma_{G0}/2 = 2.5k$, so that the corrected conductivity prefactor is about 100 Ω^{-1} cm^{-1}, but again no correction has been made for γ_T. Thus a consistent prefactor is obtained from the conductivity data, provided the temperature dependences of E_F and the band gap are properly included.

The preceding analysis accounts for the region of the Meyer–Neldel relation with conductivity prefactors in the range 1–10^4 Ω^{-1} cm^{-1}. The much larger values of σ_0 of up to 10^7 Ω^{-1} cm^{-1}, require γ in the range $10k$–$20k$, which is larger than can reasonably be expected by the mechanisms so far discussed. Some other explanations are, however, possible. The measurement of the conductivity is prone to experimental artifacts, of which the most obvious are contact resistance and surface effects. Most metals form Schottky barriers to a-Si:H, which can have a higher resistance than the sample, particularly when the defect density is low and the depletion layer is very wide. Similarly there is band bending extending into the layer when there is any space charge at the surface. The large conductivity changes that occur when gases are adsorbed on the surface of a-Si:H are described in Chapter 8. Conductivity measurements are performed in either the gap-cell or sandwich configurations depending on the placement of the electrodes. The gap-cell measurements are less affected by the contact resistance because the separation of the electrodes is large, but are more affected by surface band bending. These artifacts are greatest in low defect density undoped material because the depletion layer is large and the conductivity is low, so that other conduction mechanisms can easily dominate. The high conductivity and high defect density of doped films make them less susceptible to measurement artifacts, but there are surface adsorbate and contact resistance effects even in moderately doped layers.

Assume that the conductivity of an undoped a-Si:H sample is σ_x at the measurement temperature of T_x. If the temperature dependence of $\sigma(T)$ reveals an activated behaviour near T_x, then

$$\sigma_x = \sigma_{0x} \exp(-\Delta E_x/kT) \tag{7.21}$$

Around the temperature T_x,

$$\ln \sigma_{0x} = \ln \sigma_x + \Delta E_x/kT_x \tag{7.22}$$

Fig. 7.6. Plot of the conductivity activation energy and prefactor and the dangling bond density at different stages of annealing of electron-bombarded a-Si:H (Stuke 1987).

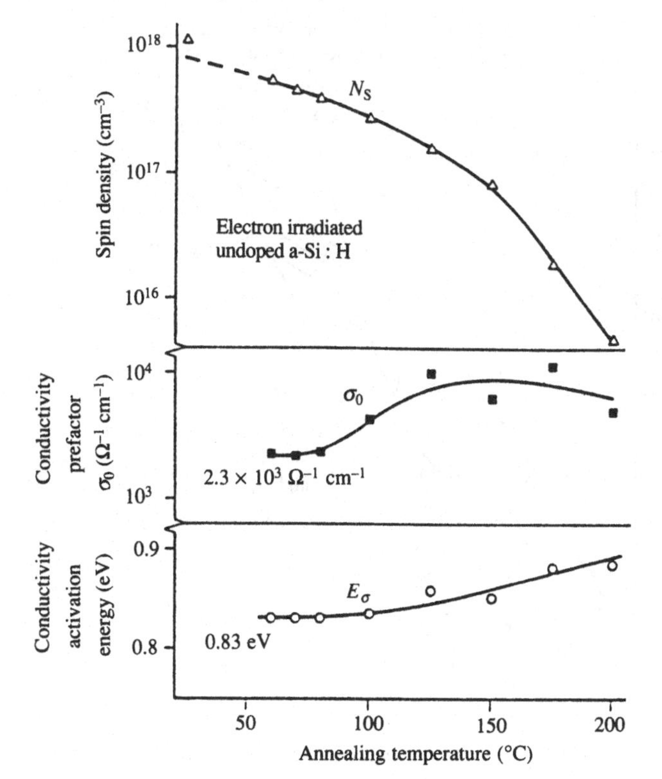

which is in the form of the Meyer–Neldel relation. Thus when the conductivity is within a limited range of values at the temperature of measurement, the prefactor follows the Meyer–Neldel relation irrespective of the actual physical origin of the activation energy. For measurements between room temperature and 550 K, with values of σ_x of 10^{-8} and $10^{-4}\ \Omega^{-1}\ \mathrm{cm}^{-1}$ respectively, the prefactor lies between the two lines in Fig. 7.7.

This discussion suggests that the Meyer–Neldel relation is actually a composite of effects illustrated schematically in Fig. 7.7. Those samples with the lowest activation energy are predominately n-type samples in the frozen state, in which the statistical shift of E_F suppresses the prefactor. p-type samples in the frozen state have a larger activation energy but a smaller statistical shift because of the broader valence band tail. The doped samples in equilibrium have higher activation energies but no contribution from the statistical shift. Undoped samples have activation energies from 0.6 eV upwards and the prefactor is raised by the temperature dependence of the band gap. Finally the

Fig. 7.7. Relation between the conductivity prefactor and activation energy for different conduction regimes described in the text.

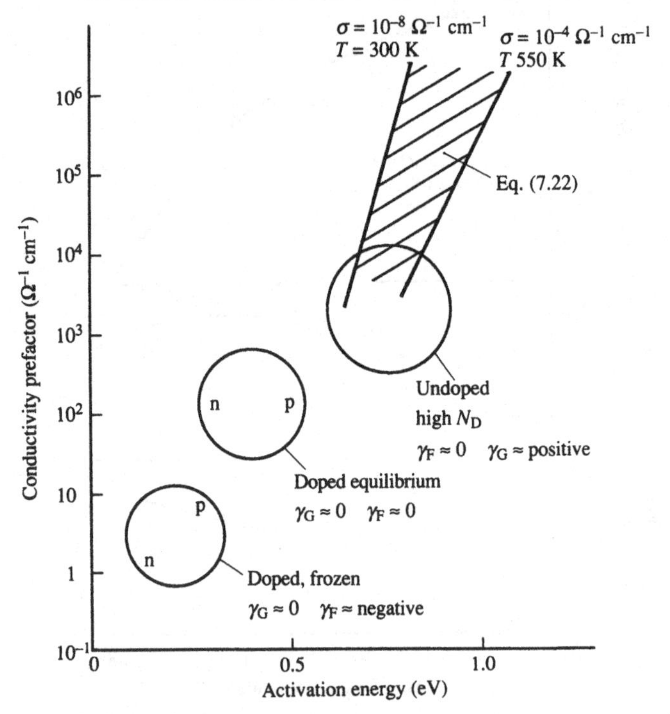

high values of activation energy and prefactor may arise from some artifacts of unspecified origin.

7.2 Carrier mobility

The carrier mobility, $\mu(E)$, is an alternative measure of the conductivity, the two being related through the density of state by Eq. (7.5). There are several techniques for measuring the mobility in a-Si:H, most notably the time-of-flight method. All the techniques measure the average motion of the carriers over a time longer than that taken to trap a carrier in the band tail states, so that the drift mobility is always measured, rather than the free carrier mobility. The drift mobility depends on the distribution of traps and the free mobility can only be extracted if the density of states distribution is known. Chapter 3 describes how the time-of-flight experiment is used to determine the shape of the band tail through the analysis of the dispersive transport process.

The drift mobility, μ_D, is

$$\mu_D = \mu_0/(1+f_{trap}) \tag{7.23}$$

where f_{trap} is the ratio of the time that the carrier spends in localized traps to that spent in mobile states. For a single trapping level,

$$f_{trap} = [N_T/N(E_{TR})kT]\exp(E_T/kT) \tag{7.24}$$

where N_T is the trap density of binding energy E_T, and $N(E_{TR})$ is the density of states at the conduction path. In general the density of traps is less than the density of conduction band states and, when T is large, $\mu_D \to \mu_0$. The data for the electron mobility shown in Fig. 7.8 indicate that the high temperature extrapolated value of μ_D is in the range 5–20 $cm^2\ V^{-1}\ s^{-1}$. The hole mobility is too low to yield a reliable extrapolated value.

The multiple trapping model of transport in an exponential band tail is described by Eq. (3.20) in Section 3.2.1 and a fit to this expression is given in Fig. 7.8. The free carrier mobilities are 13 and 1 $cm^2\ V^{-1}\ s^{-1}$ for electrons and holes respectively, with the band tail slopes of 300 °C and 450 °C (Tiedje *et al.* 1981). Implicit in the analysis is the assumption that the exponential band tail extends up to the mobility edge, but the density of states model developed in Fig. 3.16 shows that this is a poor approximation. The band tail changes slope below E_C and this may change the estimated values of the mobility.

It is possible to extract the free mobility and the shape of the band tail from the dispersive transport data without any assumptions about the form of the tail (Marshall, Berkin and Main 1987). The average trapping energy of the carriers as they cross the sample is the demarcation energy $E_D = kT\ln(\omega_0 t)$. During the transit, an average carrier is not trapped in a state deeper than E_D. The free carrier lifetime, τ_f, is therefore related to the trapping rate by

$$\tau_f^{-1} = N(E_D)\, v_c \sigma\, dE_D \tag{7.25}$$

where v_c is the carrier velocity at the mobility edge and σ is the capture cross-section. The total free carrier lifetime during the transit is given in terms of the free carrier mobility by

$$\tau_f = d/\mu_0 F \tag{7.26}$$

where d is the sample thickness and F is the applied field. Detailed balance equates the attempt frequency ω_0 to the product $N(E_T)kTv_c\sigma$, resulting in the following expression,

$$N(E_D) = [N(E_T)\mu_0 kT/\omega_0 d]\, \mathrm{d}F/\mathrm{d}E_D \tag{7.27}$$

E_D is the drift mobility activation energy whose dependence on the applied field is measured (see, for example, Fig. 3.13), allowing the quantity $N(E_D)/\mu_0 N(E_T)$ to be evaluated. The shape of the band tail together with other information about the density of states at the band edge results in estimated free mobilities of about 20 cm^2 V^{-1} s^{-1} for electrons and 10 cm^2 V^{-1} s^{-1} for holes.

The analysis of the drift mobility is based on the model of an abrupt mobility edge, in which transport only occurs above the energy E_C. The

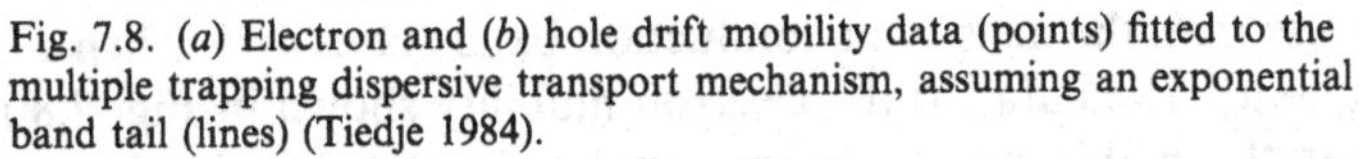

Fig. 7.8. (*a*) Electron and (*b*) hole drift mobility data (points) fitted to the multiple trapping dispersive transport mechanism, assuming an exponential band tail (lines) (Tiedje 1984).

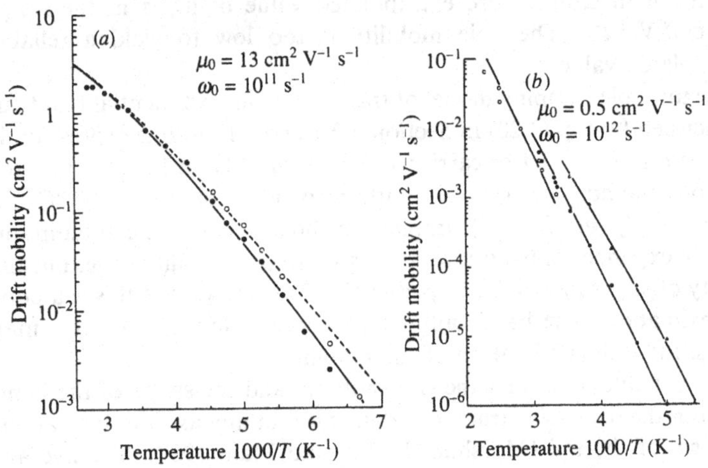

data are evidently consistent with this model but do not prove that it is correct. No attempt has been made to investigate whether the results are also consistent with a conduction energy that is temperature-dependent or with a mobility that increases slowly with energy. Such effects would be most evident at low temperatures where any contribution from transport deep in the tail would tend to dominate. It is therefore of interest that recent time-of-flight experiments have reported an unexpectedly high mobility at low temperatures (Cloude, Spear, LeComber and Hourd 1986). Other measurements have not yet corroborated these findings and further investigations are needed to confirm the results.

The time-of-flight experiments are limited to samples of low conductivity and cannot be applied to heavily doped a-Si:H. An alternative measurement of the mobility using surface acoustic waves (SAW) requires a high conductivity and is preferred in doped samples (Fritzsche 1984a). In the SAW technique, the sample is placed just above a piezoelectric crystal such as $LiNbO_3$. A SAW is propagated in the lithium niobate crystal and is accompanied by an electric field wave. The fringing field in the a-Si:H film causes a voltage between the two electrodes separated in the direction of the travelling wave. The acousto-electric voltage is

$$V_{ae} = \tfrac{1}{2}(\mu_D/v_s)\,Lk^2\,(A^2+B^2)\,\Phi_0{}^2 \tag{7.28}$$

L is the electrode spacing, v_s the wave velocity and k the wave vector, each of which are given by the experimental configuration. A and B are attenuation coefficients and Φ_0 is the potential at the crystal surface, and these three parameters can also be calculated from the geometry and the material properties. The drift mobility is measured in this experiment and is defined by

$$\mu_D = \mu_0 n_v / n_0 \tag{7.29}$$

where n_v is the density of carriers that can respond at the frequency of the acoustic wave and n_0 is the density of all the carriers that can contribute to the conductivity.

Fig. 7.9 shows some of the SAW data for electrons and holes in doped a-Si:H (Takada and Fritzsche 1987). The mobility data are not proportional to the measured V_{ae} because the attenuation coefficients depend on the dc conductivity, which is temperature-dependent. The solid lines are calculated fits to the data assuming an exponential density of tail states. The results are similar to the time-of-flight data on undoped a-Si:H and the deduced values of the free carrier mobilities are also similar. The electron mobility decreases at high doping levels.

The drift mobility in n-type a-Si:H is also obtained by combining measurements of the dc conductivity with the density of electrons in the band tail, measured by sweep out (Street *et al.* 1988b)

$$\sigma = n_{\mathrm{BT}} e \mu_{\mathrm{D}} \tag{7.30}$$

Fig. 7.10 shows the drift mobility obtained by this technique for two different phosphorus doping levels and compared with time-of-flight data for undoped material. The different data points on each curve correspond to the different equilibrium and frozen states of the sample. The facts that these give the same mobility and that there is no change at the equilibrium temperature confirm that the equilibration effects are due only to the changes in the electron density and not in the transport path. A similar conclusion is obtained from the SAW experiments and the decrease of the drift mobility with doping in Fig. 7.10 also agrees with the SAW data.

The drift mobility defined by Eq. (7.30) is easily calculated from the density of states distribution. The position of the Fermi energy is calculated at each temperature from the measured n_{BT} and either the conductivity or the mobility is found by assuming that transport is by mobile carriers at E_{TR}. The calculated fits to the data in Fig. 7.10 are obtained with a free mobility of 15 cm^2 V^{-1} s^{-1}. The reduction in the drift mobility with doping is not consistent with a broadening of the exponential tail, but is best explained by a shift of the transport energy with doping probably due to potential fluctuations.

Virtually all the mobility data indicate that the free carrier mobility for electrons is 10–20 cm^2 V^{-1} s^{-1}. The range of values for holes is larger, with some estimates as low as 1 cm^2 V^{-1} s^{-1}. There have been some suggestions that the electron mobility might be as large as 100–1000 cm^2 V^{-1} s^{-1} (Silver *et al.* 1982), but there seems to be little experimental support for such values. When the mobility data are

Fig. 7.9. Temperature dependence of the acousto-electric voltage, V_{ae}, and drift mobility of (*a*) electrons and (*b*) holes obtained by the SAW technique. The solid lines are calculated for trapping in an exponential band tail of slope T_{C} or T_{V} (Takada and Fritzsche 1987).

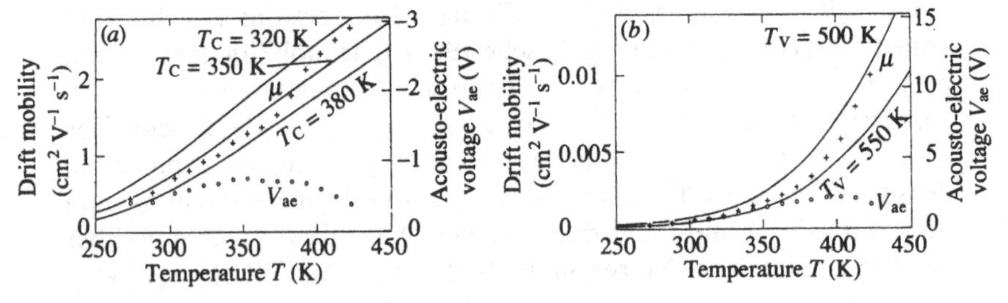

analyzed in terms of an abrupt temperature-independent mobility edge with the band edge distribution given by Fig. 3.16, the density of states at E_{TR} is deduced to be about 2×10^{21} cm^{-3} eV^{-1}. The corresponding room temperature conductivity evaluated from Eq. (7.5) is,

$$\sigma_0 = 2 \times 10^{21} \times 0.0025 \times 1.6 \times 10^{-19} \times \mu_0 = 80\text{–}160\,\Omega^{-1}\,\text{cm}^{-1} \tag{7.31}$$

This value agrees with the results of the conductivity data, so that a rather consistent picture of the transport begins to emerge. Unfortunately it is less clear whether this simple model of the transport

Fig. 7.10. Temperature dependence of the drift mobility of n-type and undoped a-Si:H obtained from Eq. (7.30) and in different thermal states. Solid lines are calculated values (Street *et al.* 1988b).

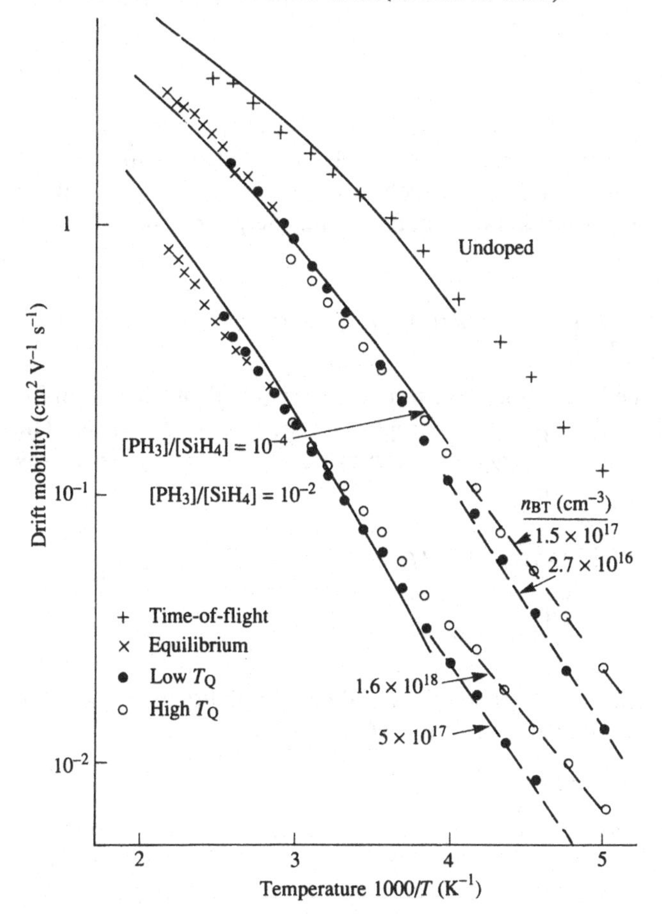

is consistent with the present models of the mobility edge. Furthermore, the thermopower data described in the next section cannot be explained by the simple mobility edge model.

7.3 Thermopower and Hall effect

The thermopower measures the average energy, with respect to E_F, which is transported by the electronic carriers and provides an alternative measure of the energy of the main conduction path. When there is a temperature difference ΔT across a conductor, there is a net flux of carriers from the hot end to the cold, which carries heat energy and charge. The charge displacement develops a voltage V_s and the thermopower or Seebeck coefficient, S_T, is V_s/T. The Peltier coefficient, Π, is the constant of proportionality relating the heat flux and the electrical current and is related to the thermopower by the Onsager relation,

$$\Pi = eS_T T \tag{7.32}$$

The sign of the thermopower is negative for electrons and positive for holes and its measurement provides a definitive identification of the type of conduction. When conduction is by a single type of carrier (either electrons or holes) far from the Fermi energy in a homogeneous material, the thermopower is

$$S_T = \frac{1}{e\sigma T}\int (E-E_F)\sigma(E)\exp[-(E-E_F)/kT]\,\mathrm{d}E \tag{7.33}$$

The integrand contains the same terms as the equivalent expression for the conductivity, Eq. (7.4), except for the factor $(E-E_F)$. The thermopower is the energy of the carriers averaged by the energy dependence of the conductivity,

$$eTS_T = \frac{\int \sigma(E)(E-E_F)\,\mathrm{d}E}{\int \sigma(E)\,\mathrm{d}E} \tag{7.34}$$

When transport occurs above an abrupt mobility edge, E_C, the thermopower is (Cutler and Mott 1969),

$$S_T = \pm(k/e)[(E_C-\mathrm{E_F})/kT+A] \tag{7.35}$$

AkT is the average energy of the conducting electrons above E_C and the magnitude of A depends on the exact form of $\sigma(E)$ above E_C. For example, A is unity when $\sigma(E)$ is constant above E_C and is greater than unity if $\sigma(E)$ increases with energy. Irrespective of the exact form of $\sigma(E)$, the significant result is that the temperature dependence of S_T is given by $(E_C - E_F)$, when $\sigma(E)$ drops abruptly to zero at E_C.

The statistical shift of E_F and the possible temperature dependence of the conduction energy, which complicate the analysis of the conductivity, also enter into the thermopower equations. However, $\sigma(T)$ and $S(T)$ can be combined into a function $Q(T)$ which eliminates the statistical shift, since the Fermi energy position drops out of the combined expression (Beyer and Overhof 1979),

$$Q(T) \equiv \ln \sigma(T) + eS_T(T)/k \tag{7.36}$$

When σ and S are given by Eqs. (7.6) and (7.35), then

$$Q(T) = \ln \sigma_{\min} + A \tag{7.37}$$

The assumption of an abrupt mobility edge at E_C predicts that neither $\sigma_{\min}$ nor A is temperature dependent, in which case $Q(T)$ is a constant; the conductivity activation energy exactly cancels the thermopower energy.

In practice, this predicted behavior is almost never observed. The examples of $Q(T)$ for doped and undoped a-Si:H in Fig. 7.11 show that its temperature dependence is generally described by

$$Q(T) = C - E_\Delta/kT \tag{7.38}$$

Fig. 7.11. Temperature dependence of the Q-function defined in Eq. (7.37), for (*a*) n-type and (*b*) p-type a-Si:H. The gas-phase doping levels range from 10^{-6} to 2×10^{-2} (Beyer and Overhof 1984).

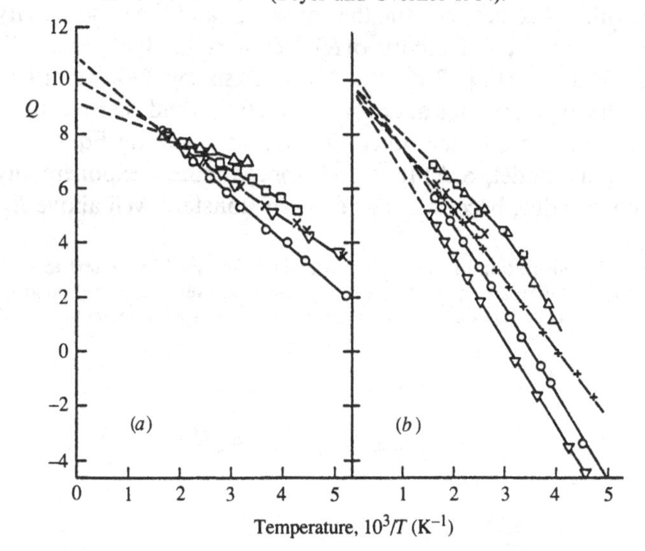

where C is a constant about 10 and E_Δ is an energy of 50–200 meV. The thermopower energy is always slightly less than the conductivity energy and the difference gets larger as the doping level increases.

There are several possible reasons for the energy difference between the thermopower and the conductivity. For example, when there is both electron and hole conduction, these contribute to the thermopower with opposite sign, but have the same sign in the conductivity. This effect may be important in undoped a-Si:H where the Fermi energy is near the middle of the band gap, but does not contribute in doped material when one sign of carrier completely dominates.

Three other mechanisms have been proposed to explain E_Δ when conduction is by a single carrier and these are illustrated in Fig. 7.12. The energy difference may be due to polaron transport (Emin, Seager and Quinn 1972). Polarons occur when there is strong lattice relaxation at the electron site. The analysis of lattice relaxation in Section 4.1.1 demonstrates that strong coupling increases in the binding energy after the electron is added to an undistorted site. Polaron conduction occurs by hopping from site to site as illustrated in Fig. 7.12(*a*). In order for the electron to jump from one site to the next (A to B), there must be thermal excitation of the local network which raises the energy of the filled site at A and lowers the energy at B so that the electron can transfer. The energy of this lattice distortion contributes to the energy required for the electron to hop from site to site, but does not get transported with the carrier. The energy therefore appears in the conductivity but not in the thermopower, accounting for the term E_Δ.

An energy difference between the thermopower and the conductivity also occurs when the conductivity $\sigma(E)$ increases gradually near E_C rather than abruptly (Fig. 7.12(*b*)). An analysis by Döhler (1979) showed that the experimental data for conductivity and thermopower can be transformed to give the function $\sigma(E)$, by inverting Eq. (7.33). According to this model, $\sigma(E)$ increases approximately exponentially near the mobility edge, becoming more nearly constant well above E_C.

Fig. 7.12. Illustration of the three mechanisms described in the text to explain the difference in conductivity and thermopower energy: (*a*) polaron hopping; (*b*) conduction over an extended energy range; (*c*) potential fluctuations.

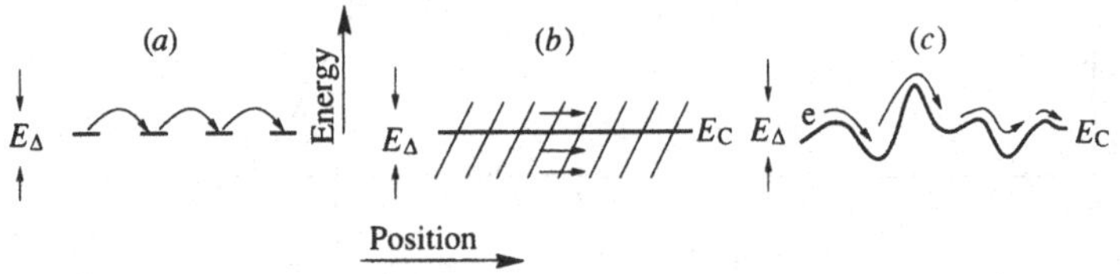

The deduced value of $\sigma(E)$ changes by about three orders of magnitude over an energy range of about 100 meV and the slope of $\sigma(E)$ is smaller in doped samples to account for the larger value of E_Δ.

The third possible explanation of the energy difference is that the conductivity is spatially inhomogeneous. For example, there might be long range fluctuations of the energy bands and of the associated mobility edge, as illustrated in Fig. 7.12(*c*) (Overhof and Beyer 1981). The difference in the thermopower and conductivity energy arises from the different spatial averaging. The exponential Boltzmann factor in Eq. (7.4) for the conductivity means that the spatial average is dominated by the highest energy regions of the potential fluctuations where the conductivity is lowest. On the other hand, the thermopower measures the average energy over the complete transport path and the low and high energy regions contribute more equally. The difference in conductivity and thermopower energy E_Δ is therefore roughly equal to the magnitude of the potential fluctuations. Expressions for the spatial averaging are given by Herring (1960). The possible origins of potential fluctuations in a-Si:H and their distribution are discussed in Section 7.4.5.

Perhaps the most important conclusion to be drawn here is that the model of an abrupt change in conductivity at the mobility edge of a homogeneous material is unable to account for the difference between the thermopower and conductivity energies. The polaron model does not receive much support in a-Si:H although it may be important in chalcogenides glasses where a similar value of E_Δ is observed. There is evidence for lattice relaxation of band tail holes (see Section 8.3.2) but the electron–phonon coupling is expected to be small near the mobility edge, where the localization radius is large. The model of a slowly varying $\sigma(E)$ is pertinent to the present theories of conduction near the mobility edge, which indicate that the mobility edge is not sharp. Long range potential fluctuations have frequently been proposed for amorphous semiconductors and are certainly important in doped a-Si:H in which there are charged dopant and defect states (see Section 7.4.5).

7.3.1 *The Hall effect*

The final transport measurement to be considered is the Hall effect. This is the most intriguing transport property, but has been so difficult to understand that it has not contributed much towards the elucidation of the conduction mechanisms. The reason is that the Hall effect is anomalous and has the opposite sign from that which is normally expected. Thus holes give a negative Hall voltage and electrons a positive one. The sign reversal is characteristic of amorphous semiconductors and is observed in all material so far investigated, but the correct sign is found in microcrystalline silicon in which the grain size is as little as 30–50 Å. The sign reversal therefore seems to be a result of the very small scattering distances in the amorphous films. Fig. 7.13 shows that the Hall mobility of a-Si:H lies between 0.01 and 0.2 $cm^2\ V^{-1}\ s^{-1}$ and has a small activation energy.

Two theories have been proposed to explain the Hall effect in amorphous semiconductors. Both treat the strong scattering by expressing the result in terms of a transfer integral J between specific sites. Friedman's calculation is based on the random phase approximation in which there is assumed to be no phase correlation between the electron or hole wavefunction on adjacent sites (Friedman 1971). The calculation yields

$$\mu_H = 4\xi(ea^2/\hbar)\,[a^3 J N(E_C)]\,(z/z_0) \tag{7.39}$$

where z is the atomic coordination, z_0 is the number of sites that can interact, a is the interatomic spacing and ξ is a factor of about $\frac{1}{3}$. The free carrier mobility μ_0 is also calculated within the same model, with the result that

$$\mu_H = \mu_0 kT/J \tag{7.40}$$

The transfer integral is estimated to be about 1 eV, so that the Hall mobility is predicted to be 10–100 times smaller than μ_0 at room

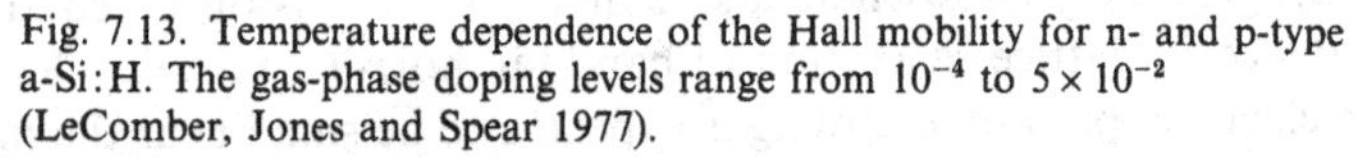
Fig. 7.13. Temperature dependence of the Hall mobility for n- and p-type a-Si:H. The gas-phase doping levels range from 10^{-4} to 5×10^{-2} (LeComber, Jones and Spear 1977).

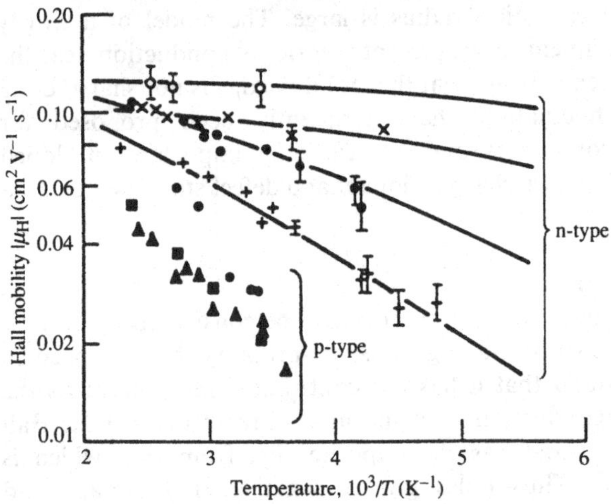

temperature. The measured values in Fig. 7.13 are consistent with this estimate. In the Friedman model, the sign of the Hall mobility is the same for electrons and holes and so is unable to explain the double sign anomaly which is observed.

Emin (1977) proposed an alternative model based on polaron hopping, in which carriers transfer between adjacent sites. The Hall effect results from the interference between the different paths between an initial and final state and involves at least three sites. The sign of the Hall coefficient depends on the number of such sites and on the sign of the transfer integral. When there are an odd number of interconnecting sites and they are connected by anti-bonding orbitals, the sign of the Hall coefficient is positive, but bonding or lone pair orbitals give a negative Hall effect. Since the conduction and valence bands of a-Si:H and the chalcogenide glasses are of these types, the double sign reversal is explained.

The difficulty with this model is that it requires hopping between nearest neighbor atoms, to preserve the correlation between the wavefunctions and to cause interference effects in the transfer integrals. Such a model might apply to a narrow band material or when the mobility edge is at the center of the band, but in a-Si:H conduction is near the band edge. The localization length is about 10 Å which is larger than the interatomic separation. Conduction should then occur by transfer between states that are farther apart than nearest neighbors and there is no obvious reason why there should be a correlation between these wavefunctions.

7.4 Theories of electronic conduction

The theories of conductivity are concerned with the calculation of the energy-dependent conductivity $\sigma(E)$. There is no conductivity within the localized states at zero temperature, and Sections 7.4.1 and 7.4.2 describe the evaluation of $\sigma(E)$ at and above the mobility edge. The comparison of theory and experiment is complicated by the fact that all measurements are performed at non-zero temperatures. Thermal effects modify the theory of extended state conduction, as described in Section 7.4.3, and also allow thermally activated hopping conduction within the localized states. Hopping conductivity dominates when the Fermi occupation factor gives the states nearest to the Fermi energy a sufficiently strongly weighted contribution to the total conductivity. The three energy ranges where conduction is expected are illustrated in Fig. 7.14. Conduction at the Fermi energy, which is invariably within the localized state distribution, is by hopping. This mechanism dominates in unhydrogenated amorphous silicon in which there is a very large defect density, but is less important in a-Si:H because of its lower defect density. Hopping conduction can also take place in the band tails where the density of states again becomes large, but the carrier concentration is low. Extended state conduction occurs at the lowest energies at which such conduction can take place, so that the properties of the mobility edge are crucial. Hopping at the Fermi energy is described briefly in Section 7.4.4, but most of the discussion concerns the conduction near the mobility edge. All three mechanisms are observed in a-Si:H (LeComber and Spear 1970), although extended state conduction is the most important.

A further consideration in applying the conductivity theories to a-Si:H is the possibility of long range inhomogeneities in the material. The conductivity is a spatial average over the complete conduction path of the carrier, which may be different from the local conductivity at any particular site. The role of inhomogeneities are considered in Section 7.4.5.

Fig. 7.14. Illustration of the three main conduction paths in amorphous semiconductors.

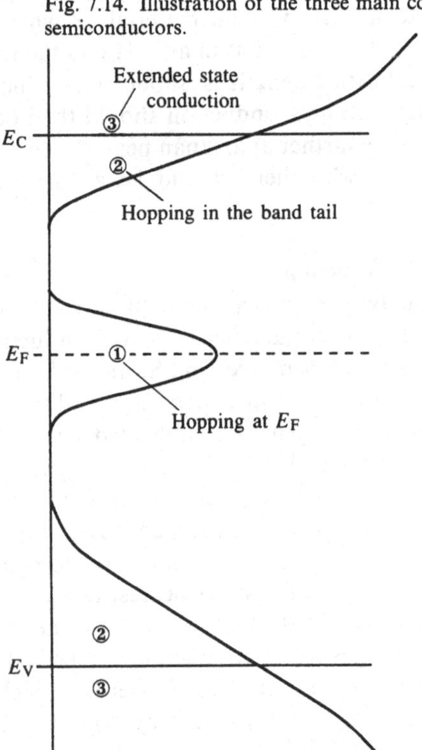

7.4.1 *Early models of extended state conduction*

The ideas about conduction near the mobility edge have changed substantially in recent years and are still in a state of flux, with several different models being developed. It is helpful to begin with the first model developed by Mott around 1965–70 (see Mott and Davis (1979) Chapter 2) to see how the ideas have evolved and to understand the underlying physical concepts.

Mott proposed that an amorphous semiconductor has a well-defined mobility edge near the band gap, where there is a discontinuous change in the conductivity. The theory combines the concept of Anderson localization with the reasoning by Ioffe and Regel (1960), that scattering lengths less than the interatomic spacing are impossible. The localization model is used to obtain the energy of the mobility edge with respect to the density of states distribution and the conductivity at E_C is calculated from the minimum scattering length. The Anderson model, which is illustrated in Fig. 7.15 and also described in Chapter 1, gives as the criterion for localization of the electronic states,

$$V_0/J > 3 \tag{7.41}$$

where V_0 is the disorder potential and J is the band width in the absence of disorder. (Anderson estimated the constant to be 5, but recent work seems to have converged on 3 for a four-fold coordinated network.) In the presence of disorder, the band width increases to $(J^2 + V_0^2)^{\frac{1}{2}}$, which is approximately $3J$ at the localization transition. The density of states in the middle of the band is consequently reduced by the same factor 3, since the total number of states in the band is fixed.

The disorder potential of the amorphous semiconductors is not large

Fig. 7.15. Illustration of the Anderson localization model showing atomic potentials and the shape of the band, with and without the disorder.

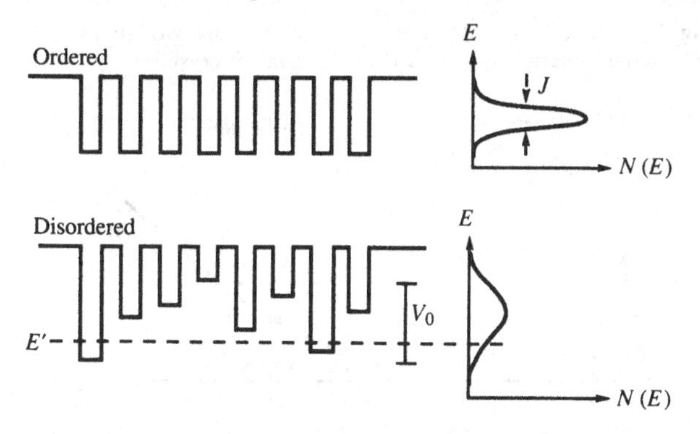

enough to satisfy the Anderson criterion, so that complete localization of all the states does not occur. For example, the photoemission data in Fig. 3.3 show that the conduction and valence band widths of a-Si:H are hardly broadened compared to crystalline silicon. The disorder potential is less than 0.5 eV, compared to a band width of 5–10 eV. Mott showed that under these circumstances the states at the center of the band remain extended, while those at the edge of the band are localized, and introduced the idea of a mobility edge separating the two types of state. Only a few atomic sites contribute to the electronic states near the band edge, as is illustrated by the dashed line in Fig. 7.15. The effective band width J' at this energy E' is greatly reduced because of the smaller wavefunction overlap between these fewer states and the mobility edge occurs when the Anderson criterion, Eq. (7.41), applies to V_0/J'; J' decreases near the band edge, giving extended states in the middle of the band and localized states at the band edge.

A parameter needed to locate the mobility edge and to calculate σ_{min} is

$$g \equiv N(E_F)/N_{xt}(E_F) \tag{7.42}$$

g is the factor by which the density of states at E_F is reduced compared to the equivalent crystalline material and is illustrated in Fig. 7.16. The same number of electrons are assumed to be present in both the amorphous and crystalline material and fill the states up to E_F. Because the band edge is broadened by the disorder, $N(E_F)$ in the disordered material is smaller than $N_{xt}(E_F)$. The Anderson criterion of Eq. (7.41) implies that the complete band becomes localized when $g \approx \frac{1}{3}$. Mott reasoned that the mobility edge occurred when the same criterion applies to the reduced band width J', so that

$$g(E_C) = \tfrac{1}{3} = J'(E_C)/V_0 = N(E_C)/N_{xt}(E_C) \tag{7.43}$$

Fig. 7.16. Illustration of the reduction of the density of states at E_F in a disordered material compared to the equivalent crystal.

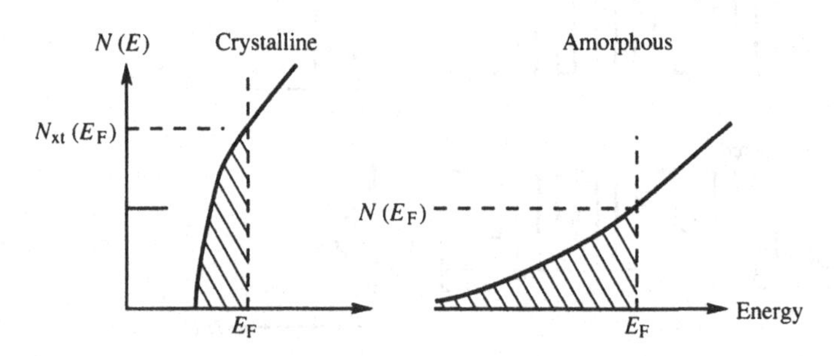

Fig. 7.17 estimates the parameter g for the actual conduction band density of states distribution of a-Si:H in Fig. 3.16. The integral of the density of states up to energy E is plotted against $N(E)$. The equivalent ordered state is taken to be a parabolic band with the density of states of crystalline silicon. The parameter g decreases from the middle of the band to the band edge as expected and the results indicate that the mobility edge should occur near $N(E_C) = 10^{21}$ cm^{-3} eV^{-1}, which is quite close to the value indicated by experiment. Unfortunately, this does not provide an accurate procedure for measuring E_C, because there is not an exactly equivalent crystal with which to compare the density of states. Nevertheless it illustrates the principle.

The next step in the theory is to calculate the conductivity above and below the mobility edge. In the Anderson model, localized states are defined by a decreasing probability that the electron diffuses a larger distance from its starting point. Mott and Davis (1979 Chapter 1) prove that the dc conductivity in the localized states is zero at $T = 0$ K. They use the Kubo–Greenwood formula for the conductivity,

$$\sigma(E) = (2e^2\hbar^3\Omega/m^2) < D_{ij}^{\,2} > [N(E)]^2 \tag{7.44}$$

Ω is the volume and D_{ij} is the matrix element of the two wavefunctions Ψ_i and Ψ_j,

$$D_{ij} = \int \Psi_i{}^* \frac{\mathrm{d}}{\mathrm{d}x} \Psi_i{}^* \, \mathrm{d}^3x \tag{7.45}$$

The matrix elements D_{ij} vanish when the states are localized and the conductivity below E_C is zero.

The conductivity above E_C is determined by the electron scattering. Ioffe and Regel (1960) proposed that scattering lengths less than the interatomic spacing, a_0, are impossible. An extended state wavefunction which changes phase randomly from site to site is the most disordered situation possible, given the short range order of an amorphous semiconductor, because this is the extent of the spatial frequencies in the disorder potential.

The scattering length only equals the interatomic spacing when there is complete localization of the band. When the disorder potential is smaller than is given by the Anderson criterion and E_C is near the band edges, then the minimum scattering length, a_E, is larger, because only a subset of the sites with the low energy of the transport path will be able to scatter carriers. The scattering length is then given by,

$$a_E{}^3 = a^3 N_{max}/N(E_C) \tag{7.46}$$

where N_{max} is the density of states at the center of the band. In the model of Fig. 7.15, a_E is the average separation of only those wells which contribute to the particular energy E' at the edge of the band. Eq. (7.46) scales the minimum scattering length to the density of states at the mobility edge. An assumed band width of 10 eV, and $N(E_C) = 2 \times 10^{21}$ cm^{-3} eV^{-1} results in an estimate for a_E of,

$$a_E = (2\text{–}3)a_0 = 5\text{–}7 \text{ Å} \tag{7.47}$$

Several different approaches have been used to calculate the conductivity at the mobility edge, each giving the same result that,

$$\sigma(E_C) = Ce^2/\hbar a_E \tag{7.48}$$

where C is a constant for which the various estimates converge on a value of 0.03. One way to obtain Eq. (7.48) is from the Boltzmann expression for the conductivity, σ_B, of a free electron metal,

$$\sigma_B = ne^2\tau/m_e = ne^2L/m_e v_c \tag{7.49}$$

where τ is the carrier lifetime, L the mean free path, and v_c the electron velocity. The relations

$$v_c = \hbar k/m_e = \pi\hbar/a_E m_e \text{ and } n = a_E^{-3} \tag{7.50}$$

give

$$\sigma_B = e^2/\pi\hbar a_E \tag{7.51}$$

Fig. 7.17. The calculated reduction in the density of states of a-Si:H compared to the equivalent crystal. The mobility edge is expected to occur when $g \simeq 1/3$.

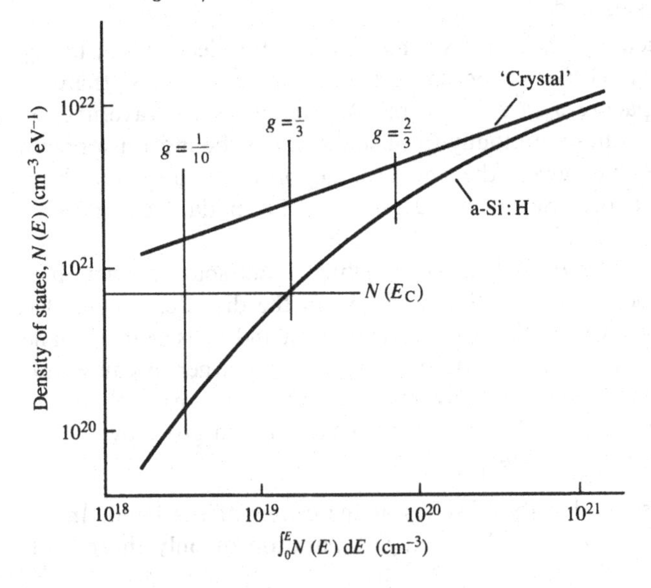

It is also necessary to include a factor g^2 in Eq. (7.51) because the conductivity varies as $N^2(E)$ (see Eq. (7.44)), and the disorder reduces $N(E)$ by the factor g at the mobility edge. Hence,

$$\sigma_{\text{min}} = (g^2/\pi)\, e^2/\hbar a_{\text{E}} = 0.03\, e^2/\hbar a_{\text{E}} \tag{7.52}$$

The idea of a lower limit on the scattering length leads to the concept of a minimum metallic conductivity $\sigma_{\text{min}} = \sigma(E_{\text{C}})$. Conduction below this value was thought to be impossible in extended states. In this model the conductivity drops discontinuously at $T = 0$ K because there is no conduction in the localized states below E_{C}. We shall see shortly that more recent ideas have changed this conclusion.

One aspect of the model which seems counter-intuitive is that σ_{min} is inversely proportional to the scattering length a_{E}. Thus, longer scattering lengths result in a smaller minimum conductivity, which is opposite to the normal conductivity mechanisms. The theory calculates σ_{min} at the transition, when the material contains just enough electrons for E_{F} to be at E_{C}. A change in the disorder causes E_{C} to move to a different energy, so that the density of electrons also changes. In effect, the weaker disorder allows the mobility edge to drop further down the band edge density of states and so the weaker scattering reduces σ_{min} rather than increasing it.

This model of a mobility edge at which $\sigma(E)$ changes discontinuously contains several implicit assumptions. For example, the scattering of the electrons is assumed to be elastic. Any inelastic scattering with a time constant τ_{i}, smears out the mobility edge by $\Delta E = \hbar/\tau_{\text{i}}$, which is the uncertainty in the electron energy. In contrast, there is no broadening of the mobility edge for elastic scattering because the electron does not change energy. The conduction model also assumes that the material is macroscopically homogeneous. In the presence of long range inhomogeneity, the conductivity should be described by a percolation model and it is expected that $\sigma(E)$ decreases continuously to zero near E_{C}. Percolation and the effects of long range potential fluctuations are described in Section 7.4.5.

The model also assumes that the disorder is uniform. The disorder is introduced at the beginning of the calculations by the Anderson localization criterion, but the effects of disorder on the transfer of the electron from site to site is not considered at a microscopic level. Scaling theory, which is described next, considers the microscopic disorder and reaches some different conclusions.

7.4.2 *Scaling theory*

The central idea of the scaling theory, as proposed by Abrahams and coworkers (Abrahams, Anderson, Licciardello and Ramakrishman 1979, Abrahams, Anderson, and Ramakrishman 1980), is to relate the conductivity $\sigma(E)$ to the size of the sample, taking into account the microscopic disorder. As an electron moves from one microscopic region to another, the specific electron states that it can occupy may change in energy because of disorder. The scaling argument starts by considering adjacent small volumes of material as indicated schematically in Fig. 7.18. The average separation W between the energy levels within a volume of size L^3 is

$$W(L) = [N(E)L^3]^{-1} \tag{7.53}$$

where $N(E)$ is the density of states. If τ is the time taken for the electron to diffuse from one volume to the next then,

$$\tau = L^2/D \tag{7.54}$$

where D is the electron diffusion coefficient. The transition time causes the electron energy to be broadened by ΔE, due to the uncertainty principle, so that

$$\Delta E = \hbar/\tau = \hbar D/L^2 \tag{7.55}$$

The conductivity $\sigma(E)$ is obtained from Eqs (7.5), (7.53) and (7.55), using the Einstein relation, $\mu = eD/kT$, to relate the mobility and diffusion,

$$\sigma(E) = N(E)e\mu_0 kT = (e^2/\hbar)\,\Delta E/LW(L) \tag{7.56}$$

Fig. 7.18. Illustration of the scaling theory model, showing energy levels associated with a small volume of material. W is the energy level splitting and ΔE is the uncertainty principle broadening.

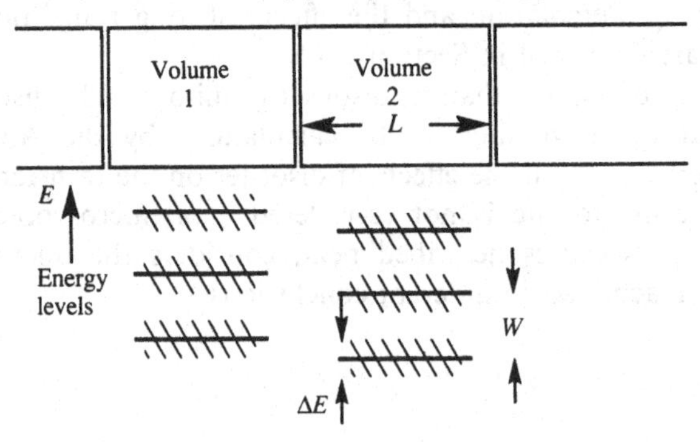

A dimensionless conductance $G(L)$ is next defined by

$$G(L) = L\sigma(E)\hbar/e^2 = \Delta E/W(L) \tag{7.57}$$

and is exactly the ratio of the two energies defined by Eqs (7.53) and (7.55).

In order for conduction to take place at zero temperature, the electron must be able to transfer from one volume to another without change in energy. The transfer cannot occur when $\Delta E \ll W$ because the energy levels generally do not coincide. Correspondingly, a small value of ΔE means that the time taken to cross the volume is long, implying that the states are localized and the conductivity is low. When $\Delta E \gg W$, the carriers can diffuse easily because there are always energy states to transfer into which lie within the uncertainty principle broadening and so the conductivity is high. Thus $G(L)$ has the correct physical characteristics of the conductance.

The next step in the argument is to consider how $G(L)$ changes with the size of the sample. Abrahams *et al.* (1979) make the scaling hypothesis that when small volumes are combined into a larger one of size bL, $G(L)$ is the only quantity needed to calculate the new $G(bL)$, so that,

$$G(bL) = f^n(b, G(L)) \tag{7.58}$$

It then follows that when the scaling is considered as a continuous function of L

$$\mathrm{d}\ln G(L)/\mathrm{d}\ln L = \beta(G) \tag{7.59}$$

where $\beta(G)$ is a universal, one-parameter scaling function. $\beta(G)$ describes how the conductivity scales with sample size at different values of G, covering both the extended and the localized states.

The form of $\beta(G)$ is constructed from its asymptotic values. When the conductivity is large, for example, well above the mobility edge, then the normal macroscopic transport theory must apply, for which σ is independent of the size of the sample, so that from Eq. (7.57), β is unity. Generalized to d dimensions, this gives

$$\lim_{G\to\infty} \beta(G) = d-2 \tag{7.60}$$

At the other extreme of low conductivity, the electrons are localized with an exponential decay of the wavefunction, which represents the conductivity,

$$G(L) = G_0 \exp(-L/L_0) \tag{7.61}$$

where L_0 is the localization length. Eq. (7.59) gives for this case,

$$\lim_{G\to 0} \beta(G) = \exp(-G/G_0) \tag{7.62}$$

Abrahams *et al.* (1980) give arguments that $\beta(G)$ is a continuous function of G and must have the form shown in Fig. 7.19 for two and three dimensions. More detailed calculations are used to show that the shape of $\beta(G)$ is as given in the figure.

The principal result of these scaling arguments is that the conductivity, in general, depends on the size of the sample. The theory has been at least partially confirmed by experiment in two- and three-dimensional systems (e.g. Thomas (1985)). The three-dimensional case is the only one for which $\beta(G)$ crosses zero. When β is positive, then Eq. (7.58) shows that $G(L)$ is an increasing function of the size L, but when β is negative, $G(L)$ vanishes at large sizes. The crossing point therefore represents the mobility edge, being the separation between energy ranges which have finite or zero macroscopic conductivity. Integration of Eq. (7.59) at the zero crossing ($\beta = 0$) gives

$$\sigma(E_C) = \text{const. } e^2/\hbar L \tag{7.63}$$

The conductivity at the mobility edge therefore decreases to zero as the

Fig. 7.19. The form of the scaling function $\beta(G)$ for two- and three-dimensional conduction (Abrahams *et al.* 1979).

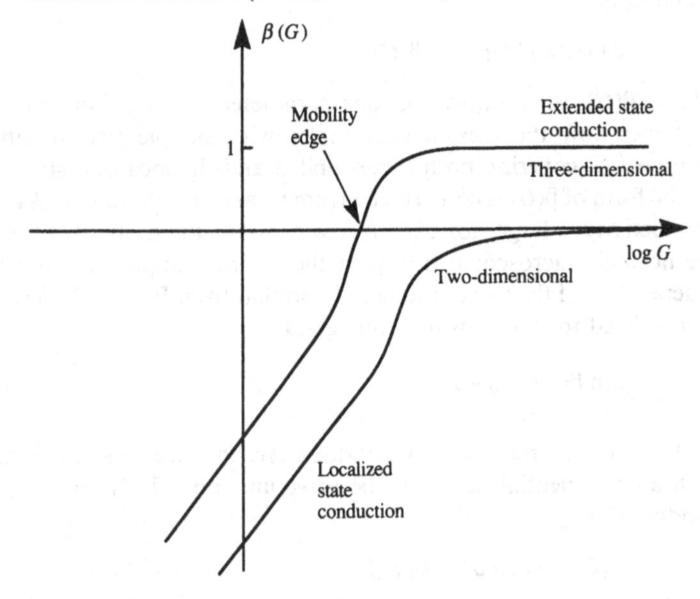

size of the sample increases. Thus it is concluded that there is no minimum metallic conductivity at E_C. Abrahams *et al.* (1980) went on to show that the conductivity goes to zero near E_C as

$$\sigma(E) \approx (E-E_C)^{\nu} \tag{7.64}$$

where $\nu = 1$, although other values of ν have since been proposed, depending on the details of the model.

There is one more step in the scaling argument as it applies to amorphous semiconductors. Thouless (1977) points out that the inelastic scattering length gives a cut-off to the scaling arguments. In the scaling theory, the uncertainty in energy, ΔE, is related to the time taken to cross the volume of size L. Inelastic scattering causes an electron transition to a different energy state and the scaling argument fails because the energy uncertainty is determined by the scattering time rather than the time to cross the volume. The inelastic scattering length, L_i, is therefore equivalent to the size of the sample and the macroscopic conductivity for larger samples behaves normally without any further size dependence. Thus,

$$\sigma(E_C) = \text{const. } e^2/\hbar L_i = \sigma_{min} \tag{7.65}$$

and the minimum metallic conductivity is once again restored.

Some different situations now need to be considered. At zero temperature and with an equilibrium distribution of electrons, there can be no inelastic scattering, because there are no empty states for the electron to scatter into and so the theory predicts that $\sigma(E)$ goes continuously to zero at E_C. However, this situation never applies to amorphous semiconductors because the Fermi energy always lies within the localized states. There are (so far) only two experimentally accessible conditions. At zero temperature, non-equilibrium electrons or holes may be excited above the mobility edge, for example, by illumination and there is inelastic scattering into the empty states between E_C and E_F. Equilibrium conductivity can take place only at non-zero temperatures and again there is inelastic scattering into the unoccupied localized states. Scaling theory predicts that there is a minimum conductivity in both situations but with the value given by Eq. (7.65) rather than Eq. (7.52). The relevant length in the expression for σ_{min} is the inelastic, rather than the elastic, scattering length.

The inelastic scattering greatly complicates the understanding of the transport. The mobility edge is ill defined by the uncertainty principle broadening of the energy. A further complication is that the experimentally accessible situations of optically excited and thermally excited carriers lead to an additional contribution to the conductivity from localized states below E_C. The next section discusses further the effects of inelastic scattering.

7.4.3 *Inelastic scattering and phonon effects*

Scaling theory leads to the conclusion that the conductivity goes continuously to zero at the mobility edge as the sample length increases. However, inelastic scattering limits the length scale and restores the minimum conductivity. In a-Si:H the experimental conditions are such that inelastic scattering by phonons is always present. There have been a few theories that attempt to deal with the effects of inelastic scattering and the electron–phonon interaction. Three of these are described briefly to illustrate the assumptions that are made and the conclusions that follow.

The first model is by Mott and Kaveh (1985), who obtain the scaling theory result by a different approach. A feature of the model is that it starts with a specific physical mechanism, that of multiple scattering, rather than a scaling argument. There is also a direct connection to the earlier model for σ_{min}. The theory uses a modified form of the scattering equation given by Kawabata (1981),

$$\sigma = \sigma_B g^2[1 - C(1 - a_E/L)/g^2(k_F a_E)^2] \tag{7.66}$$

σ_B is the Boltzmann conductivity given by Eq. (7.51), k_F is the momentum at the Fermi energy and C is a constant estimated to be near unity. This equation is derived as a description of the effects of weak disorder on the conductivity well above E_C and is extended to describe the conductivity at the mobility edge. The first term on the right hand side is the usual Boltzmann conductivity to which a factor g^2 is added for the same reason as in Eq. (7.52). The second term in the bracket describes the effects of multiple scattering on the electron. Briefly, the amplitude of the wavefunction contains a sum over the scattering terms a_i, such that $\Sigma a_i^2 = 1$. Only the first order term a_0 contributes to the conductivity so that σ is proportional to,

$$a_0^2 = 1 - \sum a_i^2 \tag{7.67}$$

The sum on the right hand side of Eq. (7.67) corresponds to the second term in the bracket of Eq. (7.66). Kawabata's calculation did not include the size of the sample, which Mott and Kaveh (1985) add as the factor $(1 - a_E/L)$, reasoning that when the elastic scattering length is comparable to L, then the multiple scattering is reduced. Furthermore, in the presence of inelastic scattering, the term L is replaced by L_i, for the same reason as the substitution was made in the scaling theory, that the coherence of the wavefunction is lost.

The next step in the argument is to extrapolate the conductivity to the mobility edge. In the absence of inelastic scattering (or for an infinite sample size), the conductivity of Eq. (7.66) goes to zero when

$$C/g^2(k_F a_E)^2 = 1 \tag{7.68}$$

This expression defines the mobility edge. When there is a finite value of the inelastic scattering length, L_i, the conductivity at the mobility edge is no longer zero but from Eq. (7.66) has the value

$$\sigma_{min} = \sigma_B g^2 a_E / L_i = 0.03\, e^2/h L_i \tag{7.69}$$

The second expression results when σ_B is substituted from Eq. (7.51) and $g = \frac{1}{3}$.

The multiple scattering model therefore reproduces the scaling theory results. The prefactor is the same as was given by Mott's early theories and only the length scale is changed from the elastic scattering length, a_E, to the inelastic length L_i. The above argument is not a rigorous way of extrapolating Eq. (7.66) to the transition and a more detailed discussion of this point is given by Mott and Kaveh (1985) (also Mott (1988)).

Two models for the effects of the electron–phonon coupling on the conductivity at elevated temperatures are now described. The first combines the extended state conduction above E_C with hopping conduction below E_C into a single theoretical analysis (Müller and Thomas 1984). The theory uses a tight binding model with interactions between nearest neighbors, described by a transfer matrix element. The calculation does not include the effects of inelastic scattering and the conductivity is obtained from the spatial correlation function between neighboring states given a specific density of states distribution. The calculation of the conductivity at zero temperature gives,

$$\sigma_e = \sigma_{min}(1 - \delta) \tag{7.70}$$

where $\sigma_{min} = 0.03\, e^2/\hbar a_E$. The factor δ is similar to the multiple scattering correction term in the Kawabata formula of Eq. (7.66). In the theory, δ is given in terms of the transfer matrix element and the density of states. At the mobility edge, $\delta = 1$, again predicting that the conductivity goes smoothly to zero. Above E_C, δ decreases continuously to zero so that the conductivity is given by the early model of the minimum metallic conductivity. This model therefore also reproduces the results of scaling theory and of the model just described, apart from the effects of inelastic scattering.

The model predicts that below E_C and at elevated temperatures, the conductivity is given by the usual thermally activated hopping rate, which depends on the density of states and increases with temperature (see Section 7.4.4). An important result of the theory is that near E_C the conductivity is not just a simple sum of the $T = 0$ K extended state conductivity and the hopping term. Fig. 7.20 shows schematically that the conductivity changes smoothly from one extreme to the other and is larger than both terms near E_C. The origin of the extra conductivity is termed 'phonon-induced delocalization'. The effect is described by the expression for hopping below E_C,

$$\sigma(E < E_C) = \sigma^h \delta/(\delta - 1) \tag{7.71}$$

and above E_C

$$\sigma(E > E_C) = \sigma_e + \sigma^h/(1 - \delta) \tag{7.72}$$

where δ is the same parameter as in Eq. (7.70). δ is large at energies well below E_C, so that $\sigma \to \sigma^h$ which is the normal hopping term. Near E_C, δ approaches unity so that the hopping conductivity is enhanced by the large factor $\delta/(\delta - 1)$, which describes the delocalization. This has the effect of broadening the mobility edge at elevated temperatures and lowering the energy at which conduction takes place. The magnitude of the energy shift depends on the shape of the density of states as well as on the other parameters of the model. Fenz, Müller, Overhof and Thomas (1985) give some examples of how different assumed density of states distributions change the predicted temperature dependence of the conductivity

Cohen, Economou and Soukoulis (1983) have proposed a different model of the electron–phonon coupling near the mobility edge, but one which also considers how two different conduction mechanisms merge together near E_C. In this case the mechanisms are small polaron hopping resulting from a strong electron–phonon coupling and conduction in extended states in the absence of strong coupling. In the standard small polaron model, self-trapping of the carrier occurs at every site, which results in a narrow band width and a hopping conductivity. Such a model does not apply well to a-Si:H because transport above E_C is not described by a hopping model and there is no obvious reason why extended states should couple to phonons more strongly in a-Si:H than in crystalline silicon. On the other hand the effect of phonon coupling gets stronger in localized states because the electron wavefunction is more confined and interacts more strongly with specific atomic bonding electrons. Thus lattice relaxation in the band tail states is a real possibility and, indeed, is indicated in photoluminescence experiments for the valence band tail (see Section 8.3.2). The model described by Cohen *et al.* includes the disorder, so that the phonon coupling at different sites is not treated identically and some sites are allowed to have stronger coupling than others.

The calculation is particularly concerned with the polaron effects in the region near E_C where scaling theory arguments apply and, in the absence of inelastic scattering, the phase coherence length or localization length is large. The theory predicts that there is no polaron formation far above E_C and that the conductivity is unaffected by the electron–phonon interaction. However, there is a critical energy near E_C at which polaron formation takes place. The conductivity drops discontinuously to zero at this energy which takes the place of the mobility edge. A sharp mobility edge is thus restored, but the conductivity at the new mobility edge is not the original Mott value, but depends on the strength of the electron–phonon coupling. Zero coupling strength gives the scaling theory result of a continuous change in σ_0 and a very large coupling gives Mott's original value of $0.03\, e^2/\hbar a_E$.

These three models illustrate the different effects that can influence the conductivity near E_C. The results are summarized as follows, with the additional comment that the conclusions apply only to material without macroscopic inhomogeneities.

(1) At zero temperature and in the absence of inelastic scattering or phonon coupling, it is generally agreed that the conductivity goes continuously to zero at E_C in large samples and that the early model of a discontinuous change is not correct.

Fig. 7.20. The model of conduction near a mobility edge at elevated temperature, showing the phonon-induced delocalization which enhances the conductivity near E_C (Fenz *et al.* 1985).

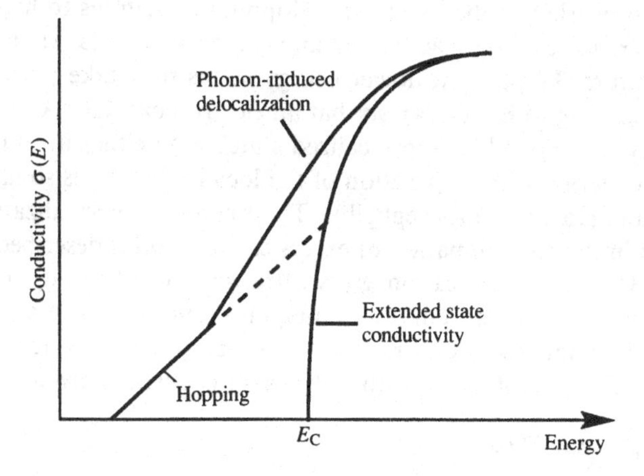

(2) The effect of inelastic scattering is to give a minimum conductivity at E_C of $0.03\,e^2/\hbar L_i$. The inelastic scattering time, $\tau_i = L_i^2/D$ also results in an uncertainty principle broadening of the mobility edge by an energy of $\hbar/\tau_i$.

(3) When the electron–phonon interaction is included, there may be a 'phonon-induced delocalization' which further broadens the mobility edge and allows conductivity to occur below E_C. However, sufficiently strong electron–phonon coupling results in polaron formation in the band tails which suppresses the conductivity in the localized states and restores the discontinuous change in σ_0.

(4) Conductivity sufficiently far down the band tails is by hopping and goes to zero at $T = 0$ K. The hopping may or may not have a polaron behavior depending on the size of the phonon interaction.

7.4.4 *Hopping conductivity*

A fundamental result of Anderson localization is that there is no conductivity in localized states at zero temperature. However, transitions between localized states can take place at elevated temperatures and result in a hopping conductivity. The transition probability, p_{hop}, between two states separated by distance R and energy W is (see Section 1.2.5),

$$p_{\text{hop}} = \omega_0 \exp(-2R/R_0 - W/kT) \qquad W > 0 \tag{7.73}$$

$$p_{\text{hop}} = \omega_0 \exp(-2R/R_0) \qquad W < 0 \tag{7.74}$$

R_0 is the localization length and the factor $\exp(-2R/R_0)$ is the wavefunction overlap of the two states. Hopping transitions to higher energy states require thermal activation and can only take place at non-zero temperature. Hopping to lower energy states may take place at $T = 0$ K, according to Eq. (7.74), so that an electron excited above the Fermi energy can move between localized states until either it returns to the Fermi energy or the separation of the localized states is so large that the hopping rate becomes negligible. This process of thermalization is important in the recombination of excess carriers and is described in Section 8.1.3. The thermalization gives a transient conductivity after the excitation of the carrier, but is a non-equilibrium process. A steady state hopping conductivity occurs only at non-zero temperature.

Carrier hopping is diffusive, with a diffusion coefficient given by

$$D_{\text{hop}} = p_{\text{hop}} R^2/6 \tag{7.75}$$

where R is the average hopping distance. When p_{hop} is given by Eq. (7.73) and the conductivity is defined by Eq. (7.5), then

$$\sigma_{hop}(E) = \tfrac{1}{6}N(E)\omega_0 e^2 R^2 \exp(-2R/R_0 - W/kT)$$
$$= \sigma_{oh} \exp[-W/kT] \quad (7.76)$$

When $N(E)$ is in the range 10^{19}–10^{20} cm^{-3} eV^{-1}, the hopping prefactor σ_{oh} is of order 10^{-1}–10^{-2} Ω^{-1} cm^{-1}, even without the $\exp(-2R/R_0)$ term. Hopping conductivity is therefore generally characterized by a low conductivity prefactor, unless the density of states is very large.

Eq. (7.76), however, takes no account of the microscopic process of hopping which depends on the local distribution of sites. The hopping conductivity is derived from the transition rate between the different localized states, which can be written as

$$\mathrm{d}n_i/\mathrm{d}t = \sum p_{ji}n_j - \sum p_{ij}n_i \quad (7.77)$$

where n_i is the occupation number of site i. The first term in Eq. (7.77) describes the hop of an electron to site i from all the nearby sites j and the second term is the return transition rate onto site j. Hopping transport is another situation in which the random distribution of states is crucially important and most of the interesting physical effects arise from the specifics of the distribution. The calculation of the conductivity is difficult because an electron only contributes to the dc conductivity if it has a path completely through the sample. The hopping rate between two sites is largest for the small fraction of pairs which are particularly close, but such events do not contribute to the conductivity if there are no other close neighbors. The average conductivity is dominated by the least probable hop within a path rather than the most probable. Also if the band of localized states is wider than kT, as is usually the case, then there is a trade-off between hopping to a close state with high energy or a more distant state with a smaller energy difference.

Although there have been several detailed calculations of the hopping conductivity based on Eq. (7.77), the physical mechanisms are best illustrated using the model proposed by Mott (1968) for variable range hopping. The model assumes a uniform density of states, $N(E_F)$, extending much more than kT, around the Fermi energy. The conductivity is dominated by hops between states at the average separation and also by those hops which require thermal excitation. Within a range of energy W from the Fermi energy, the density of states is $N(E_F)W$ and the average separation, R, is $(N(E_F)W)^{-\frac{1}{3}}$. Larger values of W result in a smaller average separation between the states, so that the first term in the hopping rate of Eq. (7.73) is reduced at the expense of the second. The largest contribution to the conductivity is when the hopping rate exponent in Eq. (7.73), $(2R/R_0 + W/kT)$, is minimized. This is easily shown to occur when the average hop distance is

$$R_{av} \approx [R_0/N(E_F)kT]^{\frac{1}{4}} \tag{7.78}$$

The conductivity is given by,

$$\sigma = \sigma_{oh} \exp(-A/T^{\frac{1}{4}}) \tag{7.79}$$

where

$$A = 1.7[1/R_0{}^3kN(E_F)]^{\frac{1}{4}} \tag{7.80}$$

and using Eq. (7.76)

$$\sigma_{oh} = \tfrac{1}{6}e^2N[1/R_0{}^3kN(E_F)]^{\frac{1}{2}} \tag{7.81}$$

This form of the conductivity is termed variable range hopping because the average hopping distance is not constant, but decreases as the temperature is raised. The conductivity follows a characteristic $T^{\frac{1}{4}}$ law rather than an Arrhenius behavior. More detailed analysis of the hopping mechanism confirms Eq. (7.79), but gives a substantially different value for the prefactor σ_{oh}.

Variable range hopping dominates the conductivity of unhydrogenated amorphous silicon because $N(E_F)$ is large and examples are shown in Fig. 7.21. These data show that ion bombardment increases the defect density and the hopping conductivity, while annealing recovers the lower defect density. The evaporated material typically has a defect density of about 10^{19} cm^{-3} eV^{-1}, so that the average hopping distance given by Eq. (7.78) is about 20 Å. The defect density is much lower in undoped a-Si:H and the contribution from variable range hopping is much smaller. The defect density has to be larger than 10^{18} cm^{-3} eV^{-1} for there to be any real possibility of observing hopping at E_F.

Hopping conduction can also occur between band tail localized states. The conductivity increases with the density of states at energies higher up the tail. The relative contribution of the hopping conductivity or conduction at the mobility edge depends on whether the lower hopping conductivity is more than offset by the larger carrier concentration at the lower energy due to the Fermi occupation factor. Early calculations considered a linear band tail and predicted that hopping dominated the conductivity at low temperature, with a conductivity activation energy $E_A - E_F$, where E_A is the energy of the bottom of the tail. A calculation of variable range hopping in a linear band tail was reported by Grant and Davis (1974).

More recently, hopping in an exponential band tail has been considered (Monroe 1985). An electron at the mobility edge thermalizes down the tail by hopping between localized states, until it reaches an energy at which further thermalization is balanced by thermal excitation to higher energy, where there is a higher density of states. This is the transport energy at which conduction takes place and is obtained as follows. The band tail density of states is defined as the exponential $N_0\exp(-E/kT_C)$. At an energy E the rate of hopping down the tail is given by Eq. (7.74) substituting for the density of states below E,

$$p_H(E) = \omega_0 \exp[-2R_0^{-1}N_0^{-\frac{1}{3}}\exp(-E/3kT_C)] \tag{7.82}$$

Fig. 7.21. Examples of the $T^{\frac{1}{4}}$ conductivity law for evaporated unhydrogenated amorphous silicon. After deposition, the films are bombarded by Si^+ ions at low temperature and then annealed at the indicated temperatures (Apsley, Davis, Troup and Yoffe 1977).

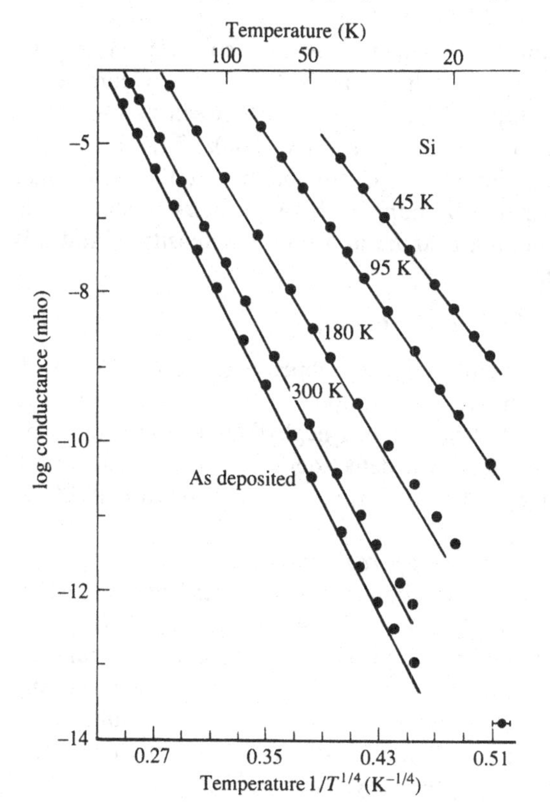

The concept of a time-dependent demarcation energy, $E_D(t)$, defined by $p_H(E_D)\, t = 1$, from the theory of dispersive transport gives

$$E_D(t) = kT_C \ln(8/R_0 3N_0) - 3kT_C \ln\ln(\omega_0 t) \qquad (7.83)$$

The rate of energy loss by thermalization is compared with the energy gain by thermal excitation. The two rates balance at a time t_s, where

$$\omega_0 t_s = \exp(3T_C/T) \qquad (7.84)$$

When this time is substituted into Eq. (7.83) the transport energy is

$$E_t = kT_C \ln(8/27R_0^3 N_0) - 3kT_C \ln(T_C/T) \qquad (7.85)$$

The first term is an energy which is estimated to be close to the mobility edge, so that the second term represents the shift of the transport path below E_C. Band tail hopping is most significant when the band tail density of states distribution is broad and at low temperatures such that $T < T_C$. The transport energy moves up to the mobility edge when the temperature is above T_C and the contribution from the band tail hopping is small.

Conductivity at non-zero frequency is not constrained by the requirement that carriers must have a conducting path completely through the material. Hopping back and forth between two localized states contributes to the ac conductivity $\sigma(\omega)$, but makes no contribution to the dc conductivity. Consequently $\sigma(\omega)$ is larger than $\sigma(0)$ and is often dominated by hopping between pairs of states. The conductivity due to hopping near E_F in a uniform density of states is (Austin and Mott 1969).

$$\sigma(\omega) = e^2 N(E_F)^2 R_0^5 kT\omega[\ln(\omega_0/\omega)]^4 \qquad (7.86)$$

$\sigma(\omega)$ has a slightly sublinear frequency dependence, $\sigma(\omega) \sim \omega^s$, with $s \approx 0.8$. Eq. (7.86) is observed in amorphous semiconductors with large defect densities. However, a-Si:H has a negligible ac conductivity by this mechanism because of the low defect density. The ac conductivity of extended states is independent of frequency at least up to 1–10 MHz.

7.4.5 *Potential fluctuations and percolation*

All of the conduction mechanisms so far described assume that the material is macroscopically uniform. Both the hopping conductivity and the models of transport at the mobility edge take into account the randomness of an amorphous material, but this is done without any consideration of large scale inhomogeneity. Structural non-uniformity and Coulomb potentials are examples of inhomogeneity in a-Si:H which may influence the conductivity. Investigations of the a-Si:H structure described in Chapter 2 show that the bonded hydrogen is distributed non-uniformly, in the form of hydrogenated voids in the best material and as a large scale columnar microstructure in material characterized by PVD growth. The small voids have an average separation of perhaps 20–30 Å and the columns have dimensions of 100 Å or greater. The Coulomb potential of random distributions of point charges also causes long range inhomogeneity. Potential fluctuations with a length scale of order 100 Å certainly exist in doped a-Si:H, caused by the charged dopant and defect states.

In order to be considered macroscopic, the size of an inhomogeneity must be larger than the wavefunction of a localized state or the coherence length of an extended state, both of which are about 10 Å. The larger scale inhomogeneities can be considered classically. The issues to be addressed are how the inhomogeneity affects the local conductivity and how the transport is averaged over the different regions. Both density and potential fluctuations affect the local energy of the mobility edge. The potential fluctuations are easier to analyze because the physical model of a random distribution of point charges is well defined. The density fluctuations are more difficult to deal with because not much is known about the exact structure and distribution of the voids, so that it is hard to calculate their effect on the mobility edge. Presumably, the high hydrogen concentration around the void causes the band gap to be locally larger than in the bulk of the a-Si:H network. Thus the voids probably represent regions where the electrons and holes are excluded. The problem of transport in a medium which contains excluded regions is one of percolation and is described briefly below.

The potential fluctuations due to a random distribution of point charges are calculated as follows. Near a single charge the potential falls with distance r as

$$V_c = e/4\pi\varepsilon\varepsilon_0 r \tag{7.87}$$

which gives a potential of 25 meV at a distance of 50 Å for the dielectric constant of silicon which is 12. In addition to this short range potential there is a long range fluctuation from the random distribution. Within a volume of dimension L the average number of charges is L^3N_c, where N_c is the charge density. There is a statistical deviation from the average number of order

$$\Delta N = (L^3N_c)^{\frac{1}{2}} \tag{7.88}$$

Within the volume there is an average charge $e\Delta N$, which contributes a potential fluctuation of magnitude

$$\Delta V = \Delta Ne/4\pi\varepsilon\varepsilon_0 L = (LN_c)^{\frac{1}{2}} e/4\pi\varepsilon\varepsilon_0 \tag{7.89}$$

The potential apparently increases indefinitely with the size of the volume considered which is, of course, unphysical. The fluctuations are, in fact, limited by screening from mobile carriers, which sets the length scale equal to the screening length,

$$L_{\text{screen}} = (8\pi\varepsilon\varepsilon_0/n_f e)^{\frac{1}{2}} \tag{7.90}$$

where n_f is the density of mobile carriers. Combining Eqs. (7.89) and (7.90) gives

$$\Delta V = \left[\frac{2N_c{}^2 e^3}{(4\pi\varepsilon\varepsilon_0)^3 n_f}\right]^{\frac{1}{4}} \tag{7.91}$$

The potential fluctuations therefore increase with the density of fixed charges, but decrease with the density of mobile charge. Overhof and Beyer (1981) calculate the potential fluctuation distribution numerically from a lattice on which there is a random distribution of positive charges and a uniform background negative charge. They confirm the predicted dependence on $N_c^{\frac{1}{2}}$ as the size of the model and the assumed density of charges are varied.

Long range density and potential fluctuations cause a spatial modulation of the mobility edge as illustrated schematically in Fig. 7.12(*c*). Within such a model, transport depends on the ability of an electron or hole to find a path through the material, which is a percolation problem. At zero temperature the conductivity $\sigma(E)$ is zero until the energy is high enough to reach the percolation threshold. Even if there is a sharp mobility edge locally, the conductivity rises smoothly from zero at the percolation threshold energy, E_{pc}, because the effective volume of conducting material is small near the threshold. Percolation theory gives a conductivity of the form,

$$\sigma(E) \approx \sigma_0(E - E_{pc})^p \tag{7.92}$$

The exponent p in Eq. (7.91) depends on the physical situation and is typically calculated to be in the range 1–2. At elevated temperatures the carriers are thermally excited over the potential fluctuations and the application of percolation theory is less clear.

The potential fluctuations cause a difference in the energy of conductivity and thermopower. Fig. 7.22 shows numerical calculations of the two quantities for the same lattice of sites with random charges that was used to calculate the potential distribution (Overhof and Beyer 1981). The curves correspond to different densities of point charges. In the model the Fermi energy is chosen to be at the average energy of the mobility edge, which is assumed to be sharp. The results can be easily adapted to the situation when the Fermi energy is below E_C, but the quantity $Q(T)$ is unaffected (see Section 7.3 for a definition of Q). The potential fluctuations give a difference in the transport energies E_Δ which is roughly equal to the half width of the distribution of the fluctuations and is approximately ΔV in Eq. (7.91).

The amplitude and spatial dimensions of the potential fluctuations are different in undoped, doped, and compensated a-Si:H. The band tail electrons and holes provide the mobile screening charge in doped a-Si:H and the doping dependence of the potential fluctuations can be estimated from the experimental data. The magnitude of n_{BT} depends on the thermal equilibrium state, but is usually about 10% of the donor and defect density. According to the square root law for doping,

$$N_c = N_{don} = aC_g^{\frac{1}{2}} \qquad a = 3\times10^{19}\,\text{cm}^{-3} \tag{7.93}$$

Fig. 7.22. Calculated temperature dependence of (*a*) the conductance and (*b*) thermoelectric power (lower) of a network containing potential fluctuations. The curves give results for different charge densities (Overhof and Beyer 1981).

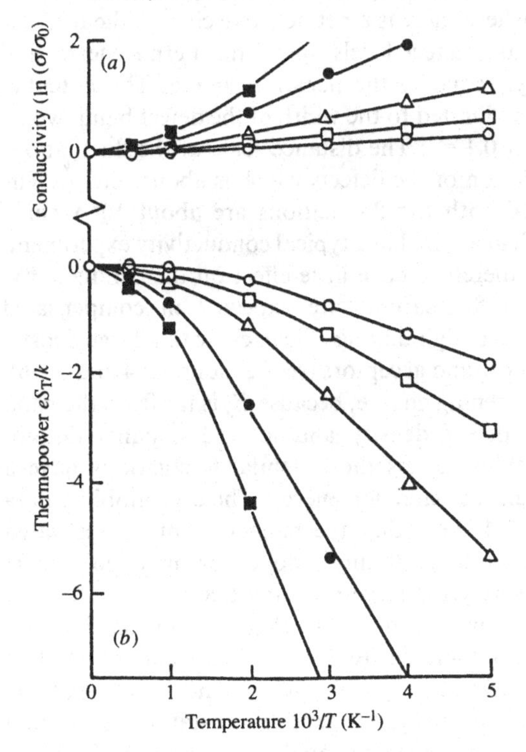

where C_g is the gas-phase dopant concentration. Combining these relations with Eqs. (7.88)–(7.91), gives for the fluctuations

$$\Delta V = 0.19\, C_g^{\frac{1}{8}}\ \mathrm{eV} \tag{7.94}$$

The potential fluctuations therefore increase slowly with the doping level because the dependence on the density of charges is partially offset by the increased screening by the larger value of n_{BT}. The constant in Eq. (7.94) is also obtained from the numerical calculations (Overhof and Beyer 1981).

It is harder to estimate whether there are significant Coulombic fluctuations in undoped a-Si:H. The defect density is much lower than in the doped material and the defects are predominately uncharged. There might, however, be about $10^{15}\ \mathrm{cm^{-3}}$ charged dangling bonds because the one- and two-electron defect bands overlap. The free carrier density is extremely low, because the Fermi energy is far from the band edges, and is only about $10^{9}\ \mathrm{cm^{-3}}$ at room temperature. A large value for the potential fluctuations is obtained when these values are substituted into Eq. (7.91), because of the weak screening. However, the fluctuations are also screened by the changes in occupancy of the defects. In any region where there is a net negative charge, the increase in the potential raises the defect levels above the Fermi energy and changes the occupancy, reducing the negative charge. The potential fluctuations are therefore limited to the width of the defect band, which is estimated to be about 0.1 eV. The distance scale of the fluctuations is of order of the separation of the defects which is about 1000 Å. The electric fields associated with the fluctuations are about $10^{4}\ \mathrm{V\,cm^{-1}}$, which is similar to the field applied in a typical conductivity experiment. Such fluctuations may therefore have little effect on the conductivity.

The largest potential fluctuations are expected in compensated a-Si:H. This material has a high doping efficiency, with a large density of ionized, charged, donors and acceptors (see Sections 5.2.4 and 6.2.6). It also has a very low screening charge, because E_F is far from the band edge. There is a low defect density and ionized dopants do not contribute to the screening unless the potential fluctuations have a magnitude of almost half the band gap energy. The drift mobility data described in Section 5.2.4 show that the band tails of compensated a-Si:H are much broader than of singly doped or undoped a-Si:H and a major reason is surely the large potential fluctuations.

Potential fluctuations have a different effect on the optical absorption. Long range Coulomb fluctuations cause a parallel shift of both band edges, so that the optical absorption is not spatially inhomogeneous. Therefore the joint density of states distribution derived from electrical measurements should not agree with the shape of the optical absorption when there are strong fluctuations. These two quantities agree well in undoped a-Si:H (see Fig. 3.21) and suggest that the fluctuations cannot be large. A difference is observed in compensated a-Si:H, indicating the presence of potential fluctuations (Howard and Street 1991).

7.4.6 *Conduction mechanisms in a-Si:H*

It remains to try to match the possible conduction mechanisms to the experimental transport data in a-Si:H. The present theories indicate that, in the absence of long range inhomogeneity, there is a minimum conductivity at the mobility edge at non-zero temperature, of magnitude $0.03\,e^2/\hbar L_i$ due to inelastic scattering. The scattering length is estimated from the diffusion coefficient of the electron at E_C and the scattering time τ_i by the relation $L_i = (D\tau_i)^{\frac{1}{2}}$. The diffusion coefficient is

$$D = \tfrac{1}{6}\omega_e a_E{}^2 \tag{7.95}$$

where ω_e is an electronic frequency, v_c/a_E. From Eq. (7.50)

$$\omega_e = \hbar k/m_e a_E = \pi\hbar/m_e a_E{}^2 \tag{7.96}$$

and k is the momentum π/a_E. The diffusion coefficient is therefore

$$D \approx \pi\hbar/6m_e = 0.3\,\text{cm}^2\,\text{s}^{-1} \tag{7.97}$$

Various estimates of the inelastic lifetime have been given by Mott (1988), all of which result in values of about the inverse phonon frequency, 10^{-13} s. The inelastic mean free path obtained by combining these quantities is 10–15 Å, which therefore gives a conductivity prefactor of

$$\sigma_{min} \approx 40\,\Omega^{-1}\,\text{cm}^{-1} \tag{7.98}$$

with at least a factor of 2 uncertainty arising from the various estimates. The free carrier mobility at E_C is estimated from Eq. (7.5) assuming that the density of states at E_C is 2×10^{21} cm^{-3} eV^{-1}. This gives a room temperature value of

$$\mu_0 \approx 6\,\text{cm}^2\,\text{V}^{-1}\,\text{s}^{-1} \tag{7.99}$$

The analysis of the experimental data in Section 7.1 found a conductivity prefactor of about $100\,\Omega^{-1}\,\text{cm}^{-1}$, but without any correction for the temperature dependence of the mobility edge. The comparison with the predicted value of σ_{min} suggests that γ_C is less than $1k$ and is positive, which corresponds to a shift of E_C into the gap with increasing temperature. Similarly the free carrier mobility found in

Section 7.2 agrees quite well with the theoretical estimates of Eq. (7.99). The conductivity and mobility data therefore seem to be consistent with a sharp mobility edge and with the theoretical calculations of σ_{min}. Furthermore, the uncertainty principle broadening of E_C is less than kT at the normal measurement temperatures and so is not a significant effect.

The thermopower data apparently do not support this model. The analysis of Section 7.3 found that a sharp mobility edge cannot explain the difference, E_Δ, of the transport energies. Long range potential fluctuations are the obvious origin of E_Δ in doped a-Si:H, where there are charged defects and dopants, and such fluctuations must be present. The energy E_Δ is observed to increase slowly with the doping level (see Fig. 7.11) as predicted by Eq. (7.94) and the magnitude of E_Δ is in reasonable agreement with the calculated amplitude of the fluctuations. Analysis of the drift mobility data in Section 7.2 concluded that the apparent position of the mobility edge moved slowly up the conduction band edge with increased doping, which is also consistent with the presence of potential fluctuations, because the conductivity is dominated by the high energy fluctuations which act as barriers to the conduction.

Fortunately, the sharp mobility edge model for conductivity and potential fluctuation model for the thermopower can be reconciled. The numerical results in Fig. 7.22 show that the conductivity is less affected by the fluctuations than the thermopower (Overhof and Beyer 1981). This explains why the conductivity prefactor is quite close to the predictions of the homogeneous mobility edge model, even in the presence of fluctuations.

A more detailed interpretation of the experiments is difficult. The difference in conductivity and thermopower energy is smallest in undoped a-Si:H, at about 50 meV. Although the larger values of E_Δ in the doped material are fairly clearly due to charge fluctuations, it is not yet known whether this is also the correct explanation in the undoped material, because the density of charged defects has not been accurately measured. The energy difference might be due to density fluctuations, but an alternative possibility is that $\sigma(E)$ does not decrease abruptly at E_C. A smoothly varying $\sigma(E)$ accounts for the energy difference and several models of such a broadening of the mobility edge are described in Section 7.4.3. E_Δ is not much larger than kT, for temperatures above 300 K, so that the conductivity data hardly distinguish a smoothly varying $\sigma(E)$ from one with an abrupt change.

The temperature dependence of the conduction energy is also uncertain. The agreement between the experimental and theoretical values of σ_{min} means that γ_T cannot be large. There are also alternative theories for the sign of γ_T. The model of phonon-induced delocalization predicts that the mobility edge moves to lower energy with increasing temperature as shown in Fig. 7.23. However, at temperatures of 100–200 K, band tail hopping may dominate the conductivity and the transport energy decreases with lower temperature, according to Monroe (1985). At these temperatures a-Si:H is in the frozen state and so the conductivity has a large statistical shift. A careful calculation, with an accurate density of states model is needed to extract γ_T from the data and this is yet to be performed. The effects of γ_T are also offset by the Boltzmann factor which tends to raise the average transport energy. For an exponential tail of slope T_C, the electron distribution is peaked at the top of the tail when the temperature exceeds T_C. Thus for electrons in a-Si:H for which T_C is about 300 K, the transport energy must be above the exponential tail above room temperature, since in all models $\sigma(E)$ increases monotonically with energy. The modeling of the transport does find that the transport path is just above the exponential region. In this temperature range γ_T must be small.

The conclusions of the transport are summarized as follows.

(1) The experimental results are consistent with the present theory of conduction. The uncertainty in the interpretation of the data is related primarily to whether the data is properly

Fig. 7.23. Schematic illustration of the temperature dependence of the conduction energy showing the effects of different transport mechanisms. The peak of the band tail electron distribution is also shown. The energy is referenced to the top of the exponential band tail.

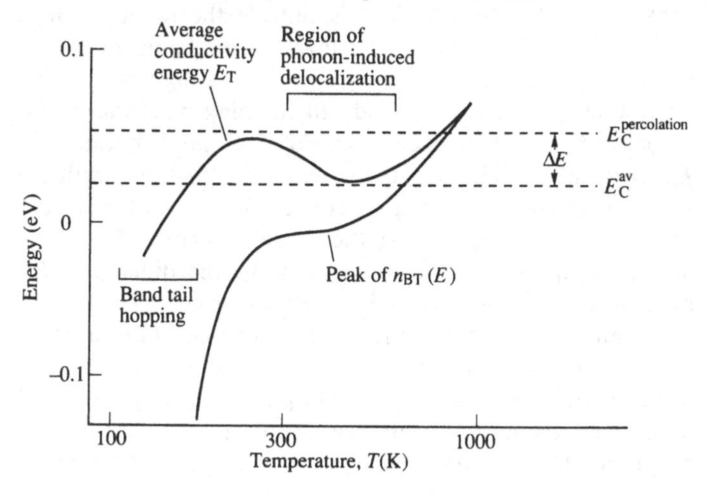

corrected for the temperature dependences of the energies and the uncertainty of which model best describes the physical situation at non-zero temperature.

(2) Potential fluctuations due to charged defects and dopants provide the most plausible explanation of the energy difference between the thermopower and the conductivity in doped material. In undoped a-Si:H either potential fluctuations or a broadened mobility edge which gives a continuous energy dependence of $\sigma(E)$ could be the explanation.

(3) The simple model of a sharp mobility edge with a conductivity prefactor of about 100 cm^2 V^{-1} s^{-1} is an adequate model to describe the conductivity at room temperature and above, even if it is not a precise description. The measured conductivity is not much changed by the potential fluctuations and the broadening of the mobility edge is small enough not to be significant above 300 K. Conductivity significantly below the mobility edge may be important below room temperature, in which case the simple mobility edge model does not apply.

(4) The temperature dependence of the transport energy, γ_T, is probably small above room temperature, but may be significant at lower temperature. The various contributions to γ_T which have been discussed are illustrated schematically in Fig. 7.23. The transport path is certainly above the energy of the peak in the electron distribution, which moves up to the top of the exponential tail near $T = T_C$, and then increases slowly with temperature, as indicated in the figure. At sufficiently high temperature, the dominant transport energy converges on the peak of $n_{BT}(E)$ although this is outside the range of most experiments. The phonon delocalization mechanism causes a reduction in the transport energy with inceasing temperature. At low temperature, the band tail hopping mechanism also reduces the transport energy. In Fig. 7.23 the two values of E_C represent the effects of the potential fluctuations, with the upper energy being that of the percolation path for the current and the lower energy being the average energy of E_C. The difference approximately corresponds to the difference, E_Δ, between the conductivity and thermopower energies.

(5) The Hall effect remains a puzzle. The theoretical explanation of the sign anomaly depends on the relative phases of the wavefunctions on adjacent sites. The phases are only correlated if the electron moves from one silicon atom to its nearest neighbor. However, the mobility edge in a-Si:H is far from the

center of the band and the distance between sites, a_E, is estimated to be about 6 Å, corresponding to third or fourth nearest neighbors. There is no obvious reason why there should be any phase correlation between the wavefunctions at this distance to give the sign anomaly.

8 The recombination of excess carriers

Illumination creates excess electrons and holes which populate the extended and localized states at the band edges and give rise to photoconductivity. The ability to sustain a large excess mobile carrier concentration is crucial for efficient solar cells and light sensors and depends on the carriers having a long recombination lifetime. The carrier lifetime is a sensitive function of the density and distribution of localized gap states, so that the study of recombination in a-Si:H gives much information about the nature of the gap states as well as about the recombination mechanisms.

The recombination process comprises two sequential steps, as illustrated in Fig. 8.1. An excited electron or hole first loses energy by many transitions within the band, in which the energy decrements are small but frequent. This process is referred to as thermalization. The thermalization rate decreases as an electron moves into the localized band tail states and the density of available states is lower. Eventually the electron completes the recombination by making a transition to a hole with the release of a large energy. Recombination lifetimes are generally much longer than the thermalization times, so that the two processes usually occur on distinctly different time scales.

Recombination is either radiative or non-radiative. The radiative process is accompanied by the emission of a photon, the detection of which is the basis of the luminescence experiment. The radiative transition is the inverse of optical absorption and the two rates are related by detailed balance. Non-radiative recombination is commonly mediated by the emission of phonons, although Auger processes are sometimes important, in which a third carrier is excited high into the band. The thermalization process occurs by the emission of single phonons and is consequently very rapid. Non-radiative electron–hole recombination over a large energy requires the cooperation of several phonons, which suppresses the transition probability.

Electron–hole correlations are an important aspect of the recombination. The recombination transition necessarily involves two particles, the electron and the hole, and so the recombination rate depends on whether they are spatially correlated or distributed randomly. The electron–hole pairing in a crystal is manifested in excitonic effects, but these are not detectable in the absorption spectra of a-Si:H (see Section 3.3.1). Correlation effects are nevertheless significant in a-Si:H and distinguish geminate from non-geminate recombination, as is explained in the Section 8.1.4.

8.1 Thermalization and recombination mechanisms

The recombination mechanisms which operate in a-Si:H are the same as in crystalline semiconductors. The presence of disorder apparently does not lead to any new processes, but does influence which mechanisms apply and their relative contribution in different measurements. The large density of band tail localized states is the most influential factor in the recombination. Recombination at low

Fig. 8.1. Illustration of electron–hole recombination, showing thermalization and different recombination mechanisms.

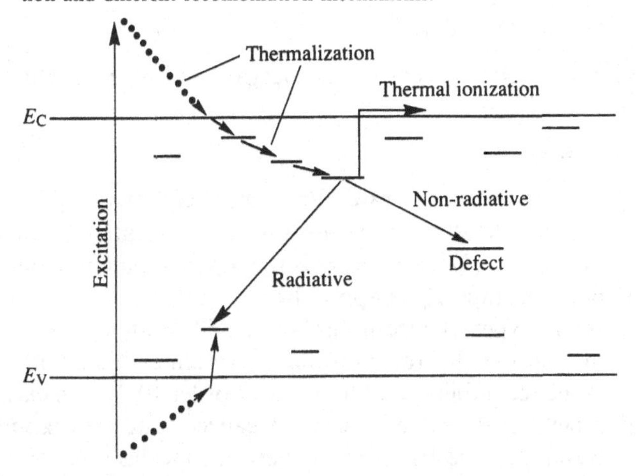

Fig. 8.2. Illustration of the dependence of the recombination rate on the separation of localized wavefunctions: (*a*) complete overlap, (*b*) weak overlap.

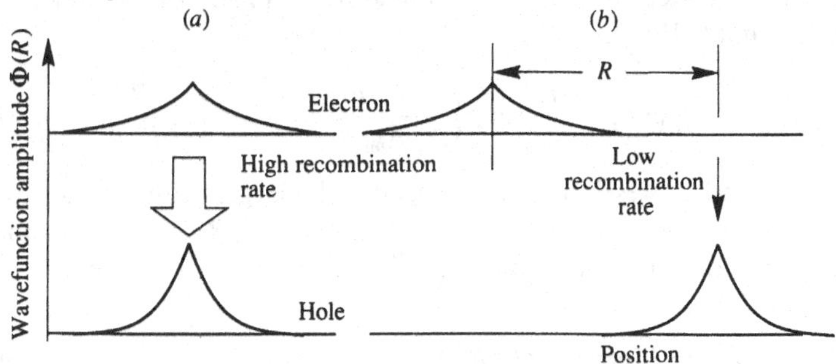

temperature invariably occurs by tunneling, because there is insufficient thermal energy to excite carriers from the band tail states to the mobility edge. Thermal excitation becomes significant above about 100 K and changes the recombination mechanisms.

8.1.1 *Radiative recombination*

The radiative recombination rate, P_{em}, for a transition between an upper and lower state emitting a photon of energy $\hbar\omega$, is given by the Fermi golden rule,

$$P_{em} = (2\pi/\hbar)|M^2|\delta(E_l - E_u + \hbar\omega) \tag{8.1}$$

where M is the matrix element of the transition and the δ-function conserves energy. Eq. (3.24) is the equivalent expression for the optical absorption. The matrix element is

$$M = M_0 J(\Phi_e, \Phi_h) \tag{8.2}$$

where J is the overlap integral of the electron and hole wavefunctions. Thus,

$$P_{em} = P_0 J^2 \tag{8.3}$$

where P_0 is the transition rate for completely overlapping wavefunctions. A dipole-allowed transition has a matrix element M proportional to $e^2 r_d{}^2/\hbar\lambda^3$, where er_d is the dipole moment, from which the recombination rate, P_0, is approximately 10^8 s^{-1}.

Two different types of recombination are illustrated in Fig. 8.2. In the first example the electron and hole have almost complete spatial overlap and the recombination lifetime is of order 10^{-8} s. An exciton or a transition between an extended and a localized state are examples. In the second example, the transition is between localized electrons and holes that are spatially separated by a distance R which is larger than the localization radii R_{0e} and R_{0h}. The wavefunction envelope for $r \gg R_{0e}$ is approximately $\exp(-r/R_{0e})$ where r is the distance from the localized state. The overlap integral is (Thomas, Hopfield and Augustiniak 1965),

$$J \approx \int \exp\left(-\frac{r}{R_{0e}}\right) \exp\left(-\frac{R-r}{R_{0h}}\right) d^3r \approx \exp\left(-\frac{R}{R_{0e}}\right) \tag{8.4}$$

The second expression follows because the largest contribution to the integral is at one or other site and applies when $R_{0e} > R_{0h}$. The matrix element contains the square of this factor, so that the transition probability is

$$P_{em} = P_0 \exp(-2R/R_{0e}) \tag{8.5}$$

A transition between states for which $R \gg R_{0e}$ has low probability and consequently a long radiative lifetime. This recombination mechanism is called radiative tunneling and occurs when there are localized states at both band edges, as in an amorphous semiconductor or a compensated crystal. It is the dominant radiative recombination mechanism in a-Si:H.

8.1.2 *Electron–phonon interactions*

Eq. (8.1) assumes that the energy conservation is maintained by the emission of a single photon. The recombination energy may instead comprise a photon and several phonons, when there is a significant electron–phonon interaction. The electron–phonon interaction is described by a configurational coordinate diagram as illustrated in Fig. 8.3. A similar description of electron capture at a defect is given in Section 4.1.1. The energies of the ground and excited states are described in terms of the configuration q by

$$\text{ground state:} \quad E_g(q) = Aq^2 \tag{8.6}$$

$$\text{excited state:} \quad E_x(q) = E^* + Aq^2 - Bq \tag{8.7}$$

where B is the linear deformation potential representing the strength of the electron–phonon interaction. The excited state has a minimum energy $E_e = E^* - W$, where W is $B^2/4A$ (see Fig. 8.3).

Fig. 8.3. Configurational coordinate diagram representing optical absorption and recombination in a material with a strong phonon coupling. A and B illustrate phonon-assisted non-radiative transitions.

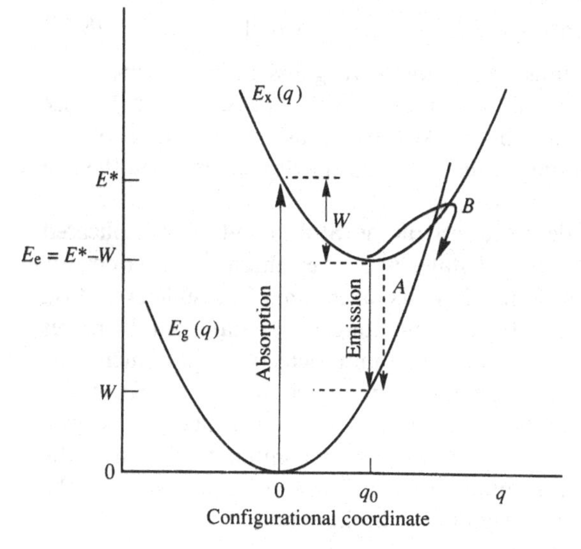

The optical transitions between the two states are described by the adiabatic approximation, in which the electronic and vibrational components of the wavefunction are treated separately. According to the Franck–Condon principle, absorption and emission take place without change in the configuration, as illustrated in Fig. 8.3. The energy of the emission is

$$\hbar\omega = E_x(q) - E_g(q) = E^* - Bq \tag{8.8}$$

At the minimum of the excited state, $q = q_0 = B/2A$ and $\hbar\omega(q_0) = E^* - 2W = E_e - W$.

The phonon coupling broadens the absorption and emission spectra. The vibrational ground state has the wavefunction of a simple harmonic oscillator with phonon frequency ω_0,

$$\Psi(q) = \text{const.}\ \exp(-Aq^2/\hbar\omega_0) \tag{8.9}$$

The absorption and emission probabilities for transitions between the two levels are proportional to the vibrational wavefunction, $\Psi(q-q_0)$. Substituting Eq. (8.8) into (8.9) gives the emission spectrum,

$$P_{em}(\hbar\omega) = \text{const.}\ \exp[-(\hbar\omega - E_e + W)^2/\sigma^2] \tag{8.10}$$

where

$$\sigma = (2W\hbar\omega_0)^{\frac{1}{2}} \tag{8.11}$$

Eq. (8.9) is the lowest vibrational wavefunction, so that the spectrum given by Eq. (8.10) applies at low temperature, $kT < h\omega_0$. The equivalent expression for the absorption spectrum is

$$P_{abs}(\hbar\omega) = \text{const.}\ \exp[-(\hbar\omega - E_e - W)^2/\sigma^2] \tag{8.12}$$

The absorption and emission bands have gaussian line shapes with their peaks separated by $2W$, as is illustrated in Fig. 8.4. The difference in energy is known as the Stokes shift; as the strength of the electron–phonon coupling increases, so do the Stokes shift and the line widths.

The experimental identification of the Stokes shift is complicated when there is a distribution of states and when thermalization occurs. Transitions between the bands have a continuum of possible excitation energies. The absorption band is no longer gaussian, but increases monotonically with energy because the joint density of states increases, as illustrated in Fig. 8.4. Thermalization allows the excited carriers to lose energy before recombination, so that in general there is always a difference in the energy of absorption and recombination even in the absence of strong phonon coupling. Section 8.3.2 discusses whether the luminescence energy has a significant Stokes shift.

8.1.3 *Thermalization and non-radiative transitions*

Non-radiative transitions invariably involve the conversion of excitation energy into phonons. Thermalization involves many inelastic transitions between states in the band or band tails. Three mechanisms of thermalization apply to a-Si:H. Carriers in extended states lose energy by the emission of single phonons as they scatter from one state to another. Transitions between localized states occur either by direct tunneling or by the multiple trapping mechanism in which the carrier is excited to the mobility edge and recaptured by a different tail state.

The rate of loss of energy of a carrier at an energy E by a single phonon is calculated to be (Mott and Davis 1979),

$$dE/dt = h\omega_0 P_{nr}(E) = h\omega_0^2[2\pi B^2 N(E)/Mv_s^2] \quad (8.13)$$

where $P_{nr}(E)$ is the non-radiative transition range, B is the deformation potential, N is the density of states, M is the atomic mass, and v_s is the velocity of sound. The term in brackets is of order unity. Thus the maximum thermalization rate is roughly $h\omega_0^2$, when one phonon energy is lost in the time of a phonon vibration. This rate is expected for electrons and holes above the mobility edge.

Thermalization in the band tail at low temperature occurs by tunneling between localized states. The low temperature only permits transitions to states of lower energy. The transition probability to a neighboring site contains the same overlap factor as for the radiative transitions,

$$P_{nr}(E) = \omega_0 \exp(-2R/R_0) \quad (8.14)$$

Fig. 8.4. Predicted line shapes for luminescence and absorption in a material with strong phonon coupling when the excited state is (*a*) discrete or (*b*) part of a continuum.

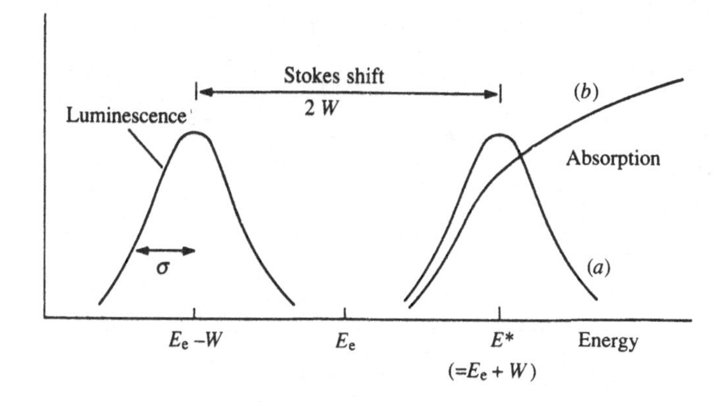

where R_0 is the localization radius. The average distance to the nearest available localized state is approximately the mean separation of the states, $N_t^{-1/3}$, where N_t is the total density of states which lie deeper into the band tail. The rate of energy loss due to tunneling in an exponential band tail, $N_0 \exp(-E/kT_0)$, is therefore (Monroe 1985)

$$\frac{dE}{dt} = kT_0 P_{nr}(E) = kT_0 \omega_0 \exp\left[-\frac{2(kT_0 N_0)^{-\frac{1}{3}} \exp(E/3kT_0)}{R_0}\right] \tag{8.15}$$

The thermalization rate of carriers from above the exponential tail is shown in Fig. 8.5 for two assumed band tail widths, corresponding to the conduction and valence bands of a-Si:H. The rate decreases very rapidly and, for practical purposes, thermalization stops at an energy $5kT_0$ below the mobility edge.

At elevated temperatures, thermalization also occurs by thermal emission out of the traps to the mobility edge, followed by retrapping. This is the familiar multiple trapping mechanism which governs the dispersive carrier drift mobility and which is analyzed in Section 3.2.1.

Fig. 8.5. Calculated thermalization rates in extended and localized states for the different mechanisms.

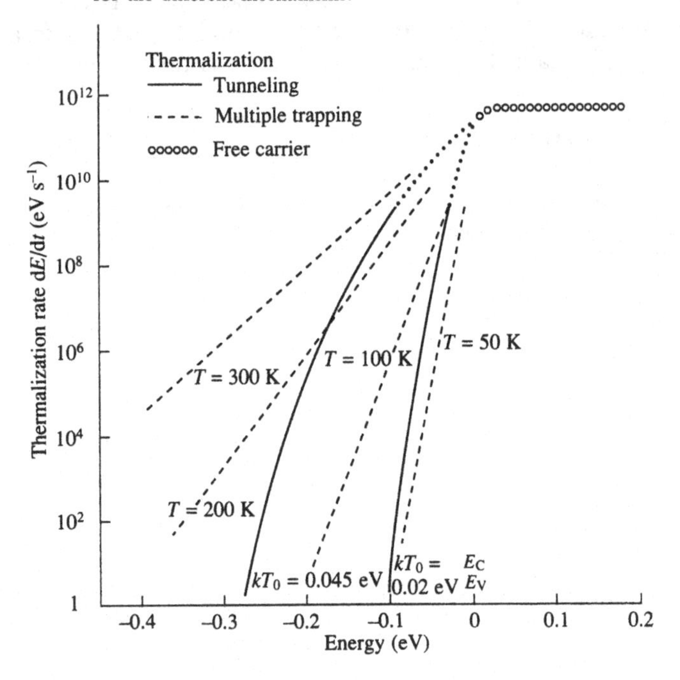

The time dependence of the average trap energy of the carriers is given by the demarcation energy

$$E_D = kT \ln(\omega_0 t) \quad (8.16)$$

The thermalization rate is temperature-dependent and given by

$$dE/dt = \omega_0 kT \exp(-E/kT) \quad (8.17)$$

This rate is compared with the tunneling thermalization rates in Fig. 8.5. The multiple trapping rate is independent of the band tail slope, unlike tunneling thermalization. Multiple trapping dominates above 200 K for holes and 50 K for electrons (see Fig. 8.5). Eq. (8.17) only applies in the dispersive regime when $T < T_0$, as the carriers equilibrate at higher temperature.

The maximum rate of the phonon-assisted transitions is 10^4–10^5 times greater than the fastest radiative rate and 10^{10} times greater than the average rate (see Fig. 8.14). Luminescence therefore occurs only after the majority of carriers have thermalized to a sufficiently low density of states that further phonon-assisted transitions are suppressed by the weak overlap to neighboring sites.

The recombination of an electron–hole pair, or the capture of a carrier into a deep trap, releases much more energy than can be taken up by a single phonon. Multiphonon recombination is represented by transition A in Fig. 8.3. The probability of the simultaneous emission of n phonons is (Stoneham 1977),

$$P_A = \omega_0 \exp(-\gamma n) = \omega_0 \exp[-\gamma(E_e - W)/\hbar\omega_0] \quad (8.18)$$

where γ is a constant in the range 1–2, depending on the strength of the electron–phonon interactions. The rate decreases rapidly as the transition energy increases. An alternative recombination mechanism when the phonon coupling is strong is indicated by transition B in Fig. 8.3. The carrier in the upper level is thermally excited to the crossing point of the configurational coordinate diagram. The carrier makes a transition to a highly vibrationally excited state of the lower energy level, from which it can return to the ground state by sequential emission of phonons. The transition probability is limited by the rate of thermal excitation to the crossing point,

$$P_B = \omega_0 \exp(-W_X/kT) \quad (8.19)$$

where

$$W_X = W(E^*/2W - 1)^2$$

The thermally activated transition probability over the barrier decreases at low temperatures, which allows the multiphonon mechanism to dominate.

The activation energy, W_X, decreases as the electron–phonon coupling gets stronger. Henry and Lang (1977) give a detailed calculation of the transition rates and apply the theory to the capture cross-sections of deep traps in crystals. The model predicts a recombination rate (or trapping cross-section) of the form shown in Fig. 8.6. Regions *A* and *B* correspond to the transition mechanisms marked *A* and *B* in Fig. 8.3. The transition rates have a large activation energy when the phonon coupling is small and vice versa. The temperature dependence is absent below $\hbar\omega_0/2k$, and at high temperature the rates all tend to ω_0, or alternatively to a capture cross-section of about 10^{-15} cm^{-2}.

8.1.4 *Geminate electron–hole pairs*

Optical absorption and recombination processes involve two or more particles and so may include correlation effects. Electron–hole pairs form excitons in a crystal, with the result that the absorption and emission spectra are not described by the one-particle density of states distributions. Although excitons can exist in an amorphous material (see Chapter 3), they are not detected in the optical spectra and the absorption is described by the convolution of the one-particle densities of valence and conduction band states. The correlation effects in

Fig. 8.6. The expected temperature dependence of the capture cross-section for multiphonon trapping at defects with different phonon coupling strength (Henry and Lang 1977).

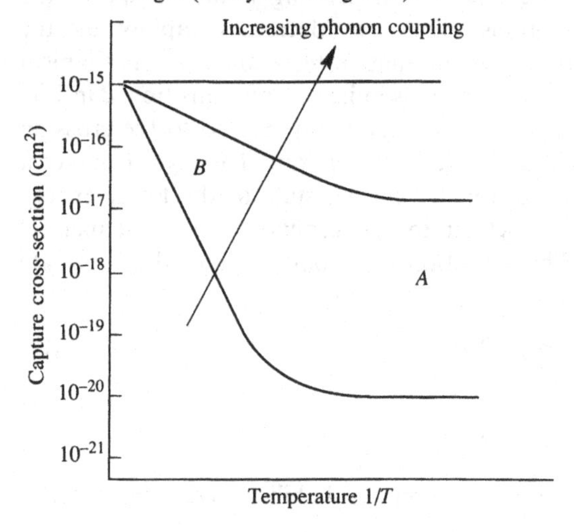

amorphous semiconductors are manifested in the form of geminate pairs, which result from the short free carrier scattering lengths and are not important in crystalline semiconductors.

The absorption of a photon creates an electron–hole pair whose wavefunctions initially overlap. After the absorption process, the electron and hole thermalize to the band edge and diffuse apart. The different thermalization mechanisms in extended and localized states are reflected in the diffusion properties. In the extended states, the thermalization time, t_{th}, required to emit the excess energy ΔE as n phonons is

$$t_{th} = n\omega_0^{-1} = \Delta E/\hbar\omega_0^2 \tag{8.20}$$

After this time, the carrier crosses the mobility edge and is trapped in localized states, so that further movement is much slower, although the distance between the sites is larger. During thermalization in extended states the carriers diffuse apart a distance,

$$L_T = (Dt_{th})^{\frac{1}{2}} = (D\Delta E/\hbar\omega_0^2)^{\frac{1}{2}} \tag{8.21}$$

where D is the diffusion coefficient. The thermalization distance is 70 Å when the energy ΔE is 0.5 eV, D is 0.5 cm^2 s^{-1} (corresponding to a free mobility of 20 cm^2 V^{-1} s^{-1}) and the phonon energy is 0.05 eV. The electron and hole therefore remain close to each other and are called a geminate pair.

Due to their close proximity, the electron–hole pair is bound by their mutual Coulomb interaction as shown in Fig. 8.7. When the potential is strong enough, the particles diffuse together, giving geminate recombination. Otherwise the electron and hole diffuse apart and any subsequent recombination is non-geminate. The Onsager (1938) model

Fig. 8.7. Onsager model of geminate recombination showing the Coulomb interaction of the photoexcited electron and hole.

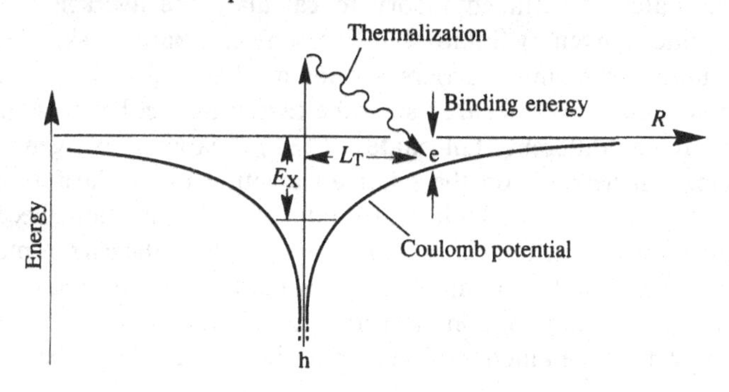

provides an estimate of the criterion for geminate recombination, which is when the Coulomb binding energy is larger than kT,

$$e^2/4\pi\varepsilon\varepsilon_0 L_T > kT \tag{8.22}$$

Combining this with Eq. (8.21) gives,

$$kT < (\omega_0 e/4\pi\varepsilon\varepsilon_0)(\hbar/D\Delta E)^{\frac{1}{2}} \tag{8.23}$$

Geminate recombination is favored at low temperature and small ΔE. A geminate pair does not give photoconductivity, because the electron and hole do not separate. This effect is observed in many chalcogenide glasses and organic polymers which have a higher photon energy threshold for photoconductivity than for optical absorption.

The Coulomb potential model in Fig. 8.7 does not take into account quantum effects. The binding potential does not increase indefinitely as the electron and hole diffuse together, but instead an exciton is formed of binding energy E_X. The exciton can either dissociate thermally with rate $\omega_0 \exp(-E_X/kT)$, or else can recombine with rate W_{em}. An additional condition for geminate recombination is therefore,

$$E_X > kT \ln(\omega_0/W_{em}) \tag{8.24}$$

and both inequalities (8.23) and (8.24) must be satisfied for geminate recombination to occur. According to expression (8.24), geminate recombination is prevented if either the exciton binding energy is small or the recombination rate is low. Both factors are strongly influenced by the magnitude of the electron–phonon coupling. Strong coupling increases both the exciton binding energy and the multiphonon recombination rate.

The Onsager model is incomplete because it does not consider tunneling between the localized states in either the thermalization or the recombination processes. Both types of transition occur over distances of about 50 Å, which is similar to the diffusion distance in extended states. A detailed theory to calculate the average recombination rate is given by Shklovskii, Fritzsche and Baranovskii (1989). In the tunneling regime, carriers do not need to diffuse together to recombine, but at any localized state the carrier may either recombine or hop to a different tail state. The probability of geminate recombination depends on the relative transition rates. Carriers near the mobility edge have a high probability of thermalization, because there are many deeper tail states. However, each time the carrier moves down the band tail, the number of available sites decreases and eventually the recombination transition dominates. Shklovskii *et al.* (1989) show that the kinetics of recombination are largely independent of the detailed shape of the band tail and that the most probable recombination lifetime, τ_m, only depends on the prefactors for radiative and non-radiative tunneling (see Eqs. (8.5) and (8.14)),

$$\tau_m = \omega_0 P_0^{-2} \tag{8.25}$$

This simple expression accounts well for the measurements of luminescence decay described in Section 8.3.3. This model only applies at low temperature because the multiple trapping mechanism is not considered.

Geminate recombination is suppressed when the density of excited electron–hole pairs is large. For example, a pair density of 10^{17} cm^{-3} results in an average separation of the carriers of 50 Å. The geminate pairs overlap when this distance is less than the thermalization length, L_T, and non-geminate recombination between carriers from different pairs occurs. Geminate recombination is therefore most likely at low temperatures and weak excitation intensities and in a-Si:H it is only observed under these conditions.

The recombination kinetics depends on whether the recombination is geminate or non-geminate. Simple rate equations describing the two situations are,

$$\left.\begin{array}{ll}\text{geminate:} & dN_{gp}/dt = G - \alpha_1 N_{gp} \\ \text{non-geminate:} & dN_{ng}/dt = G - \alpha_2 N_{ng}^2\end{array}\right\} \tag{8.26}$$

In the first case, pairs are isolated from each other and the recombination is monomolecular, with a rate which is independent of illumination intensity. The non-geminate electrons and holes act as independent particles and the recombination rate is proportional to the product of the two densities (here assumed equal). The observable difference is a recombination lifetime which is independent of the excitation intensity for geminate recombination, but which decreases with increasing illumination intensity for non-geminate recombination. The simple rate equations also predict a different form of the time dependence, but a more realistic model must also include a distribution of recombination rates due to the tunneling recombination.

A spin-dependent recombination rate is another consequence of the electron–hole correlation. The conservation of spin selection rule is preserved in amorphous materials. The final state of the recombination process has zero spin and both radiative and phonon-assisted non-radiative transitions occur without change in spin, so that recombination can only proceed from an initial state of zero spin. A weakly interacting electron–hole pair forms four possible spin states, one singlet and one triplet. Of the four states, only the singlet and one triplet state have allowed optical transitions. Several possible spin-dependent effects can be anticipated in the recombination. For example, geminate pairs are created from an absorption process which selects the spin-allowed transitions. Thus, provided there is no spin relaxation during thermalization, all geminate pairs can recombine, whereas non-geminate pairs, which have random spin alignments, have some forbidden configurations. The spin alignment can be deliberately perturbed by microwave-induced transitions of the Zeeman split levels when a magnetic field is applied.

8.2 Carrier thermalization

8.2.1 *Thermalization in extended states*

Carrier thermalization in the extended states is expected to take of the order of 10^{-13} s and the only experiment with this time resolution is photoinduced absorption using the pump-and-probe method (Tauc 1982). A mode locked laser provides the very short duration light pulses, with minimum pulse times that are constantly being reduced, but are presently in the range 10^{-14} s. The beam is split into two and one pulse is delayed with respect to the other, by using different path lengths. The two beams coincide on the sample. The first pulse excites carriers out of the ground state and the second pulse records the difference in the absorption due to these excess carriers as they thermalize and recombine back to the ground state. The transmitted or reflected intensity of the probe beam contains the information about the transient response of the carriers. A nice feature of the experiment is that the measurement does not involve fast timing electronics.

The induced absorption is

$$\Delta\alpha(t) = \sum_i n_i(t)\Delta\sigma_i(t) \tag{8.27}$$

where $n_i(t)$ is the time-dependent density of excited carriers of one type and $\Delta\sigma_i$ is the difference in absorption cross-section between the ground and excited state of the carriers. The total induced absorption is the sum over contributions from electrons and holes in different excited states. A problem with the interpretation of the experiment is that the time dependence of the induced absorption can be due to the recombination of the carriers or to the thermalization of carriers into new states which have different absorption cross-sections. There is no unambiguous method of separating the two contributions.

The result of one of the first experiments to measure the thermalization times is shown in Fig. 8.8 (Vardeny and Tauc 1981). The pump and probe both have energies of 2 eV, which is above the band gap and excites extended state electrons and holes. The data illustrate some of the complexities of the experiment. There is a constant induced absorption at long times, which implies that recombination is slow. Near time $t = 0$, there is some structure that depends on the polarization of the pump and probe beams and is due to coherence effects. When this term is subtracted, the thermalization time is deduced to be 1–2 ps, corresponding to an energy dissipation rate of 0.5–1.0 eV ps^{-1}. The thermalization is difficult to detect, as it corresponds to a small change in the absorption because the cross-section is weakly energy-dependent and is also masked by the coherence artifact. In undoped a-Si:H there is no particular reason to expect a strongly energy-dependent cross-section.

More recent measurements vary the energy of the pump and probe independently (Fauchet *et al.* 1986). Some data are shown in Fig. 8.9 for a 2.0 eV pump and different probe energies. The transient response of the induced absorption is almost constant when the probe energy is above 2 eV as found by Vardeny and Tauc, but there is a fast decay of the absorption at lower probe energies. The decay is interpreted in terms of the bleaching of the absorption transitions which occurs when a carrier occupies a conduction band state and so inhibits the excitation of electrons from the valence band. The transient absorption is the

Fig. 8.8. Induced absorption of a-Si:H indicating a thermalization time of about 1 ps. Curves (*a*) and (*b*) are for different polarizations of pump and probe (Vardeny and Tauc 1981).

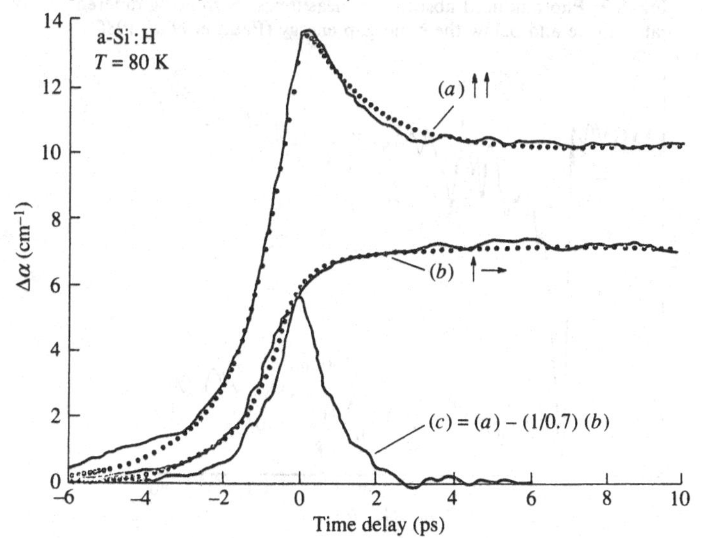

combined effect of an approximately constant induced absorption of the excited carrier, reduced by a time-dependent bleaching as the carriers thermalize to the band edges. The thermalization only bleaches the lower energy transitions but not those of higher energy. The thermalization time depends on the induced carrier density and ranges from 1–10 ps.

The transient induced absorption is different in doped or compensated a-Si:H. The compensated material has fewer deep defects to cause rapid recombination so that the time dependence arises only from thermalization. The absorption transients are shown in Fig. 8.10 for a pump and probe energy of 2 eV (Thomsen *et al.* 1986). In contrast to the almost constant absorption of the undoped material (Fig. 8.8), there is now a rapid change from absorption to induced bleaching after 1–2 ps. There has been considerable debate about the origin of this effect, but the present consensus seems to be that the bleaching is caused by trapping in donor or acceptor states which have a lower absorption cross-section than the intrinsic band tail states. Thus the time constant for bleaching represents the thermalization time into the dopant states which are estimated to be at least 50 meV beyond the mobility edges.

The evidence of these experiments is therefore that thermalization in the extended states is faster than 10^{-12} s and is barely resolvable by present experiments. The thermalization which is observed in a time scale of 1–10 ps corresponds to trapping in shallow localized band tail

Fig. 8.9. Photoinduced absorption measurements showing different decay rates above and below the band gap energy (Fauchet *et al.* 1986).

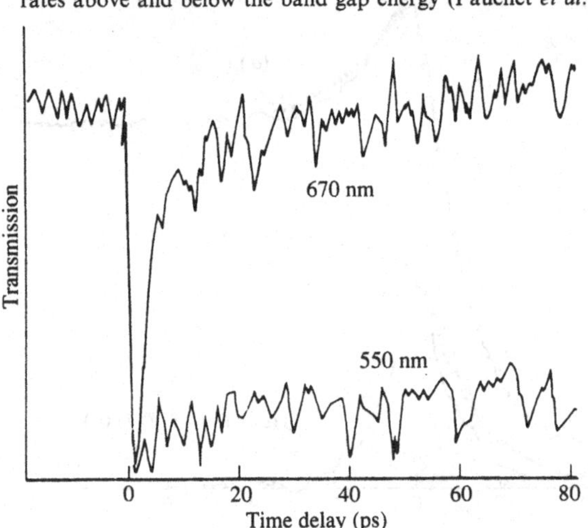

states, or into deep gap states when these have a very high concentration. The thermalization times generally agree with the estimates in Fig. 8.5, which predict that extended state thermalization over 0.5 eV takes only 1 ps, but that transitions in the localized tail states are much slower.

8.2.2 *Thermalization in localized states*

Low temperature tunneling thermalization in the band tail is observed by time resolved luminescence, which is a transition between band tail electrons and holes, as described in Section 8.3.2. Fig. 8.11 (*a*) shows the change in energy of the luminescence peak with increasing time after the excitation pulse, measured at 12 K. The decrease of the energy at times up to 10^{-5} s is caused by carrier thermalization – the changes at

Fig. 8.10. Induced absorption of compensated a-Si:H showing absorption at short times and induced transparency at longer times (Thomsen *et al.* 1986).

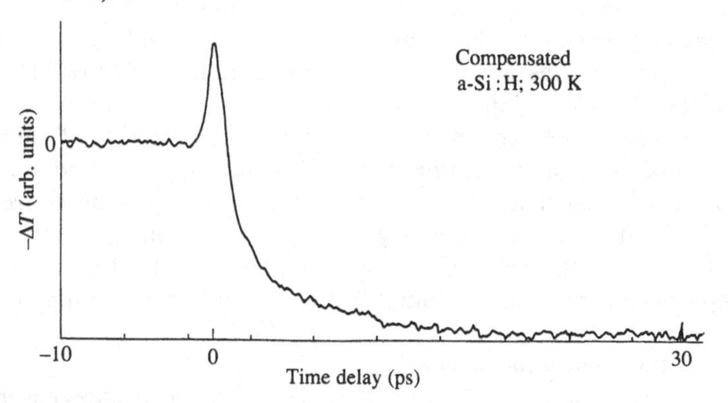

Fig. 8.11. (*a*) Time dependence of the luminescence peak energy, which reflects the thermalization of carriers at times less than 10^{-4} s. (*b*) Temperature dependence of the luminescence peak and band gap energy showing thermalization by multiple trapping (Tsang and Street 1979).

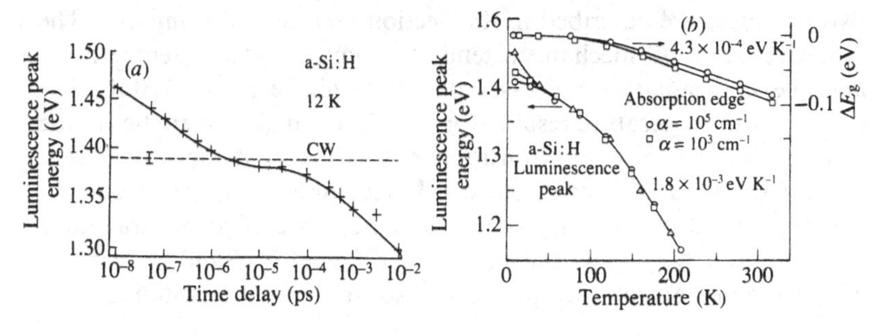

longer times have another origin. The energy change of 50–100 meV between 10^{-8} s and 10^{-5} s is consistent with the predictions in Fig. 8.5. The luminescence experiments do not distinguish between electron and hole thermalization, and the induced absorption data is similarly unclear. Nevertheless the models of low temperature thermalization are well supported by the data.

Thermalization by multiple trapping dominates over direct tunneling above 50–100 K. The luminescence energy reflects the carrier distribution and so is expected to have a temperature-dependent shift of the peak energy, given from Eq. (8.16) by

$$E_P(T) = E_P(0) - kT \ln(\omega_0 \tau_R) \tag{8.28}$$

where E_P is the energy of the luminescence peak and τ_R is the recombination lifetime. There is a large temperature shift of E_P, as shown in Fig. 8.11(*b*), and the temperature coefficient of about $20k$ is much larger than that of the band gap energy. The recombination lifetime is about 10^{-3} s (see Section 8.3.3), although it decreases with increasing temperature. Inserting this time constant and $\omega_0 = 10^{12}$ s^{-1} into Eq. (8.28) gives a temperature dependence for $E_P(T)$ of $21k$, in excellent agreement with the data.

In summary, the experimental data confirm the models of carrier thermalization. Thermalization in extended states is very rapid and is completed in less than 10^{-12} s. Thermalization by tunneling between localized states becomes increasingly slow as the carriers move into the band tail and at high temperatures is overtaken by the multiple trapping mechanism of sequential thermal excitation and trapping.

8.3 Band tail recombination

The rapid thermalization of carriers in extended states ensures that virtually all of the recombination occurs after the carriers are trapped into the band tail states. The two dominant recombination mechanisms in a-Si:H are radiative transitions between band tail states and non-radiative transitions from the band edge to defect states. These two processes are described in this section and the following one. The radiative band tail mechanism tends to dominate at low temperature and the non-radiative processes dominate above about 100 K. The change with temperature results from the different characteristics of the transitions. The radiative transition rate is low, but there is a large density of band tail states at which recombination can occur. In contrast, the defect density is low but there is a high non-radiative transition rate for a band tail carrier near the defect. Band tail carriers are immobile at low temperatures, so that the recombination is dominated by the higher probability that a carrier thermalizes into a band tail state rather than to a defect. Carriers diffuse rapidly from site to site at high temperatures and the faster transition rate at the defect dominates the recombination.

8.3.1 *Photoluminescence*

Photoluminescence is the radiation emitted by the recombination process and as such is a direct measure of the radiative transition. Information about non-radiative recombination can often be inferred from the luminescence intensity, which is reduced by the competing processes (Street 1981a). The most useful feature of the luminescence experiment is the ability to measure the emission spectrum to obtain information about the energy levels of the recombination centers. The transition rates are found by measuring the transient response of the luminescence intensity using a pulsed excitation source. Time resolution to about 10^{-8} s is relatively easy to obtain and is about the maximum radiative recombination rate. The actual recombination times of a-Si:H extend over a wide range, from 10^{-8} s up to at least 10^{-2} s.

The absolute luminescence efficiency is difficult to measure precisely and can typically be obtained only within a factor 2. The relative efficiency in two samples can be measured much more accurately, provided that care is taken with the sample configuration. There are strong optical interference effects in the luminescence of thin a-Si:H

Fig. 8.12. (*a*) Examples of the luminescence spectra of undoped, doped and compensated a-Si:H. (*b*) Comparison of the absorption and luminescence spectra of undoped a-Si:H.

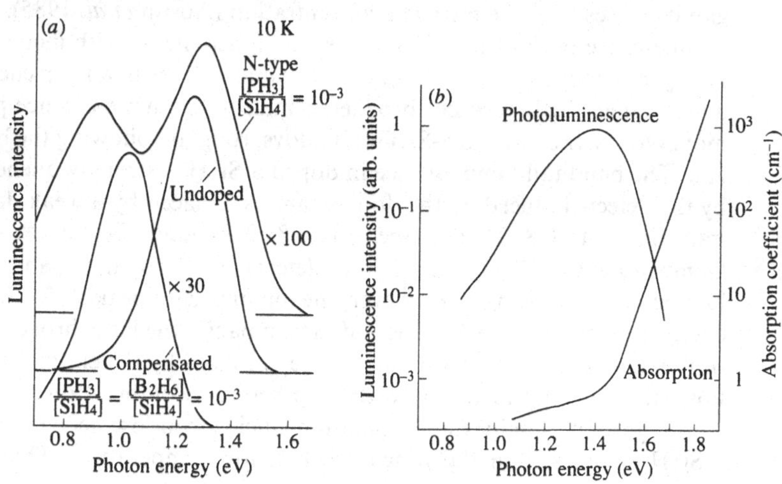

films on smooth substrates, which modulate the spectrum. Much of the luminescence emitted isotropically in the films is totally internally reflected at the surface. Samples are usually deposited on ground glass substrates to avoid both effects.

Luminescence measurements are usually made at low temperature because the competing non-radiative transitions are enhanced at elevated temperatures. At room temperature the luminescence intensity is low and almost undetectable. The intensity is almost constant below about 50 K, so that a low temperature of 10–20 K is adequate for the measurements.

8.3.2 *Luminescence spectra*

Examples of the low temperature luminescence spectra are shown in Fig. 8.12. The luminescence intensity is highest in samples with the lowest defect density and so we concentrate on this material. The role of the defects is discussed in Section 8.4. The luminescence spectrum is featureless and broad, with a peak at 1.3–1.4 eV and a half width of 0.25–0.3 eV. It is generally accepted that the transition is between conduction and valence band tail states, with three main reasons for the assignment. First, the energy is in the correct range for the band tails, as the spectrum lies at the foot of the Urbach tail (Fig. 8.12(*b*)). Second, the luminescence intensity is highest when the defect density is lowest, so that the luminescence cannot be a transition to a defect. Third, the long recombination decay time indicates that the carriers are in localized rather than extended states (see Section 8.3.3).

Equivalent luminescence bands are seen in the alloys of a-Si:H. Fig. 8.13 shows examples of the spectra of a-Si:N:H in which the band gap increases with the nitrogen concentration (Austin *et al.* 1985). The luminescence peak of the alloy moves to higher energy with increasing nitrogen concentration, although not as rapidly as the band gap energy, and the line width becomes broader. Similarly, the luminescence peak moves to lower energy in a-Si:Ge:H alloys, roughly following the band gap. The band tail luminescence in doped a-Si:H is strongly quenched by the defects induced by the doping, and is replaced by a weak defect transition at 0.8–0.9 eV (see Fig. 8.12(*a*) and Section 8.4.1). Compensated a-Si:H has fewer defects and so the band tail luminescence is stronger. However, the luminescence peak shifts to low energy as seen in Fig. 8.12. The shift accompanies the large broadening of the band edge (see Section 5.2.4) and is caused by the electrons and holes thermalizing further into the gap before recombination.

The absolute luminescence quantum efficiency in low defect density a-Si:H samples is in the range 0.3–1 at low temperature. Thus the luminescence measures the dominant recombination mechanism. An accurate value of the efficiency is hard to measure from the light emission, because there are many corrections for the collection optics. Jackson and Nemanich (1983) estimate the efficiency as about 0.3 using photothermal deflection spectroscopy, which distinguishes radiative and non-radiative processes by the heat generated in the sample.

The origin of the luminescence band width is controversial. The two possible broadening mechanisms are disorder broadening, due to the energy distribution of band tail states, and phonon broadening associated with the Stokes shift. Clearly there must be a component of disorder broadening, but the Stokes shift may also be large. When both mechanisms are present the total linewidth, ΔE, is

$$\Delta E^2 = \Delta E_{\text{disorder}}{}^2 + \Delta E_{\text{Stokes}}{}^2 \tag{8.29}$$

The identification of a Stokes shift is based on a comparison of the absorption and luminescence line shapes (Street 1978) (also see Section 8.1.2). The comparison is difficult because there is a continuum of electronic states and there is no peak in the absorption spectrum which can be associated directly with the luminescence. Furthermore the

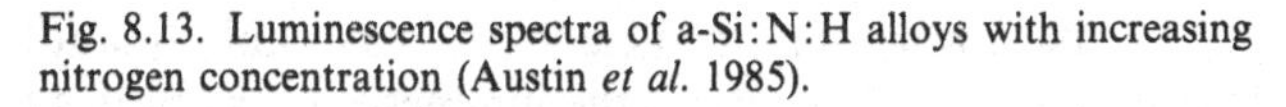

Fig. 8.13. Luminescence spectra of a-Si:N:H alloys with increasing nitrogen concentration (Austin *et al.* 1985).

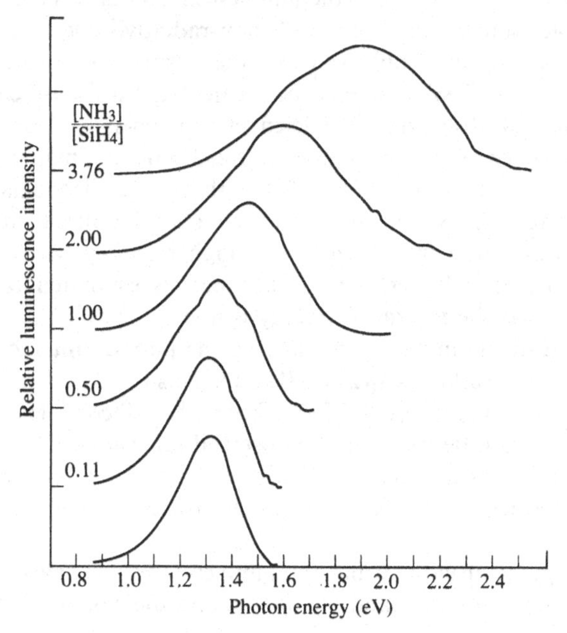

thermalization of the carriers is not a part of the Stokes shift, but can easily be mistaken for the relaxation energy. However, any emission process has an inverse absorption process and the two are related by the detailed balance expression,

$$\frac{N}{\tau} = \left(\frac{8\pi n^2}{\lambda^2}\right)\frac{g_1}{g_2}\int \alpha\, d\nu \tag{8.30}$$

N is the density of states, ν the frequency, n the refractive index, g_1 and g_2 the degeneracies of the ground and excited states, and λ is the wavelength. The absorption coefficient increases with the number of states and with the transition probability, τ^{-1}.

The total luminescence intensity is the product of the number of excited electron–hole pairs and the recombination rate, N_{occ}/τ. If η_L is the luminescence quantum efficiency and G is the excitation intensity, then by definition, $\eta_L G$ is the luminescence intensity. N_{occ} is the density of states which are occupied by electron–hole pairs. Since N_{occ} cannot be greater than the available density of states at the emission energy, then it follows from Eq. (8.30) that

$$\eta_L G = N_{occ}/\tau \leqslant \left(\frac{8\pi n^2}{\lambda^2}\right)\frac{g_1}{g_2}\int \alpha\, d\nu \tag{8.31}$$

which relates the absorption coefficient to measurable quantities. As the excitation intensity G increases, the luminescence saturates because all the available states are filled and non-radiative mechanisms dominate. The saturation is observed and the absorption coefficient corresponding to the 1.4 eV luminescence is calculated to be 10^2 cm^{-1}. The actual band tail absorption coefficient at this energy, measured from the luminescence excitation spectrum, is 2–3 orders of magnitude lower (Street 1978), seemingly in conflict with the detailed balance relation. The discrepancy is removed by a Stokes shift, because then the corresponding absorption transition is at a higher energy where the absorption coefficient is larger. The Stokes shift is estimated to be about 0.4 eV, so that the relaxation energy is about 0.2 eV.

The half width of the luminescence line by the phonon interaction mechanism, from Eq. (8.11), is $2[(2\ln 2)W\hbar\omega_0]^{\frac{1}{2}}$. This is 0.25 eV for the maximum phonon energy of 0.05 eV from the silicon network vibrations, which is a little less than the observed line width. Thus the phonon model indicates that the luminescence spectrum is dominated by the phonon interaction and that the disorder broadening contributes less.

The opposite point of view is that the luminescence line shape and energy are determined by the band tail shapes and the convolution of the one-electron density of states, without significant relaxation effects (Dunstan and Boulitrop 1981, Searle and Jackson 1989). The luminescence peak energy is given by the energy separation of electrons and holes after thermalization

$$E_{PL} = E_C - E_V - E_{te} - E_{th} \tag{8.32}$$

E_{th} and E_{th} are the energies to which the electrons and holes thermalize from the mobility edges. The data in Fig. 8.5 give estimates that E_{te} is about 0.10 eV and E_{th} is 0.25 eV, so that a mobility gap of 1.75 eV is needed to account for the 1.4 eV luminescence peak energy. The spectrum is described by

$$A(E_L) = \text{const.}\, f(E_L) \int N_V(E)\, N_C(E+E_L)\, \mathrm{d}E \tag{8.33}$$

where $f(E_L)$ is a function describing the effects of thermalization on the shape of the spectrum. There is no thermalization on the low energy side of the spectrum, so that $f(E_L) = 1$, and the spectrum follows the joint density of states. On the high energy side of the spectrum, $f(E_L) < 1$ and decreases monotonically to zero where the thermalization to lower energy states is complete. Dunstan and Boulitrop (1984) have analyzed the shape function in more detail and give an explicit form for $f(E)$. The joint density of states is dominated by the valence band tail which has an exponential slope of $kT_V \approx 500$ K. A difficulty with the interpretation is that the low energy side of the luminescence has a broader tail which is hard to explain in the absence of phonon coupling (see Fig. 8.12(*b*)).

The relative contributions of disorder and phonon broadening in the luminescence spectrum remain to be resolved. The disorder broadening evidently dominates in alloy materials in which the band tails are much broader. For example, the luminescence line width increases with nitrogen content in a-Si:H:N alloys, with a width which is predicted by the measured band tail slopes (Searle and Jackson 1989).

8.3.3 *Recombination kinetics*

One of the characteristic features of the luminescence in a-Si:H is the broad distribution of recombination times. Fig. 8.14 shows the luminescence decay extending from 10^{-8} s to 10^{-2} s (Tsang and Street 1979). The data are inverted in the lower part of the figure to give the distribution of lifetimes, which has its peak at 10^{-3}–10^{-4} s at low temperature and is 2–3 orders of magnitude wide. The shape of the distribution is sensitive to the excitation intensity for reasons discussed shortly and the time constants are even longer at very low intensity.

Radiative tunneling is the only recombination mechanism which can explain the decay data satisfactorily. The electron and hole are localized at different sites separated by a distance R, and the recombination time is, from Eq. (8.5)

$$\tau_R = \tau_0 \exp(2R/R_0) \tag{8.34}$$

where R_0 is the localization length and τ_0 is about 10^{-8} s. The distribution of lifetimes arises from the random variations of electron–hole pair separations. A radiative lifetime of 10^{-3} s requires $R \approx 6R_0$, which is 60 Å for a localization length of 10 Å. This pair separation is close to the 70 Å predicted for geminate pairs after thermalization.

Light induced ESR measures the density of band tail electrons and holes, and provides a different method of measuring the recombination

Fig. 8.14. The decay of the luminescence of low defect density a-Si:H following a short excitation pulse. The lower curves are the distribution of lifetimes calculated from the data (Tsang and Street 1979).

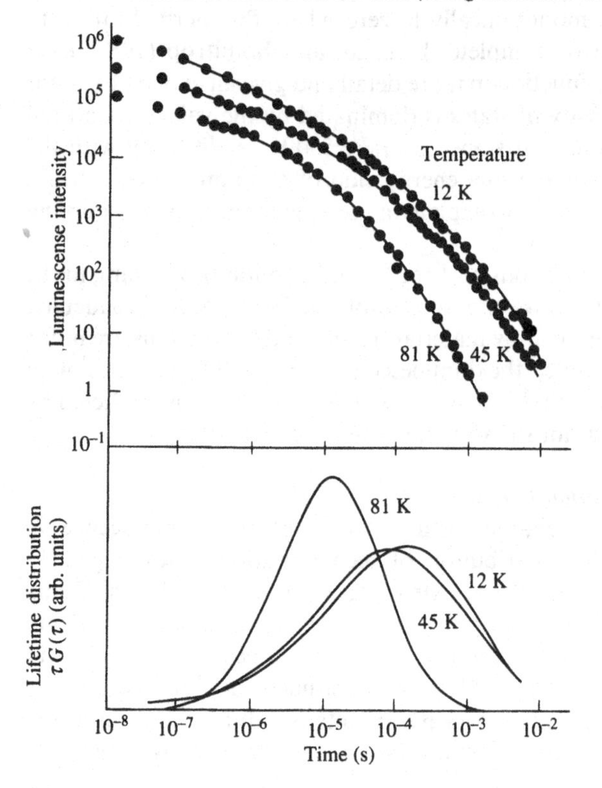

decay times (Street and Biegelsen 1982). Examples are shown in Fig. 8.15 for illumination conditions typical of the luminescence. The decay continues out to 10^3 s and even then a substantial density of unrecombined carriers remain. At first sight this result seems to conflict with the luminescence decay which does not extend beyond 10^{-2} s. The difference is a consequence of the range of electron–hole pair separations.

An electron–hole pair is created with a separation R which has a broad distribution centered at the average value, R_P, of the thermalization distance. One model for the distribution function, $f(R)$, of pair separations is illustrated in Fig. 8.16 (Biegelsen, Street and Jackson 1983) and assumes random pairing. An alternative distribution proposed by Shklovskii *et al.* (1989) based on the analysis of tunneling, has a similar general shape. The luminescence intensity is proportional to the rate of radiative recombination events and is therefore dominated by the close electron–hole pairs for which the rate is large. LESR measures the density of carriers remaining in the band tails, and so is most sensitive to the long lived states, which contribute little to the luminescence. From the results of the luminescence decay, $R_P \approx 6R_0$,

Fig. 8.15. The decay of the LESR signal of band tail electrons and holes measured at 30 K. The inset shows the intensity dependence of the spin signal (Street and Biegelsen 1982).

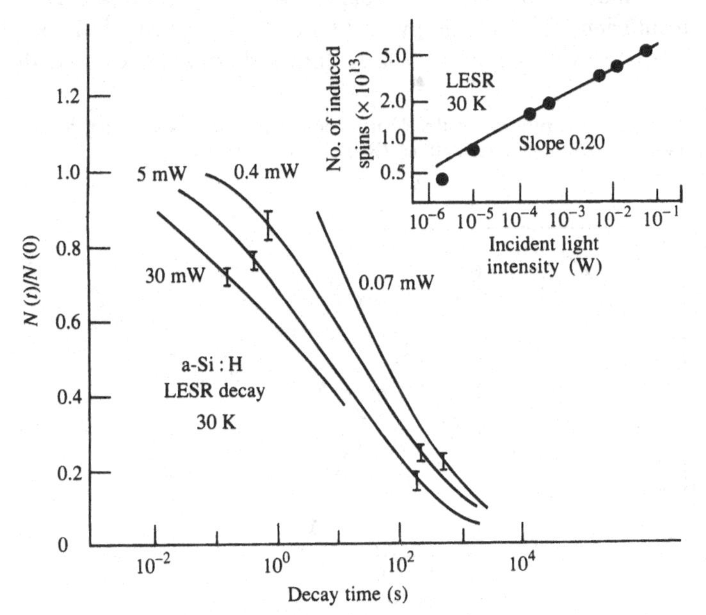

and the decay time at the peak of the distribution is 10^{-3} s (see Fig. 8.14). A distance, R_C, is defined from Eq. (8.34) by,

$$R_C = (R_0/2)\ln(t/\tau_0) \tag{8.35}$$

At time t, all pairs for which $R < R_C$ have recombined, while the more distant pairs remain. From the shape of the distribution in Fig. 8.16, about 20% of the pairs remain after 1 s, reducing to about 5% after 1000 s, and this is consistent with the LESR data. The measurable range of the recombination over which the radiative tunneling mechanism applies is therefore more than 10 orders of magnitude, from 10^{-8} to 10^3 s.

The tunneling mechanism only applies at low temperatures when the electrons and hole are immobile. The luminescence and LESR decay more rapidly above about 50 K. Hong, Noolandi and Street (1981) solved the complicated time-dependent diffusion equation for geminate recombination when there is a distribution of thermalization distances and temperature-dependent multiple hopping of carriers. The asymptotic solution for the luminescence intensity is

$$I(t) \approx t^{-\frac{3}{2}} \tag{8.36}$$

which agrees quite well with the measured decay data above 100 K.

8.3.4 *Geminate and non-geminate recombination*

Geminate recombination occurs at low temperatures because there is insufficient thermal energy to dissociate an electron–hole pair. The pair separates by about 50–100 Å during thermalization and the

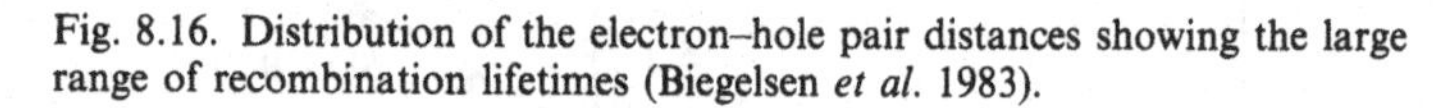

Fig. 8.16. Distribution of the electron–hole pair distances showing the large range of recombination lifetimes (Biegelsen *et al.* 1983).

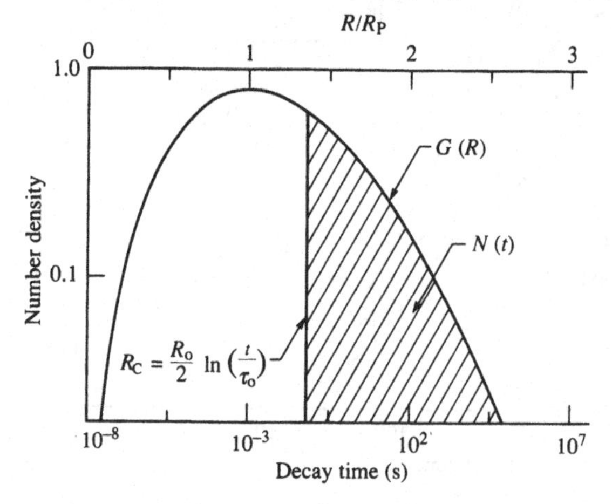

carriers are trapped in band tail states until a recombination event takes place. Recombination is geminate when the density of pairs is sufficiently low to ensure that there is negligible overlap between neighboring pairs. When the pairs overlap, an electron from one pair can recombine with a hole from another and recombination is no longer geminate. These situations are illustrated in Fig. 8.17. The transition from geminate to non-geminate recombination is seen in the luminescence data as the onset of an intensity-dependent recombination time, which signals non-geminate kinetics. The transition takes place at an electron–hole pair density of 10^{18} cm^{-3}, which corresponds to an average pair separation of 50 Å. This is consistent with the predicted range of thermalization distances and also agrees with the estimate obtained from the luminescence lifetime distribution.

Surprisingly, the transition from geminate to non-geminate recombination is not observed in LESR. Fig. 8.15 shows that the LESR spin density increases very slowly with excitation intensity, over the full range of the measurements. The sublinear behavior is characteristic of non-geminate recombination, because the geminate process has a linear intensity dependence.

The reconciliation of these two apparently conflicting results is quite interesting and is connected to the broad distribution of recombination lifetimes. Those electron–hole pairs which are created with small separations are more likely to result in geminate recombination than the more distant pairs. The close pairs are also more likely to contribute to the luminescence and the distant pairs to LESR. Thus the two experiments are selectively measuring different parts of the distribution. The density of geminate pairs of separation, R, is given by,

$$N_P(R) = Gf(R)\tau_R(R) \tag{8.37}$$

where G is the generation rate, $f(R)$ is the distribution function shown in Fig. 8.16, and the recombination time τ_R is given by Eq. (8.34). The rate of recombination of the pairs, assuming geminate recombination, is

$$S_P(R) = N_P(R)/\tau_R(R) = Gf(R) \tag{8.38}$$

The LESR experiment measures N_P and the luminescence S_P. The ratio of these quantities is τ_R, which varies by many orders of magnitude for the different pair separations. Hence the pairs with the longest lifetimes contribute most to the LESR but little to the luminescence, and vice versa.

The transition from geminate to non-geminate recombination occurs when the pairs overlap, which is at a separation, R_g, given approximately by,

$$R_g = [N_P(R_g)]^{-\frac{1}{3}} = [Gf(R_g)\tau_R(R_g)]^{-\frac{1}{3}} \tag{8.39}$$

Both R and $f(R)$ are slowly varying functions of R compared to $\tau_R(R)$. Therefore

$$G\tau_R(R_g) \approx \text{const.} \tag{8.40}$$

is the approximate solution of Eq. (8.39). From Eq. (8.34),

$$R_g \sim (R_0/2)\ln(\text{const.}/\tau_0 G) \tag{8.41}$$

Fig. 8.17 shows schematically that the transition occurs at a smaller pair separation as the excitation intensity increases. The LESR is dominated by the large separation pairs which are always non-geminate, while the luminescence is dominated by the close pairs which are mostly geminate. Geminate recombination is therefore seen in luminescence, but not in LESR. Furthermore, the excitation energy at which the transition occurs in the luminescence depends on the decay time being measured.

8.3.5 *Thermal quenching of luminescence*

The quantum efficiency of the band tail luminescence is largest at low temperature, low excitation intensity, and in samples of low defect density. Other conditions cause competing non-radiative processes which quench the luminescence intensity. Direct recombination to defect states in samples of high defect density is discussed in Section 8.4.1. The other main non-radiative mechanism is thermal

Fig. 8.17. Illustration of the conditions for geminate and non-geminate recombination. The lines connect electrons and holes created by the same photon absorption. Wavy lines indicate recombination between nearest neighbor pairs.

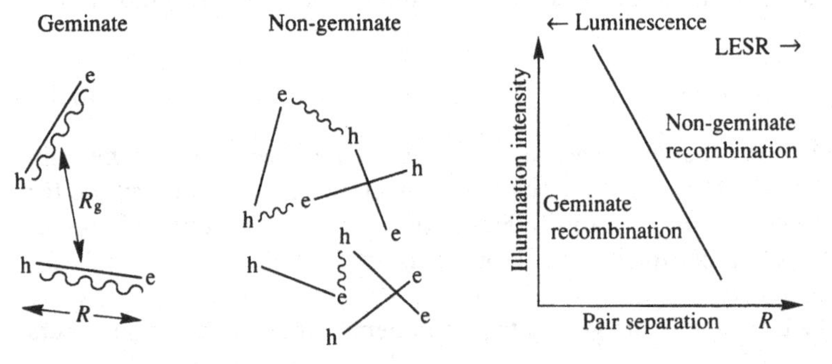

quenching; at room temperature, the luminescence efficiency is reduced by about four orders of magnitude and is barely detectable. The weaker non-radiative mechanisms of surface and Auger recombination are described in Section 8.3.6.

The competition between the radiative and non-radiative rates, P_r and P_{nr}, defines the luminescence quantum efficiency,

$$y_L = P_r/(P_r + P_{nr}) \tag{8.42}$$

and the recombination lifetime,

$$\tau_r = (P_r + P_{nr})^{-1} \tag{8.43}$$

Non-radiative quenching therefore usually reduces both the intensity and the decay time. These relations apply only when the different recombination processes compete directly. The actual processes are usually more complicated, with some competing and non-competing paths and a distribution of rates.

The increased mobility of the carriers is the primary reason for thermal quenching. The band tail luminescence has a weak transition probability, with an average radiative rate of about 10^3 s^{-1}, compared with the 10^{13} s^{-1} rate for a phonon-assisted recombination process. The large difference in rates means that a few non-radiative centers can completely overwhelm the radiative recombination, provided that the carriers are sufficiently mobile to reach them. Thus, thermal quenching is anticipated at temperatures above which there is significant carrier transport. Accordingly a correlation between carrier mobility and thermal quenching is expected. Electrical conduction is observed above about 100 K in a-Si:H, and thermal quenching begins at around this temperature. In alloy materials, the band tails are broader and both the mobilities and the thermal quenching are suppressed.

The conventional theory of thermal quenching by excitation of a carrier out of a shallow trap predicts a thermally activated process. No single activation energy is observed in a-Si:H, but it is found that the luminescence efficiency, y_L, follows the relation (Collins, Paesler and Paul 1980),

$$(1/y_L - 1)^{-1} = y_0 \exp(-T/T_L) \tag{8.44}$$

Fig 8.18 shows the thermal quenching of a-$Si_{1-x}C_x$:H alloys for different x (Liedke *et al.* 1989). The parameter T_L is about 25 K in a-Si:H and increases with the carbon concentration. It should not be too surprising that the temperature dependence originates from the exponential distribution of band tail states, or that the derivation of Eq. (8.44) follows the multiple trapping approach. Thermal quenching occurs when the rate of thermal excitation to the band edge exceeds the recombination rate, which we assume for simplicity to have a single average value τ_R. Thermal quenching therefore occurs for all carriers with binding energy less than E_D, where,

$$E_D(T) = kT \ln(\omega_0 \tau_R) \tag{8.45}$$

E_D is a demarcation energy, similar to that defined in the analysis of dispersive transport (see Section 3.2.1). It is assumed that all carriers which are thermally excited recombine non-radiatively, but the same result is obtained if some fraction are subsequently retrapped and recombine radiatively. The luminescence efficiency is given by the fraction of carriers deeper than E_D. An exponential band tail density of states proportional to $\exp(E/kT_v)$ results in a quantum efficiency of

$$y_L = y_0 \exp(-E_D/kT_v) = y_0 \exp\left[-\left(\frac{T}{T_v}\right)\ln(\omega_0 \tau_R)\right] \tag{8.46}$$

Fig. 8.18. The temperature dependence of the luminescence intensity for a-$Si_{1-x}C_x$:H alloys at different compositions x (Liedke *et al.* 1989).

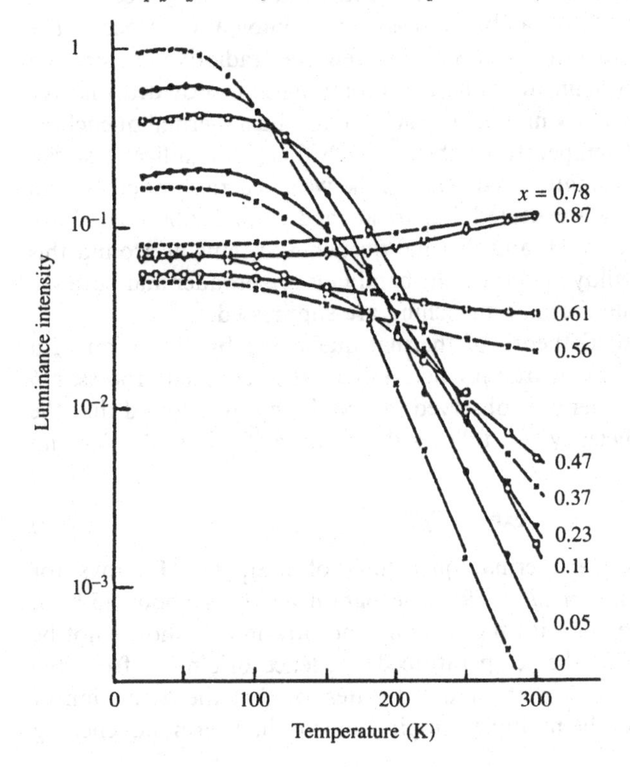

which is of the form in Eq. (8.44) when $y_L \ll 1$ with,

$$T_L = T_v/\ln(\omega_0 \tau_R) \tag{8.47}$$

The average τ_R of 10^{-3} s gives $T_L \approx T_0/23$, so that the measured value of 20–25 K is of the correct magnitude for the band tail slope in a-Si:H. The larger T_L in the alloys is because they have broader band tails. The derivation of Eq. (8.46) makes several approximations, of which the most severe is the neglect of the distribution of lifetimes, so that one should not expect an accurate fit. Also it is tacitly assumed that only one type of carrier is thermally excited. The non-radiative process obviously involves both carriers. The question of what is the rate-limiting step in the process is complicated and, in fact, the thermal quenching of the luminescence depends on the defect density and on the excitation intensity.

There is little thermal quenching below 50 K. According to the model just described, the luminescence efficiency is unity when the demarcation energy is shallower than the energy to which carriers thermalize, E_{th}, because there are no carriers to be excited to the band edge. The onset of quenching at 50 K implies that E_{th} coincides with the demarcation energy at this temperature. An estimate of the thermalization energy is therefore $E_D(50)$, which is about 0.1 eV from Eq. (8.45).

8.3.6 *Auger and surface recombination*

These two non-radiative mechanisms have a relatively minor effect on the recombination. Auger recombination occurs when the recombination of an electron–hole pair excites a third electron or hole up into the band, where it can subsequently thermalize to the band edge (Landsberg 1970). The need for the third particle means that Auger recombination is only important at high excitation intensities, although there are situations when three particles can be localized at the same site, for example, an exciton bound to a neutral donor or acceptor. The Auger transition rate in a crystalline semiconductor is constrained by the need to conserve both energy and momentum, which limits the density of final states to which the third particle can be excited. The momentum conservation requirement is relaxed in amorphous semiconductors, because of the disorder, and the Auger rate may therefore be higher than in the equivalent crystal. However, there have been no detailed theoretical calculations of the rate for any amorphous semiconductor and few experiments have been performed which allow the rate to be evaluated.

Evidence for the Auger process is contained in the low temperature luminescence data in Fig. 8.19 (Street 1981b). The thermal quenching model predicts that the luminescence intensity saturates to a constant value below 50 K, but the experiments find this only when the excitation intensity is weak. Evidently there is another non-radiative mechanism which is most effective at the higher excitation intensities. The dependence on intensity is a characteristic feature of Auger processes and the effect is observed at intensities above the transition to non-geminate recombination. Auger effects must be associated with non-geminate recombination because of the requirement that a third particle participates in the recombination.

The usual characteristic of Auger recombination is a recombination rate which varies as the second or third power of the carrier concentration and a highly non-linear luminescence intensity. In contrast the data in Fig. 8.19 show a weak effect with the quantum efficiency changing by no more than a factor 2–3 for two orders of magnitude change in the excitation intensity. The effect is one more consequence of the distribution of recombination times. The luminescence efficiency is $P_r/(P_r+P_a)$ when there is direct competition with an Auger process of rate P_a. Thus pairs for which P_r is large are hardly affected by the Auger process. Furthermore an increase in excitation intensity raises both P_r and P_a, so that the intensity dependence of y_L is weak.

Surface recombination occurs when there is a high density of centers at the surface to provide an efficient recombination path which

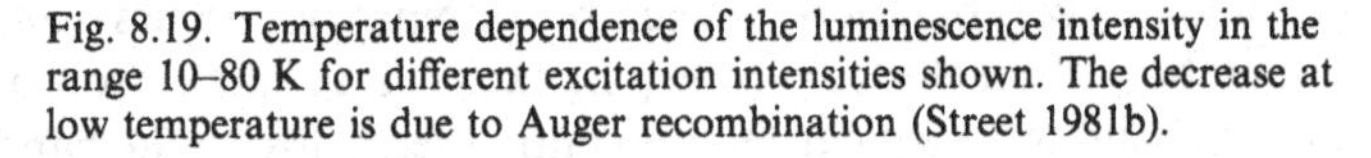

Fig. 8.19. Temperature dependence of the luminescence intensity in the range 10–80 K for different excitation intensities shown. The decrease at low temperature is due to Auger recombination (Street 1981b).

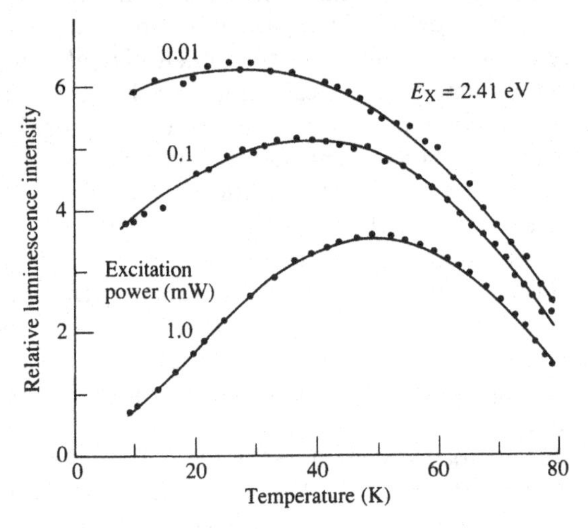

is usually non-radiative. The rate is defined by a surface recombination velocity, S_n, which is the product of the carrier velocity and the probability that recombination occurs when the particle reaches the surface. The recombination suppresses the surface carrier concentration and the rate is governed by the flux of carriers diffusing to the surface

$$D\,\mathrm{d}n/\mathrm{d}x = -S_n n \tag{8.48}$$

Even when S_n is large, the concentration gradient only extends for about the diffusion length L_D. Surface recombination is a weak effect in a-Si:H, because L_D is small and is confined to carriers created close to the surface. This occurs particularly at low temperature when the diffusion length is less than 100 Å. This contrasts with the situation in a crystal, in which the diffusion length is large and can lead to a high recombination rate even when S_n is small.

Surface recombination reduces the luminescence intensity when the excitation light has a very short absorption length. The effect is weak because of the low carrier mobility and diffusion length, so that although there are many defects at the surface of a-Si:H, the effect on the luminescence intensity is only 10–20% (Dunstan 1981). A measure of the surface recombination is the product, αL_D, of the optical absorption coefficient and the diffusion length. In most situations in a-Si:H, and particularly at low temperature, αL_D is less than unity, whereas in crystals it is often much greater.

The surface recombination of a-Si:H is greatly enhanced in multilayer structures which constrain the electron–hole pair to be near an interface. For example a thin a-Si:H film sandwiched between a-Si_3N_4 layers, has a luminescence intensity which drops rapidly when the layer thickness is less than 200 Å. These results are discussed further in the next chapter (Section 9.4.1) and show that surface recombination effects extend 100–200 Å into the film, and that electron–hole pairs created farther from the surface are not influenced by the surface at low temperature. As the carriers become more mobile at higher temperatures, the diffusion length increases and surface recombination is more significant.

8.4 Recombination at defects

Defects provide the dominant recombination path when their density is above about 10^{17} cm^{-3} or when the temperature is higher than about 100 K. The recombination mechanism depends on the temperature and on the mobility of the carrier. The low temperature mechanism is discussed first.

8.4.1 *Low temperature non-radiative tunneling*

Fig. 8.20 shows the dependence of the band tail luminescence intensity on the defect density as measured by the $g = 2.0055$ ESR resonance in undoped a-Si:H. The luminescence intensity drops rapidly when the defect density is above 10^{17} cm^{-3}, becoming unobservable at defect densities above 10^{18} cm^{-3} (Street *et al.* 1978). These data establish that the defect provides an alternative recombination path competing with the radiative band tail transition. At the low temperatures of the measurements, the electrons and holes are trapped in the band tails and are immobile. Both the radiative and the non-radiative transitions must therefore occur by tunneling. Section 8.3.3 describes the radiative tunneling process which results in a wide distribution of recombination times. A similar mechanism accounts for the non-radiative quenching.

Fig. 8.21 illustrates the recombination of an electron–hole pair near a defect. The tunneling rate of the electron to the defect is given by the usual expression

$$P_{nr} = \omega_0 \exp(-2R/R_0) \tag{8.49}$$

where R_0 is the localization radius of the band tail electron. (It is

Fig. 8.20. The dependence of the band tail luminescence intensity on the defect density measured by ESR. The solid lines are fits to Eq. (8.52) (Street *et al.* 1978).

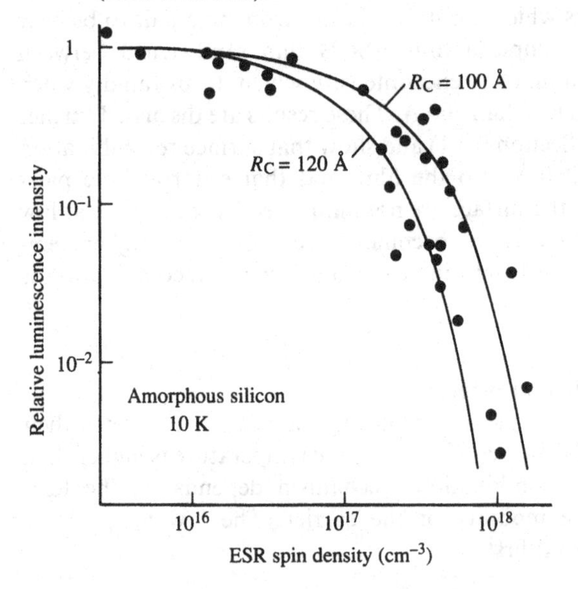

assumed that the electron is the particle which tunnels to the defect.) The critical transfer distance R_C, is defined such that if the defect is closer to the electron than R_C then non-radiative tunneling is more probable than the radiative recombination. On the other hand, the radiative transition dominates when the defect is farther away than R_C. Equating P_{NR} with the radiative recombination rate, τ_R^{-1}, in Eq. (8.49) leads to

$$R_C = (R_0/2)\ln(\omega_0\tau_R) \tag{8.50}$$

The luminescence efficiency is given by the fraction of electron–hole pairs which are created farther than R_C from the nearest defect (Street *et al.* 1978). The distribution of distances is the nearest neighbor distribution function, $G(R)$, for randomly dispersed defects, which is (Williams 1968)

$$G(R) = 4\pi R^2 N \exp(-4\pi R^3 N/3) \tag{8.51}$$

The luminescence efficiency, y_L, is therefore

$$y_L = \int_{R_C}^{\infty} G(R)\,dR = \exp\left(-\frac{4\pi R_C{}^2 N}{3}\right) \tag{8.52}$$

Eq. (8.52) gives a good fit to the data in Fig. 8.20, for values of R_C of 100–120 Å. The fit assumes that there is a single value of the critical transfer radius R_C, which is a poor approximation because τ_R is widely distributed with a corresponding varying value of R_C. Nevertheless, the average radiative lifetime of 10^{-3} s from Fig. 8.14 and $\omega_0 = 10^{-12}$ s gives

$$R_C \approx 10R_0 \tag{8.53}$$

Fig. 8.21. Illustration of the competition between band tail radiative recombination and defect recombination either by tunneling (solid line) or by thermal excitation to the band edge (dashed line).

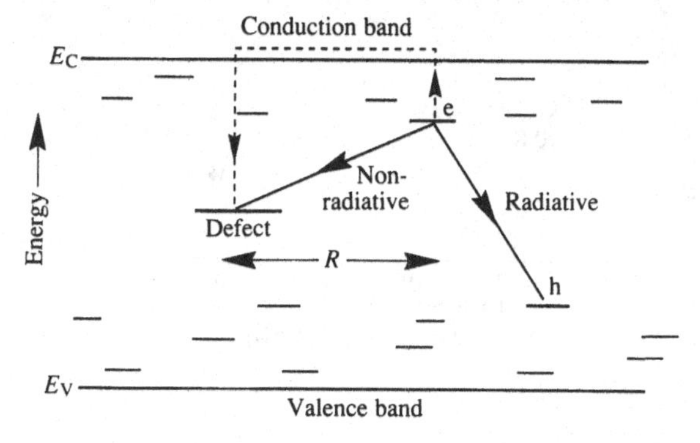

so that R_0 is 10–12 Å. This value of the localization distance agrees with the estimate obtained from the transition from geminate to non-geminate recombination and also equals the localization radius of the phosphorus donor (Section 5.2.2).

The defect tunneling mechanism is suppressed when the localization length is decreased, as happens in a-Si:Ge alloys (Street, Tsai, Stutzmann and Kakalios 1987c) and probably in other alloys. Although the dependence on defect density has the same form as for a-Si:H, a larger defect density is needed to quench the band tail luminescence and is explained by a smaller localization length.

Charged defects in doped a-Si:H quench the luminescence by the same mechanism. Fig. 8.22 shows that quenching begins at doping levels of 10^{-5} $[PH_3]/[SiH_4]$ or $[B_2H_6]/[SiH_4]$ and is complete at doping levels of 10^{-2}. The defect density is about 10^{17} cm^{-3} when the doping is 10^{-5} (see Fig. 5.9) and is the same defect density at which quenching is observed in undoped material. This quenching effect was the first indication that doping induces defects and was confirmed in several subsequent experiments. At the intermediate doping levels the band tail luminescence is replaced by a lower energy emission band at 0.8–0.9 eV, which is attributed to a weak radiative transition to defect states and is discussed further in Section 8.4.3.

Fig. 8.22. Doping dependence of the luminescence intensity of the band edge and defect transitions (Street *et al.* 1981).

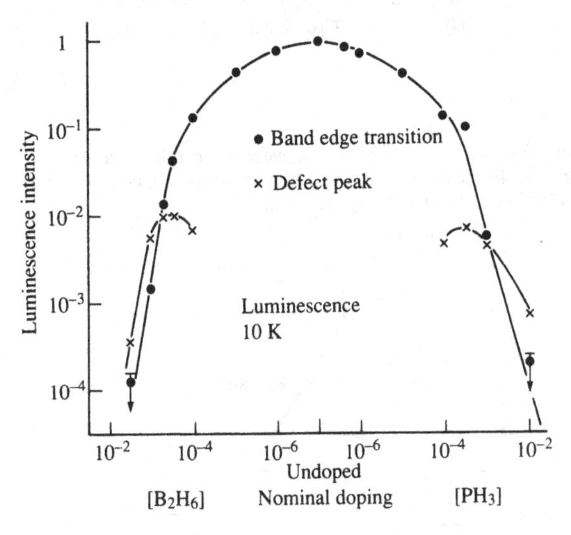

8.4.2 *High temperature trapping at defects*

At temperatures above 100 K, the defect recombination mechanism changes gradually from tunneling to direct capture of a mobile electron or hole at a defect. The capture rate defines the capture cross-section σ_c, such that the free carrier lifetime is given by

$$\tau_F^{-1} = \sigma_c v_c N_D \tag{8.54}$$

where v_c is the velocity of the carriers. This expression applies to ballistic motion, in which the capture length, d ($\sigma_c = \pi d^2/4$), is smaller than the scattering mean free path of the carriers, a_E. The capture is diffusive when this condition is not valid and the lifetime is given by,

$$\tau_F^{-1} = 4\pi d D N_D \tag{8.55}$$

where D is the diffusion coefficient.

In the discussion of transport mechanisms in Chapter 7, the inelastic scattering length is estimated to be 10–15 Å and the elastic scattering length is about 5 Å. A typical capture cross-section of 10^{-15} cm^2 corresponds to a capture length of 3–5 Å. The similarity of the length scales suggests that neither the ballistic nor the diffusive model is completely valid, but as we shall see, the difference in the results is not large. The transport theories hold that the diffusion coefficient and the free carrier mobility are related by

$$\mu_0 = eD/kT = ea_E v_c/6kT \tag{8.56}$$

Furthermore, the free carrier mobility and lifetime are related to the drift mobility and total lifetime for deep trapping, τ, by

$$\mu_0 \tau_F = \mu_D \tau \tag{8.57}$$

Combining these expressions gives for the two mechanisms of capture (Street 1984),

$$\text{ballistic};\ \sigma_c = \left(\frac{ea_E}{6kT}\right)\frac{1}{\mu_D \tau N_D} \tag{8.58}$$

and

$$\text{diffusive};\ d = \left(\frac{e}{4\pi kT}\right)\frac{1}{\mu_D \tau N_D} \tag{8.59}$$

The value of $\mu_D\tau$ is obtained from the time-of-flight experiment which measures charge transported by the drift of optically excited carriers across a sample (see Section 3.2.1). Deep trapping causes the charge collection to be less than the charge generated by the initial light pulse. With an applied voltage, V, a carrier moves a distance $\mu_D tV/d$ in a time t, where d is the sample thickness. The fraction of carriers remaining after time t is $\exp(-t/\tau)$. The total charge collection is then

$$Q = \frac{1}{\tau}\int_{\tau_T}^{\infty} \frac{Q_0 \mu_D tV}{d^2} \exp(-t/\tau)\,dt + \frac{1}{\tau}\int_0^{\tau_T} Q_0 \exp(-t/\tau)\,dt \tag{8.60}$$

where Q_0 is the initial charge and τ_T is the transit time. Integration gives (Hecht 1932)

$$\frac{Q}{Q_0} = \frac{\mu_D \tau V}{d^2}[1-\exp(-d^2/\mu_D \tau V)] \tag{8.61}$$

$\mu_D \tau$ is extracted from the voltage dependence of Q. The exponential term is negligible when the charge collection is small and Q is proportional to the applied voltage.

Measurements of $\mu_D \tau$ for doped and undoped samples of different defect densities are shown in Fig. 8.23 (Street, Zesch and Thompson 1983). $\mu_D \tau$ is inversely proportional to the defect density, so that a single value of the capture cross-section can be inferred from Eq. (8.58) or Eq. (8.59). There are four possible trapping transitions of electron and holes into dangling bond defects, and these are illustrated in Fig. 8.24. The defects are neutral in undoped a-Si:H and can capture either

Fig. 8.23. Measurements of electron and hole $\mu_D \tau$ values in undoped, doped, and compensated a-Si:H (Street *et al.* 1983).

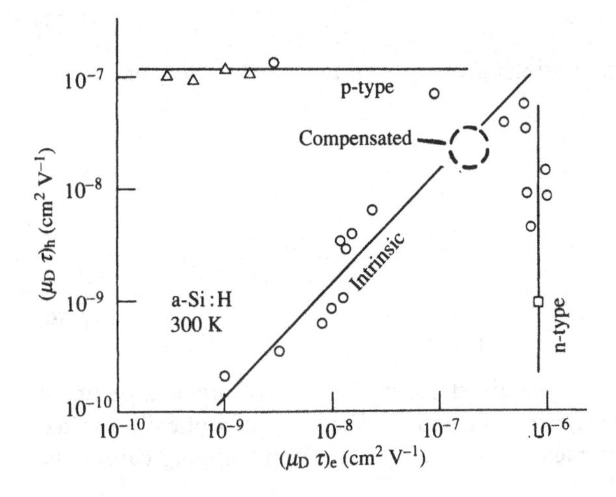

Table 8.1. *Trapping parameters for the four transitions shown in Fig. 8.24. The cross-sections are calculated from Eq. (8.58) for ballistic capture*

	Transition			
	$e \to D^0$	$e \to D^+$	$h \to D^0$	$h \to D^-$
$\mu_D \tau N_D$ ($cm^{-1}\,V^{-1}$)	2.5×10^8	5×10^7	4×10^7	1.5×10^7
σ_c (cm^2)	2.7×10^{-15}	1.3×10^{-14}	8×10^{-15}	2×10^{-14}

an electron or a hole. This is confirmed by the observation in Fig. 8.23 that $(\mu_D \tau)_e$ is proportional to $(\mu_D \tau)_h$ as the defect density changes. However, the value of $\mu_D \tau$ is smaller for holes than electrons by about a factor 7. In doped a-Si:H the defects are charged; the negative defect can capture holes but not electrons, because the defect has only two states in the gap, and vice versa for the positive defect. Again the measurements in Fig. 8.23 confirm the model, as the value of $\mu_D \tau$ drops very rapidly for the minority carrier in doped material, but the majority carrier is largely unaffected.

Table 8.1 summarizes the results for trapping of electrons and holes into the different charge states of the defects, based on measurements of $\mu_D \tau N_D$ and the assumption that a is 5 Å for electrons and 10 Å for holes (Street 1984). The capture cross-sections are in the range

Fig. 8.24. Illustration of recombination processes through traps. There are four possible transitions for a defect with two gap states.

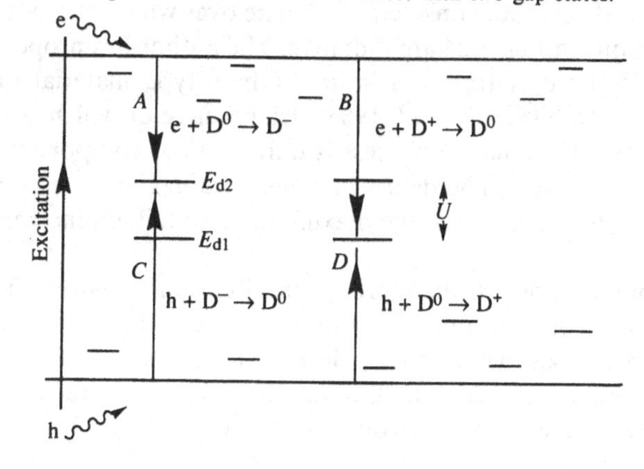

10^{-15}–10^{-14} cm^2 and differ by only a factor 2 depending on whether the ballistic or diffusive model is applied. The cross-section is a little larger for the transitions that involve a Coulomb attraction between the carrier and defect, as might be expected. The cross-sections are almost independent of temperature. One set of data found an increase at low temperature of a factor 2–3 at 150 K compared to room temperature (Street 1984). Other reports find the opposite temperature dependence, but also a weak effect (Spear, Hourd and Kinmond 1985).

The low temperature tunneling mechanism and the high temperature capture of carriers by defects merge at intermediate temperatures. Carriers move by thermal excitation to the mobility edge followed by retrapping in the tail states. The probability of tunneling to the defect is very low when the carrier is far from a defect, but increases exponentially as it gets closer. The cross-section is therefore the sum over the probabilities of the different sites, weighted by the time spent at the site. At low temperature the carrier spends most time in a single band tail trap and so the tunneling mechanism contributes most, but as the temperature increases and the carriers become more mobile, the direct capture process takes over.

8.4.3 *Radiative recombination at defects*

The defect recombination is predominantly, but not completely, non-radiative. The quenching of the band tail luminescence at high temperature or at high defect density is accompanied by the onset of a weak luminescence transition at lower energy. Some typical luminescence spectra are shown in Fig. 8.25 for doped and high defect density undoped material. The peak energy at low temperature is 0.8 eV in n-type a-Si:H and 0.9 eV in p-type and decreases slightly at elevated temperature. There is agreement that the transition is from the band edge to the defect, but considerable debate over which of the four possible transitions in Fig. 8.24 are radiative. The author has proposed that the luminescence is from transition A in n-type material and transition D in p-type (Street *et al.* 1984). Irrespective of which is the correct transition, the radiative process is only a minor component of the defect recombination. The defect luminescence has no more than 1 % of the quantum efficiency of the maximum band tail luminescence intensity.

Defect recombination is therefore primarily non-radiative. The standard theories of non-radiative transitions are based on the multiphonon processes described in Section 8.1.2. A temperature-independent capture cross-section of about 10^{-15} cm^{-2} is characteristic of a defect state with a strong electron–phonon coupling (see Fig. 8.6).

Indeed, for the activation energy of σ_c to be zero, the lattice relaxation energy should be equal to the trap depth, which is about 0.7–0.9 eV. Such a large lattice relaxation should give the defect transition a large Stokes shift, but none is observed. The optical and thermal transition energies have been measured for n-type a-Si:H and are shown in Fig. 4.24. The energies of thermal emission, observed by DLTS and by optical absorption and luminescence agree to within 0.1 eV. The Stokes shift is therefore very small and unable to account for the capture cross-section within the multiphonon model. The mechanism of non-radiative capture at defects remains puzzling.

8.5 **Photoconductivity**

Photoconductivity occurs when carriers are optically excited from non-conducting to conducting states. It is an indirect measure of the recombination and does not distinguish between radiative and non-

Fig. 8.25. Luminescence spectra of doped and undoped a-Si:H measured at 250 K, showing the band tail transition at 1.1 eV and the defect peaks at 0.81 eV and 0.91 eV (Street *et al.* 1984).

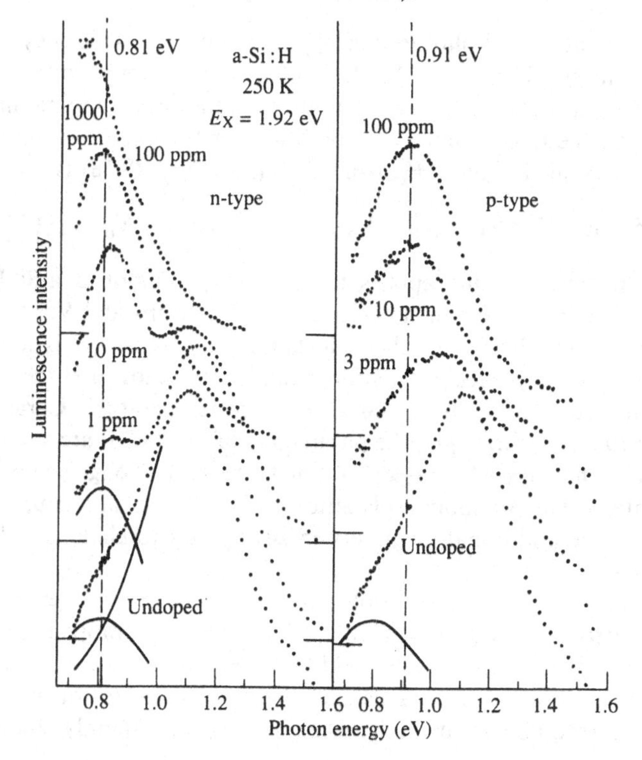

radiative mechanisms. The illumination excites electrons and holes to the band edges where they drift towards the electrodes under the applied field. When the contacts are blocking, the only contribution to the current is from the optically excited carriers and this is referred to as primary photoconductivity. Charge collection in p–i–n solar cells and the time-of-flight transient photoconductivity are examples. Secondary photoconductivity occurs when the electrons or holes which are absorbed at the contact are replaced by carriers injected from the other contact. The same number of electrons and holes are transported in a primary photoconductor by charge neutrality, but the densities are not necessarily equal in a secondary photoconductor. The nature of the contacts is therefore an important consideration in the interpretation of the photoconductivity and one that is often overlooked. Most of the common metals form blocking contacts for electrons in a-Si:H (see Chapter 9) and give secondary photoconductivity only when the contact is sufficiently leaky.

The primary photocurrent of a thin a-Si:H sample is

$$I_{\text{ph}} = \eta_G \eta_C G \tag{8.62}$$

where G is the absorbed photon flux, η_G is the quantum efficiency for generating mobile electrons and holes and η_C is the charge collection efficiency. The collection efficiency is unity when the deep trapping time is longer than the transit time. The condition for full collection of the charge at an applied voltage V_A, from Eqs. (8.61) and (8.58) is

$$d^2 < \mu_D \tau V_A \simeq e a_E V_A / 6\sigma_c N_D kT \approx 2.5 \times 10^8 \, V_A / N_D \tag{8.63}$$

where the numerical value applies to electrons in undoped a-Si:H. There is full collection for thicknesses of a few microns at 1 V.

The generation efficiency of band-to-band transitions, η_G, is unity unless there is geminate recombination. Experiments show that there is no geminate recombination in a-Si:H at room temperature (Carasco and Spear 1983). Surprisingly, the parameters of Eq. (8.23) are similar to those for chalcogenide glasses in which there is strong geminate recombination. The probable explanation is that the phonon coupling is small and the additional inequality (8.24) applies to chalcogenides but not to a-Si:H.

When there is incomplete collection of charge, the primary photoconductivity is complicated by the presence of trapped space charge which distorts the electric field (see Section 10.1.2). Although the primary photoconductor is usually the best structure for a light detector, the recombination mechanisms are more commonly studied by secondary photoconductivity, largely because the electron and hole currents are not constrained by charge neutrality and there is no space charge to distort the field.

The secondary photoconductivity is given by

$$\sigma_{ph} = n_{oe} e\mu_{oe} + n_{oh} e\mu_{oh} \tag{8.64}$$

where $n_{oe,h}$ and $\mu_{oe,h}$ are the concentrations and mobilities of free electrons and holes. It is convenient to rewrite these in terms of the band tail carrier concentrations, n_{BT}, P_{BT}, and drift mobilities

$$\sigma_{ph} = n_{BT} e\mu_{De} + P_{BT} e\mu_{Dh} \tag{8.65}$$

Eq. (8.65) assumes that electrons and holes form a quasi-equilibrium with the band tail states. This is valid for electrons near room temperature, but only approximate for holes since dispersive hole transport indicates that a quasi-equilibrium is not fully established in the band tail.

The carrier concentrations are given in terms of the optical excitation rate, G, and the recombination times, τ_e and τ_h,

$$G = n_{BT}/\tau_e = P_{BT}/\tau_h \tag{8.66}$$

The lifetimes are those of the band tail carriers, not the free carriers. Thus,

$$\sigma_{ph} = Ge(\mu_{De}\tau_e + \mu_{Dh}\tau_h) \approx Ge\mu_{De}\tau_e \tag{8.67}$$

Time-of-flight experiments find that in undoped a-Si:H, $\mu_{De}\tau_e$ is about seven times larger than $\mu_{Dh}\tau_h$, indicating that electron transport dominates. The values of $\mu_D\tau$ are not necessarily the same in photoconductivity, but there is general agreement that electrons contribute most.

The photoconductivity (from now on assumed to be only from electrons) is also expressed in terms of a quasi-Fermi energy, E_{Fn}, describing the quasi-equilibrium of band tail carriers,

$$\left.\begin{aligned} &\sigma_{ph} = \sigma_0 \exp[-(E_C - E_{Fn})/kT] \\ &\text{or} \\ &E_C - E_{Fn} \approx kT\ln(100/\sigma_{ph}) \end{aligned}\right\} \tag{8.68}$$

The second expression uses the experimental information about the conductivity prefactor derived in Eq. (7.19). The descriptions of the photoconductivity in terms of the recombination lifetimes or the quasi-Fermi energies are equivalent.

Fig. 8.26 shows some typical photoconductivity data. Below room temperature the photoconductivity far exceeds the dark conductivity and is only weakly temperature-dependent. The intensity dependence is slightly sublinear, such that $\sigma_{ph} \sim G^{\gamma}$, with $\gamma \approx 0.8$, but γ decreases to 0.5 at low temperature. High defect densities cause a reduction in σ_{ph}. The room temperature photoconductivity in low defect density material reaches $10^{-5}\ \Omega^{-1}\ cm^{-1}$ for the illumination conditions shown, which corresponds to a quasi-Fermi energy of about $E_C - 0.4$ at room temperature.

Recombination is evidently controlled by trapping into defect states, consistent with the other recombination measurements. The recombination transitions through defects with two gap states are illustrated in Fig. 8.24, with electrons and holes captured into either of the two states. This type of recombination is analyzed by the Shockley–Read–Hall approach which distinguishes between shallow traps, for which the carrier is usually thermally excited back to the band edge, and deep traps, at which the carriers recombine. A demarcation energy, which is usually close to the quasi-Fermi energy, separates the two types of states. The occupancy of the shallow states is determined by the quasi-equilibrium and that of the deep states by the recombination processes. No attempt is made here at a comprehensive analysis of the photoconductivity, which rapidly becomes complicated. Instead some approximate solutions are derived which illustrate the processes.

The simplest recombination model assumes the value of $\mu_D \tau$ obtained

Fig. 8.26. Typical measurements of photoconductivity in undoped a-Si:H showing (*a*) the sublinear dependence on excitation intensity (Wronski and Daniel 1981), and (*b*) the weak temperature dependence and approximate inverse dependence on defect density (Dersch, Schweitzer and Stuke 1983).

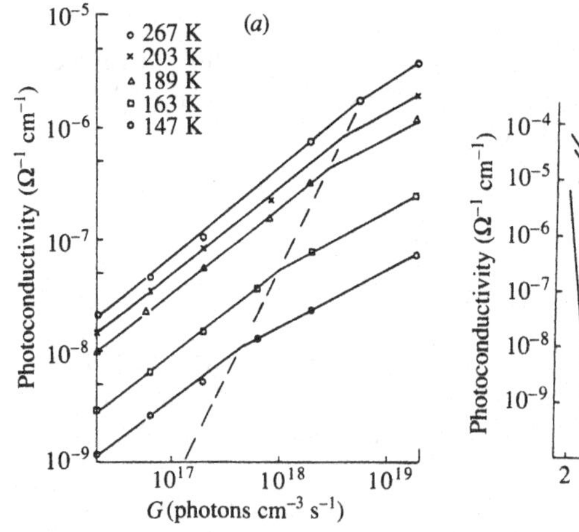

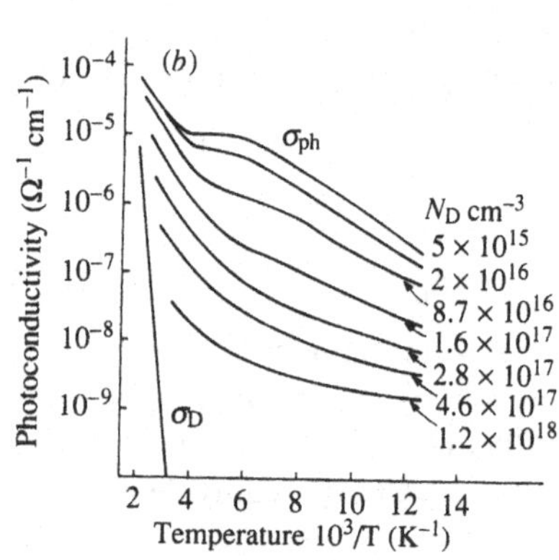

in Section 8.4.2 for the trapping of carriers. Substituting Eq. (8.58) into Eq. (8.67) gives

$$\sigma_{ph} = Ge^2 a_E/6N_D \sigma_c kT \approx 4\times10^{-11} G/N_D \,\Omega^{-1}\,cm^{-1} \tag{8.69}$$

where the numerical value is from Table 8.1 for electron capture into neutral defects. The photoconductivity is predicted to be proportional to the excitation intensity and inversely proportional to the defect density, with little temperature dependence, since the capture cross-section, σ_c, is approximately constant. The inverse dependence on N_D is found for defect densities above 10^{16} cm^{-1}, but there is a weaker dependence at lower densities. The expected linear dependence on G and independence of T are approximately, but not exactly, obeyed (see Fig. 8.26). Finally, the absolute magnitude of σ_{ph} is larger than predicted by Eq. (8.69) by about one order of magnitude.

This simple model is not exact because it takes no account of the occupancy of the defects, which is different from the equilibrium occupancy. The time-of-flight measurement of the trapping rates is performed with few excess carriers, so that the trap occupancy is the same as in equilibrium, but in a steady state photoconductivity experiment, E_{Fn} and the demarcation energies are often far from midgap. The trap occupancy is calculated from the rate equations for band tail electrons and holes

$$\frac{dn_{BT}}{dt} = G - n_{BT}(C_{e^+}\mu_{De} N_{D^+} + C_{e0}\mu_{De} N_{D0}) \tag{8.70}$$

and

$$\frac{dP_{BT}}{dt} = G - P_{BT}(C_{h^-}\mu_{Dh} N_{D^-} + C_{h0}\mu_{Dh} N_{D0}) \tag{8.71}$$

Thermal excitation out of the traps is neglected in these expressions and band-to-band recombination is also not included. The Cs are the recombination constants for the four transitions shown in Fig. 8.24. According to Eqs. (8.54) and (8.56), $C_i = 6kT\sigma_i/ea$ and the four cross-sections differ by less than a factor 10 (see Table 8.1). N_{D^+}, N_{D^0} and N_{D^-} are the concentrations of the different charge states of the defect. By charge neutrality,

$$n_{BT} + N_{D^-} = P_{BT} + N_{D^+} \tag{8.72}$$

In steady state,

$$n_{BT} = \frac{G}{\mu_{De}(C_{e^+}N_{D^+} + C_{e0}N_{D0})} \tag{8.73}$$

and

$$P_{BT} = \frac{G}{\mu_{Dh}(C_{h^-}N_{D^-} + C_{h0}N_{D0})} \tag{8.74}$$

The low hole drift mobility is a significant feature of the recombination. Experiments show that $\mu_{De} \sim 1\ cm^2\,V^{-1}\,s^{-1}$ and $\mu_{Dh} \sim 3 \times 10^{-3}\ cm^2\,V^{-1}\,s^{-1}$. Thus, if the terms in brackets in Eqs. (8.73) and (8.74) are comparable, then $P_{BT} \sim 300n_{BT}$. However, a photoconductivity of $10^{-5}\ \Omega^{-1}\,cm^{-1}$ corresponds to $n_{BT} \sim 10^{14}$–$10^{15}\ cm^{-3}$ in the temperature range of the experiments in Fig. 8.26. The density of band tail holes should then be much larger than the defect density, which violates the charge neutrality condition, Eq. (8.72). Eqs. (8.72)–(8.74) are satisfied only when most of the defects are negatively charged, which has the effect of reducing the electron recombination rate and allowing the band tail populations to be more nearly equal.

The trapped holes which recombine slowly because of their low mobility are called 'safe hole traps'. Their presence increases the electron lifetime and the photoconductivity and seems to account for the features of the photoconductivity not explained by the simple model of Eq. (8.69) (McMahon and Crandall 1989). Safe hole traps are most significant in low defect density material, when their concentration can exceed the defect density. A detailed analysis needs to take into account the full distribution of hole traps as well as the dispersive transport of holes. The role of transitions between the band edges in the recombination process also needs to be determined.

There is much still to be understood about the photoconductivity of a-Si:H. However, the measurements confirm that recombination through defects is the main mechanism, particularly when their concentration is high. Extrinsic effects further complicate the interpretation of photoconductivity. For example, surface recombination can dominate when the bulk recombination rate is low. These effects can arise from either the excess defects at the surface or from the band bending, which causes a field induced separation of the electron and hole distributions. Contacts, which are almost invariably non-ohmic, also modify the photoconductivity, in particular, the response time.

9 Contacts, interfaces and multilayers

In a thin film such as a-Si:H, surface and interface effects can exert an influence throughout the entire material. The surface has different electrical and structural properties from the bulk for various reasons. There are chemical reactions which change the composition, such as oxidation or reactions with metals. There is also transfer of electrical charge across the interface which causes band bending. Possible origins of the charge transfer are the different work functions of the materials in contact or charged species attached to the surface. This chapter describes metal contacts, the free surface, semiconductor and dielectric interfaces, and, lastly, multilayer structures in which a series of very thin layers is grown which has properties which differ markedly from those of the bulk material.

9.1 Metallic contacts

When a metal is brought into contact with a semiconductor, there is a transfer of charge across the interface to bring the two Fermi energies into alignment. The space charge in the metal remains very close to the contact, but extends much farther in the semiconductor because of the low density of states in the band gap. The resulting Schottky contact has rectifying electrical properties. A similar barrier is formed between doped and undoped a-Si:H layers. The nature of the metal contact is important in virtually all electrical measurements, as it determines whether charge can flow easily across the contact. Schottky contacts are used in transient capacitance techniques to measure the defect states (see Chapter 3) and in photosensing devices, where the blocking contact reduces the dark current and minimizes noise. The properties of the Schottky barrier on a-Si:H depend on the specific metal used, on the nature of the interface formed between the two materials and on the density of states in the a-Si:H material.

9.1.1 *Models of the Schottky barrier*

Fig. 9.1 shows a schematic diagram of a metal Schottky contact on a semiconductor. In isolation, the metal and the semiconductor generally have different work functions Φ_M and Φ_S. (The work function is the energy needed to remove an electron from the Fermi energy to the vacuum.) When electrical contact is made between the two materials, the Fermi energies must coincide, provided there is no applied electric field. The alignment is achieved by the transfer of charge from one side of the interface to the other. The charge is constrained to remain in the vicinity of the interface by the Coulomb attraction and forms a dipole layer with a potential difference equal to the difference in work function $\Phi_M - \Phi_S$. There is band bending within the space charge layer at either side of the interface. The band bending extends only a few angströms in the metal because the density of states at the Fermi energy is very high, but in the semiconductor it is far more extended. The width of the space charge depletion layer W, is determined by the gap state density and distribution, as described shortly.

The barrier is characterized by the barrier height, Φ_B, the built-in potential V_B, and the depletion layer width. An ideal Schottky contact has a barrier height and built-in potential given by,

$$\Phi_B = \Phi_M - \Psi_S \qquad eV_B = \Phi_M - \Phi_S \tag{9.1}$$

where Ψ_S is the electron affinity, defined as the energy separating the conduction band edge and the vacuum energy. Doping of the semiconductor changes the work function but not the electron affinity.

Fig. 9.1. Schematic diagram of a Schottky barrier between a metal and a semiconductor, showing the charged depletion layer extending into the semiconductor.

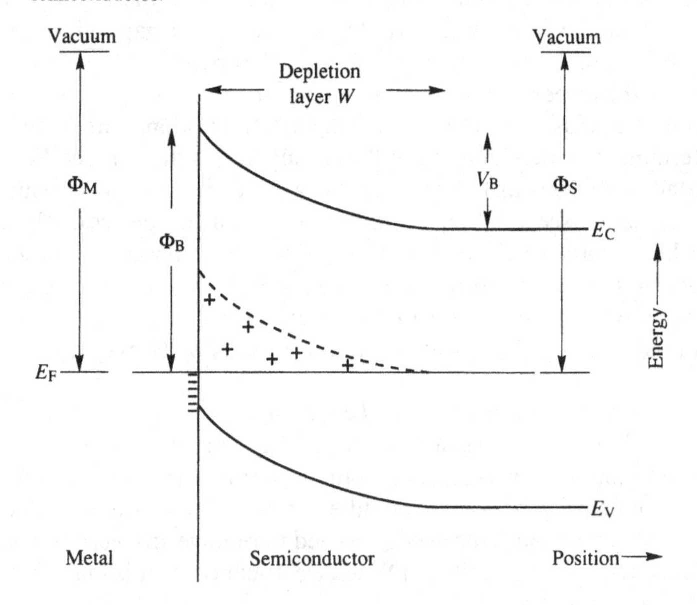

The model predicts that the barrier height is independent of the doping, but that the built-in potential varies.

The ideal Schottky barrier is rarely observed and a-Si:H is no exception. Departures from the simple model invariably depend on the nature of the semiconductor–metal contact. Any additional dipole charge at the interface adds a potential difference across the junction. A very thin barrier is transparent to electrons and causes an apparent reduction in the barrier height. The dipole layer can originate from surface states at the interface for which a density of 10^{12}–10^{13} cm^{-3} is sufficient to modify the barrier height. Alternatively there might be a dielectric layer at the interface, for example, caused by a surface oxide. The different dielectric constant of the layer causes a polarization layer at the interface. Finally there could be a dipole layer due to a configuration of polar molecules at the interface, although such a layer is probably not important in a-Si:H. When the interface effects are included, a phenomenological expression for the barrier height is

$$\Phi_B = \alpha_1(\Phi_M - \Psi_S) + \alpha_2 \quad (9.2)$$

where α_1 is reduced from unity by the interface layer and α_2 is related to the density of surface states.

The potential of the depletion layer $V(x)$ is the solution of Poisson's equation

$$\frac{d^2V(x)}{dx^2} = -\frac{\rho(x)}{\varepsilon\varepsilon_0} \quad (9.3)$$

The space charge, $\rho(x)$, arises from the ionization of band gap states which are raised above the equilibrium Fermi energy by the band bending. The charge density is related to the density of states distribution by

$$\rho(V, x) = \int_{E_F(O)}^{E_F(V(x))} eN(E)\,dE \quad (9.4)$$

Eq. (9.4) is a low temperature approximation but is easily modified to include the Fermi distribution of the electrons.

There are two simple density of states distributions which allow an analytical solution of Eq. (9.3). First is when the charge ρ is constant which, for example, arises from a discrete donor-like level in the gap with density N_D. The solution to Eq. (9.3) is

$$V(x) = \frac{eN_D(W-x)^2}{2\varepsilon\varepsilon_0} \quad (9.5)$$

The depletion layer has a parabolic voltage dependence and the width is

$$W = \left[\frac{2\varepsilon\varepsilon_0(V_A + V_B)}{eN_D}\right]^{\frac{1}{2}} \tag{9.6}$$

where V_B is the built-in potential and V_A is the applied voltage.

The second solution of Eq. (9.3) is for a uniform density of states, N, in the gap, for which the charge density is $NV(x)$, and

$$V(x) = (V_A + V_B)\exp[-x(eN/\varepsilon\varepsilon_0)^{\frac{1}{2}}] \tag{9.7}$$

The depletion potential decreases exponentially and has no obvious termination, unlike the previous case. It is usual to define the end of the depletion layer as being when the potential is kT, so that

$$W = \left(\frac{\varepsilon\varepsilon_0}{eN}\right)^{\frac{1}{2}} \ln[(V_A + V_B)/kT] \tag{9.8}$$

A-Si:H has a smoothly varying density of states, so that Eq. (9.8) applies when the band bending is small compared with the width of the defect band, which is about 0.2 eV. When the voltage is larger, the charge depends on the shape of the density of states and W cannot easily be calculated.

The simple theory depicted in Fig. 9.1 suggests that the Fermi energy intersects one of the bands when the built-in potential is more than about 1 V. In practice, this does not occur because the time for an electron to be ionized from a state below midgap is larger than the time for the same state to be reoccupied by an electron excited from the valence band. This condition is known as deep depletion, in which the Fermi energy is held at the middle of the gap by the excitation kinetics. There is a constant depletion charge, given by $N(E_F)\Delta E$, where ΔE is the shift of the Fermi energy from its equilibrium position to the deep depletion energy. The solutions for the depletion width are Eqs. (9.5) and (9.6).

The depletion layer has an associated capacitance, measurement of which is discussed in Section 4.2.3. The capacitance C, is $\varepsilon\varepsilon_0 A/W$, where A is the area, so that from Eq. (9.6),

$$\frac{1}{C^2} = \frac{2(V_A + V_B)}{\varepsilon\varepsilon_0 eN_D A^2} \tag{9.9}$$

This equation is frequently used to obtain the dopant concentration in crystalline semiconductors. The equivalent expression for the capacitance of the junction for a uniform charge distribution is

$$\frac{1}{C^2} = \frac{\ln[2(V_A + V_B)/kT]}{\varepsilon\varepsilon_0 eNA^2} \tag{9.10}$$

9.1.2 *Electrical transport across the barrier*

There are several techniques for measuring the barrier heights, of which the most common is from the current–voltage (J–V) characteristics. The different transport mechanisms across the barrier are illustrated in Fig. 9.2. Thermionic emission refers to the excitation of the carriers over the top of the barrier where the conduction band edge intersects the interface. Tunneling through the barrier reduces the apparent barrier height and is significant when the internal field is large, which occurs when the applied voltage is high or the depletion layer is narrow. Tunneling into the localized states at the band edge also causes a reduction in the barrier height. Finally there can be field emission from the Fermi energy at high reverse bias, possibly via gap states.

For simplicity, only electron transport across the interface is considered. At zero applied voltage, the equilibrium current fluxes in

Fig. 9.2. Illustration of the three main mechanisms of electronic transport across the barrier. The dashed lines illustrate the change in potential profile when a voltage, V_A, is applied.

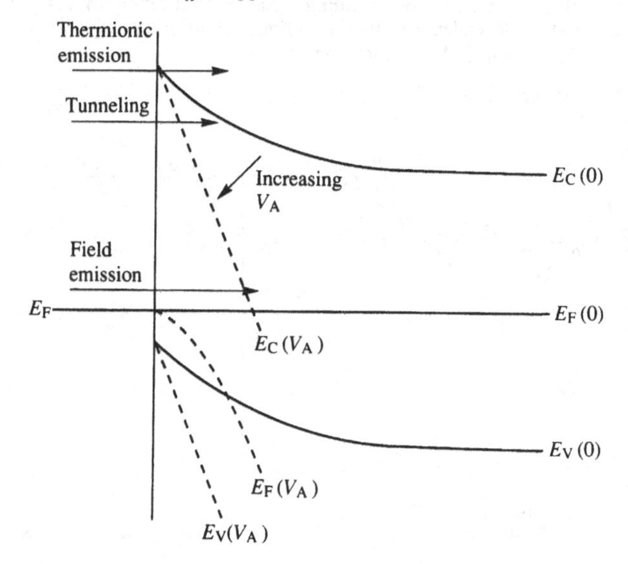

each direction are equal. The thermionic emission current density from the metal over the barrier is

$$J_0 = \text{const.} \exp(-e\Phi_B/kT) \tag{9.11}$$

The application of a bias V_A changes the current from the semiconductor by a factor $\exp(eV_A/kT)$, because the Fermi energy is raised or lowered by eV_A with respect to the metal Fermi energy. Thus,

$$J(V_A) = J_0[\exp(eV_A/kT) - 1] \tag{9.12}$$

In forward bias (positive V_A) the current increases exponentially with V_A (when $eV_A \gg kT$) and in reverse bias the current saturates at J_0. The forward bias current is more usually written as

$$J_F(V_A) = J_0 \exp(eV_A/nkT) \tag{9.13}$$

where n is called the ideality factor and is a phenomenological correction for the many possible reasons why the current differs from the ideal model. Recombination of electrons and holes in the depletion region and a non-uniform interface gives ideality factors greater than unity.

Fig. 9.3 shows the J–V characteristics of a palladium contact to a-Si:H (Thompson, Johnson, Nemanich and Tsai 1981). The exponential increase in forward bias and the saturation in reverse bias

Fig. 9.3. Electrical characteristics of a palladium/a-Si:H Schottky barrier: (*a*) the exponential forward current, the ideality factor, n, the reverse current, and the effects of annealing, described in Section 9.1.3; (*b*) the temperature dependence of the saturation current density J_0 plotted according to Eq. (9.14) (Thompson *et al.* 1981).

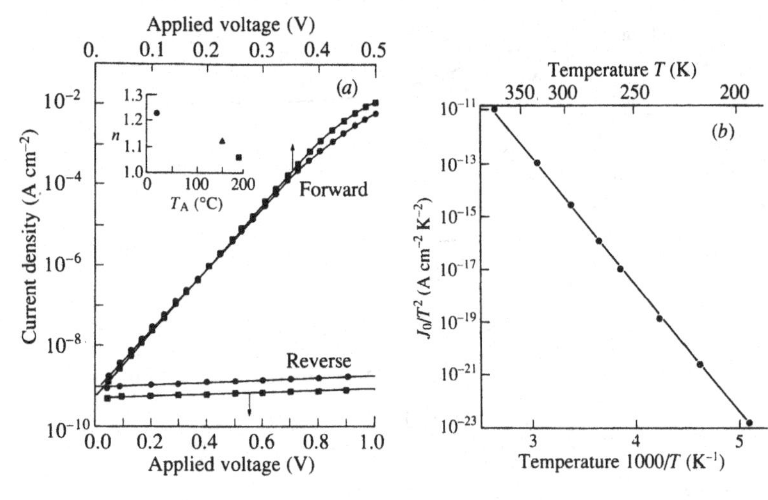

are clearly seen. Beyond about 0.5 V, the forward bias current becomes linear, indicative of a current limitation due to the bulk series resistance. The ideality factor of the data is 1.05, showing that nearly ideal contacts can be made, although there are many situations in which much larger ideality factors are observed.

The temperature dependence of J_0, measured by extrapolation of the forward J–V data, is shown is Fig. 9.3. The saturation current density is given in the thermionic emission model by

$$J_0 = A^* T^2 \exp(-e\Phi_B/kT) \tag{9.14}$$

where A^* is Richardson's constant (120 A cm^{-2} K^{-2} for free electrons in vacuum). The data in Fig. 9.3 are plotted in this form and are thermally activated with a barrier height of 0.97 eV.

Although the reverse current of an ideal Schottky barrier is J_0, in practice there are other current sources. Imperfect contacts have a leakage current which generally increases exponentially with bias. Even with an ideal contact, there is a thermal generation current caused by the excitation of electrons and holes from bulk gap states to the band edges. This mechanism determines the Fermi energy position under deep depletion conditions. The current density is the product of the density of states and the excitation rate and is approximately,

$$J_{th} = eN(E_F)kT\omega_0 \exp(-E/kT)\Omega \tag{9.15}$$

where Ω is the volume. Substituting $N(E_F) = 10^{16}$ cm^{-3} eV^{-1}, equating E to half the mobility gap, $E_m/2 = 0.9$ eV and $\omega_0 = 10^{13}$ s^{-1}, yields a current of 10^{-11} A cm^{-2} for a film thickness of 1 μm. This current is lower than J_0 for the palladium barrier data in Fig. 9.3 and therefore does not contribute much to the J–V data. However, the thermal generation current is larger than J_0 for a junction to p-type a-Si:H, which has a significantly larger barrier height (see Section 10.1.3).

Electrons are optically excited from the metal Fermi energy over the Schottky barrier by the internal photoemission mechanism. The carriers are collected at the opposite contact and detected as a photocurrent when the junction is held in reverse bias. The energy required to excite carriers is less than the band gap energy, so that the internal photoemission spectrum is distinguished from bulk band-to-band carrier generation. Models of the effect predict that the photoresponse spectrum is

$$Y(E) = A(E - \Phi_B)^2 \tag{9.16}$$

Fig. 9.4 shows some examples of the photoresponse of Schottky barriers with different metals (Wronski, Lee, Hicks and Kumar 1989).

The two sets of data correspond to barriers formed by the metals for electrons and holes. Fig. 9.4 shows that Eq. (9.16) is obeyed for some different metal contacts to a-Si:H. The extrapolation of the data gives barrier heights ranging from 0.8 to 1.1 eV and is in good agreement with the results obtained from the *J–V* characteristics. The metals with the largest barriers for electrons have the smallest barriers for holes and vice versa. The sum of the two barrier heights is the mobility gap, as can be seen from Fig. 9.1. The measurements in Fig. 9.4 indicate a mobility gap of 1.9 eV, in good agreement with the density of states model shown in Fig. 3.16.

9.1.3 *Measurement of depletion layers*

The depletion layer profile contains information about the density of states distribution and the built-in potential. The depletion layer width reduces to zero at a forward bias equal to V_B and increases in reverse bias. The voltage dependence of the junction capacitance is a common method of measuring $W(V)$. Eq. (9.9) applies to a semiconductor with a discrete donor level, and V_B is obtained from the intercept of a plot of $1/C^2$ versus voltage. The $1/C^2$ plot is not linear for a-Si:H because of the continuous distribution of gap states – an example is shown in Fig. 4.16. The alternative expression, Eq. (9.10), is also not an accurate fit, but nevertheless the data can be extrapolated reasonably well to give the built-in potential. The main limitation of the capacitance measurement is that the bulk of the sample must be conducting, so that the measurement is difficult for undoped a-Si:H.

The depletion layer profile on undoped a-Si:H can be obtained by transient photoconductivity. A pulse of light excites electron–hole pairs very near the contact. As in the time-of-flight experiment, holes are immediately collected by the contact and electrons drift down the internal field of the depletion layer, giving transient conductivity of

$$I(t) = ne\mu_D E(t) \tag{9.17}$$

where $E(t)$ is the depletion layer field at the position of the carrier packet at time t. At such time the carriers have moved a distance,

$$x = \int_0^t \mu_D E(t)\,dt \tag{9.18}$$

so that the field and voltage profiles may be calculated from the transient response. Examples of the voltage profile in Fig. 9.5 show that the zero bias depletion width in undoped a-Si:H is 1–2 μm (Street 1983). The shape of $V(x)$ is approximately exponential and implies that the density of states is reasonably uniform in the middle of the gap with a density of ~ 4×10^{15} cm^{-3} eV^{-1}. The two sets of data in Fig. 9.5 differ only in the contact metals which are chromium and platinum. Extrapolation of the data gives the built-in potentials of 0.2 and 0.5 eV respectively. Since the Fermi energy is about 0.7 eV from the conduction band in undoped a-Si:H, the resulting barrier heights are 0.9 eV and 1.2 eV, which agree reasonably well with the other measurements.

Fig. 9.4. Internal photoemission spectrum plotted according to Eq. (9.16) for different metal contacts to a-Si:H, showing the barrier for (*a*) electrons and (*b*) holes (Wronski *et al.* 1989).

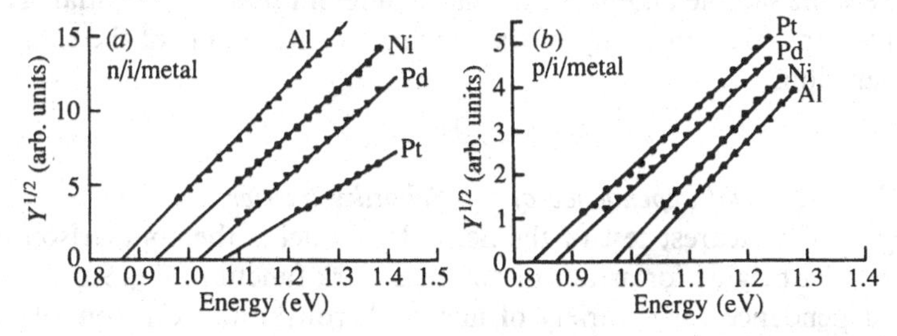

Fig. 9.5. The depletion potential profile of a-Si:H with platinum and chromium contacts measured by transient photoconductivity. The inset plots the dependence of the charge collection, Q_C, on applied bias which shows the shrinking of the depletion width in forward bias (Street 1983).

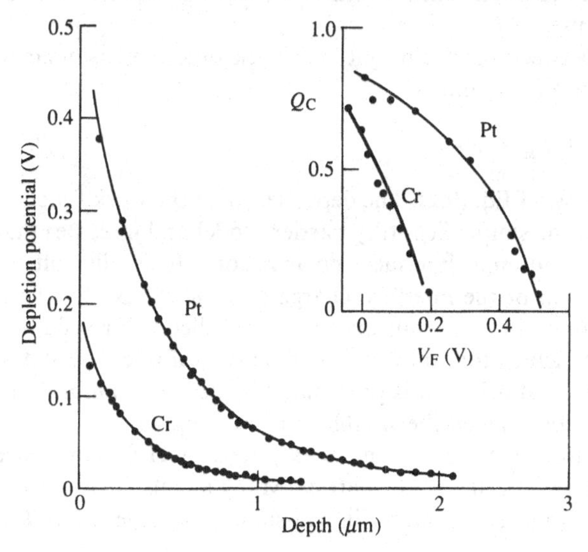

The carriers drift under the influence of the internal field to the edge of the depletion layer and then stop in the field-free region. The charge collected in the external circuit is

$$Q_C = Q_0 W/d \tag{9.19}$$

where Q_0 is the initial charge and d is the sample thickness. The measurement of $Q_C(V)$ gives a similar result as the capacitance–voltage experiment. The examples of charge collection in Fig. 9.5 give the same built-in potential as the depletion layer profiles. This type of experiment also finds that the contacts are not symmetrical and that a bottom chromium contact has a narrower depletion layer than the top contact to the same sample and probably also a different built-in potential. The results show the sensitivity of the junction to the details of the sample preparation.

9.1.4 *Material dependence of the Schottky barrier*

The clearest test of the Schottky model is the comparison of the barrier heights for metals of different work function. Fig. 9.6 shows this dependence for a variety of metals (Wronski and Carlson 1977). The values are such that most metals form a barrier for electrons to undoped a-Si:H. There are a few metals with low work functions such as samarium and ytterbium, which are blocking for holes and should give an ohmic contact for electrons (Greeb, Fuhs, Mell and Welsch 1982). Unfortunately all such metals are either uncommon or inconvenient to use.

The relation between barrier height and work function is similar to that in crystalline silicon and is

$$\Phi_B = 0.28\,\Phi_M - 0.44 \text{ eV} \tag{9.20}$$

which is of the form of Eq. (9.2). The dependence on the work function is weaker than in the simple Schottky barrier model and indicates that there is some additional interface polarization. It is difficult to determine the origin of the interface charge and the subject has been debated for many years for metals on crystalline silicon. Wronski and Carlson (1977) conclude that there is a high density of interface states. Whatever the physical origin, it is presumably the same in amorphous and crystalline silicon, given the simililar barrier heights.

Chemical reactions between the metal and the a-Si:H film influence the formation and properties of the barrier. Many metals are known to react with silicon to form silicides. Fig. 9.3 shows a change in the *J–V* characteristic of a palladium barrier upon annealing to about 200 °C, with an improvement of the ideality factor from 1.2 to 1.05. The annealing causes the growth of the silicide Pd_2Si, as is demonstrated by the Raman spectra of Fig. 9.7, measured on films only 100 Å thick (Nemanich, Tsai, Thompson and Sigmon 1981). The extra lines in the low frequency region of the Raman spectrum identify Pd_2Si and the sharpness of the lines is because the silicide layer is crystalline. Silicide formation only takes place on a clean a-Si:H surface and not on a sample which has been left in air for more than a few minutes.

A variety of surface reactions have been observed with other metals on a-Si:H (Nemanich 1984). For example, a similar silicide is formed with platinum and nickel at 200 °C and with chromium at 400 °C. Aluminum and gold form intermixed phases at low temperature, but do not form silicides. Instead both metals promote low temperature crystallization of the a-Si:H film. Dendritic crystallization occurs at 200 °C at a gold contact, giving a very non-uniform interface, and aluminum causes crystallization at 250 °C. The resulting Schottky contact for gold is surprisingly ideal, but is very poor for aluminum.

As might be anticipated, these surface reactions are suppressed when the a-Si:H surface is allowed to form a native oxide. However, the

Fig. 9.6. Measurements of the Schottky barrier height versus metal work function for various metal contacts to a-Si:H and crystalline silicon. The dashed line shows the relation for an ideal Schottky contact (Wronski and Carlson 1977).

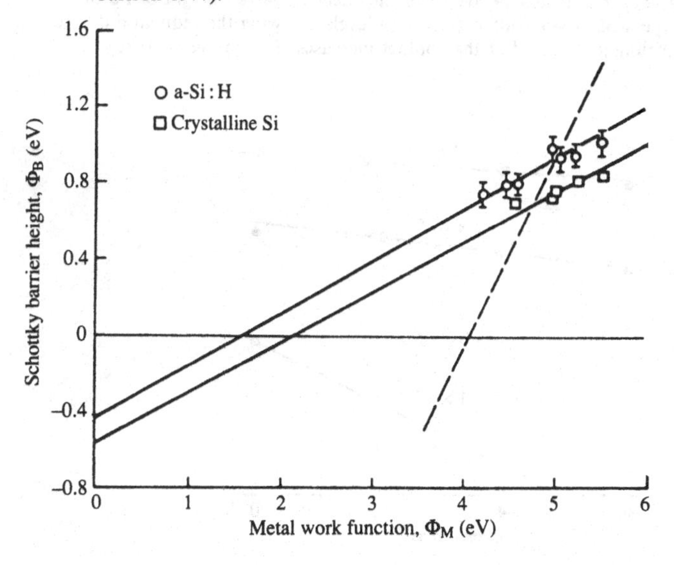

Fig. 9.7. The Raman spectra of the palladium/a-Si:H interface region: (*a*) as deposited a-Si:H; (*b*) palladium on freshly deposited a-Si:H; (*c*) sample annealed to 300 °C; (*d*) annealed to 550 °C; (*e*) palladium deposited on oxidized a-Si:H and annealed to 400 °C (Nemanich *et al.* 1981).

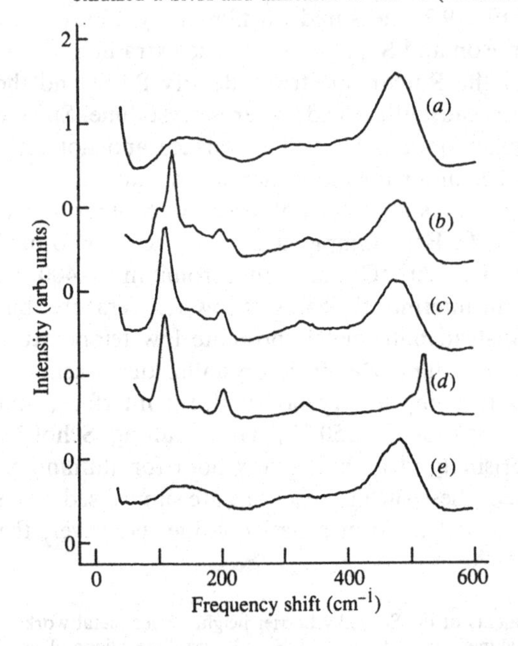

Fig. 9.8. The bias dependence of the effective barrier height of Pd on n-type a-Si:H with different doping levels, showing the reduction due to tunneling as the field at the contact increases (Jackson *et al.* 1986).

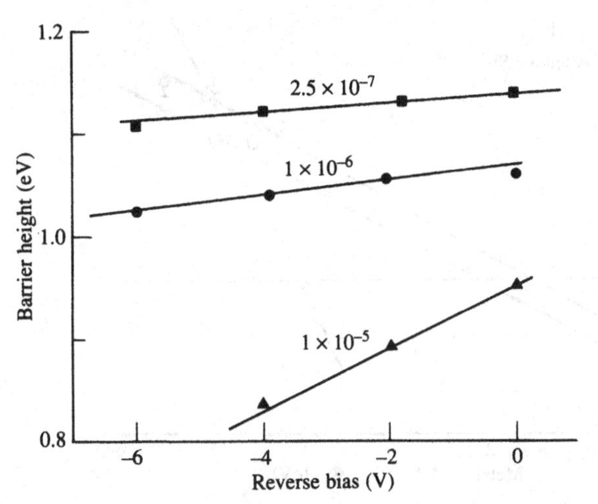

oxide layer also modifies the barrier. It is often difficult to determine exactly which factors control the barrier height. Interface chemical reactions, lateral non-uniformity, surface states and dielectric surface layers are all present under some circumstances and probably all contribute to some degree.

The barrier height decreases with n-type doping and is bias-dependent, as shown in Fig. 9.8 for a palladium contact (Jackson, Nemanich, Thompson and Wacker 1986). The reduction is largest at doping levels above 10^{-5} $[PH_3]/[SiH_4]$, even though the greatest shift in the Fermi energy is at a lower doping level. The Fermi energy does not directly change the barrier height and the origin of the reduction is the large defect density induced by doping. The calculated value of the depletion width is shown in Fig. 9.9, using Eq. (9.6) and the known values of N_D (Fig. 5.9). The narrowing of the junction allows tunneling at lower energies which reduces the apparent barrier height and is also responsible for the bias dependence of Φ_B in Fig. 9.8. The barrier collapses completely at a doping level of 10^{-2} $[PH_3]/[SiH_4]$ and the contact becomes ohmic. This property is used in many devices, such as thin film transistors.

Fig. 9.9. Calculated doping dependence of the depletion width and depletion field at different bias voltages.

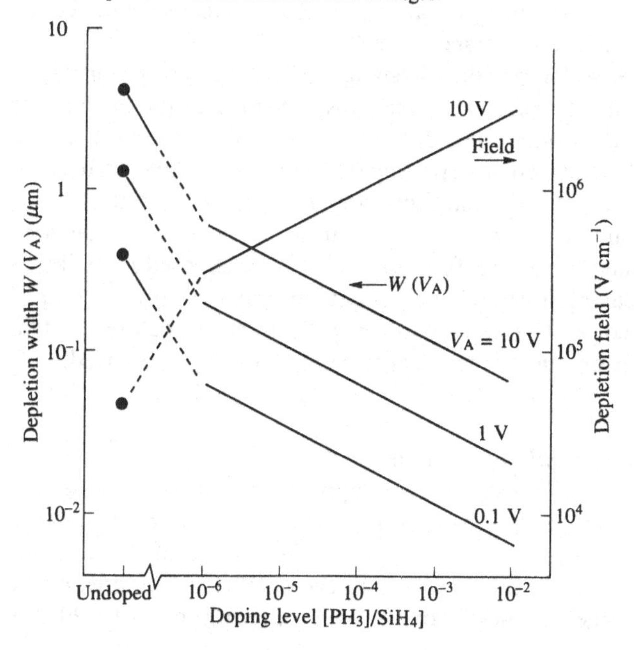

9.2 Surfaces

The surface of a semiconductor generally has different properties from the bulk. The surface of a thin film material such as a-Si:H is particularly significant because the 'bulk' is never far away. Interest in multilayer structures which are described in Section 9.4 and the development of thin film field effect transistors, which are described in Section 10.2, has focussed attention on surface and interface properties.

The surface may influence the electronic properties of the material through the presence of surface states or surface charge. Surface states act as recombination centers or traps for electrons and holes, while surface charge results in band bending extending into the bulk. The internal field causes carrier transport and separates electron–hole pairs. The band bending is important since, in general, the depletion layer width is greater than the carrier diffusion length (see Section 10.1). The direct influence of surface states extends as far as the diffusion length, which is of order 1000 Å at room temperature. Surface recombination is weak at low temperature, when the diffusion length is short (see Section 8.3.6), but is more significant at room temperature where it plays a role, for example, in the collection efficiency of photoconductivity. The band bending extends over the depletion layer width which is about 1 μm, so that the internal field tends to have a larger effect on the properties than the surface states.

There is also a compositional change at the surfaces or interface. Oxidation of a free surface is the most obvious example, but the hydrogen concentration is also different near a surface or interface and this presents difficulties in interpreting the data on multilayer structures.

Most studies of a-Si:H surfaces are made after exposure of the sample to the ambient atmosphere. So far no convenient way has been devised to obtain a fresh surface on a previously exposed material, as is done by cleaving a crystal in vacuum, so that the only method of studying a clean surface is to perform measurements before the film is exposed, but only a few such experiments have been reported.

9.2.1 *Oxidation and surface states*

Amorphous silicon oxidizes upon exposure to air, at a rate which depends on the microstructure of the film. Columnar material (see Section 2.1.1) oxidizes rapidly along the surface of the columns because the intercolumnar regions are porous. The surface area of the columns is so large that an effective oxygen concentration of 10–20 at%

is created in this way. In contrast, high quality CVD material is impermeable to oxygen and only a thin surface oxide results; its growth rate is shown in Fig. 9.10. The thickness increases as the square root of the exposure time, indicative of a diffusion limited process and the oxide thickness is 5–10 Å after a few days (Ponpon and Bourdon 1982). Initially the growth rate is lower than on crystalline silicon. It is known that hydrogen has the ability to passivate the crystalline silicon surface and so its presence on the a-Si:H film is probably the reason for the slower rate. The different oxidation properties of PVD and CVD films are a dramatic illustration of the different a-Si:H structures that can be produced.

The surface states are detected by ESR, PDS and photoemission experiments. In the first two cases, the surface state density is deduced from the thickness dependence of the defect density (Jackson, Biegelson, Nemanich and Knights 1983). When there are N_S cm^{-2} surface states and N_D cm^{-3} bulk states, then the total number of defects in a film of thickness d, is

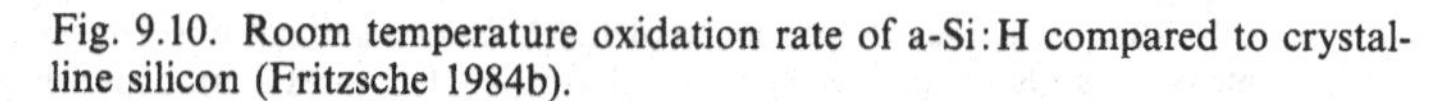

$$N_T = N_S + dN_D \tag{9.21}$$

Fig. 9.11 shows the thickness dependence of N_T for high quality undoped a-Si:H. The finite intercept upon extrapolation to $d = 0$ gives

Fig. 9.10. Room temperature oxidation rate of a-Si:H compared to crystalline silicon (Fritzsche 1984b).

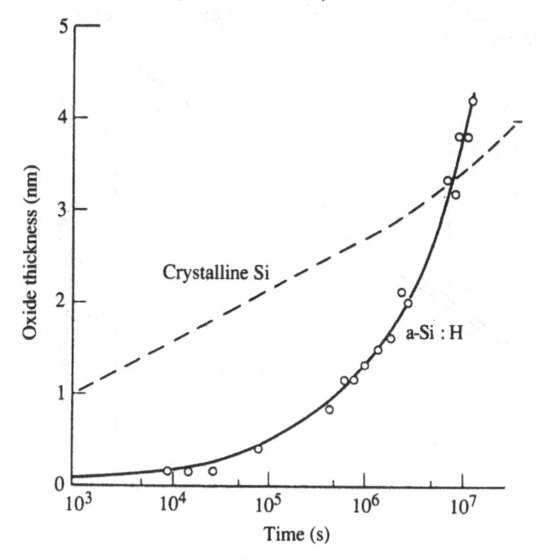

a surface states density of about 10^{12} cm^{-3}. The density increases in samples of high bulk defect density. Doped samples are interesting because the surface state density is slightly larger in n-type a-Si:H, but is an order of magnitude lower in p-type material.

Photoemission is sensitive only to the surface layer of depth 10–50 Å and so directly measures the surface region. However, this experiment alone cannot determine whether the states measured are specific to the surface region or characteristic of the bulk. The photoemission yield experiment has the sensitivity to explore the very low density of states above the valence band tail and is one of the few techniques which can be performed on clean surfaces, not exposed to air prior to measurement. Some yield spectra of the valence band tail and the defects are shown in Fig. 3.9. Clean undoped a-Si:H has an effective defect density of about 10^{17} cm^{-3}, which is attributed to surface states, since the density is higher than the known bulk defect density. A surface state density of 10^{12} cm^{-2} is inferred from the estimated emission depth of about 100 Å. Oxidation of the surface is observed to increase the surface states density by one order of magnitude or more and also to change the defect energy level by 0.2 eV (Winer and Ley 1987, 1989). Boron doping reduces the surface state density and phosphorus increases it slightly, and both results agree with PDS experiments.

Fig. 9.11. Thickness dependence of the ESR and subgap absorption measurements of defects in undoped a-Si:H. The inset is an expansion of the small thickness data. The non-zero intercept is due to surface or interface states (Jackson *et al.* 1983).

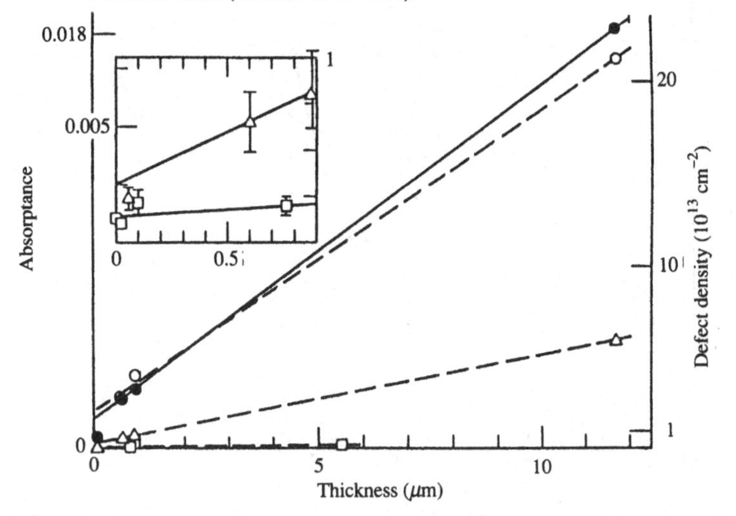

It is natural to suppose that the differing surface state densities are due to compositional or morphological changes at the surface as a result of oxidation or doping. However, accompanying the oxidation and doping are changes in the surface Fermi energy (or work function) which are shown in Fig. 9.12 for the oxidized surface of undoped and boron doped films. Large exposures of oxygen shift E_F and eventually pin its position, independent of doping level. The movement of E_F provides a different mechanism for the surface defect density, because thermal equilibration of the defects is sensitive to the position of E_F. According to the equilibrium models discussed in Chapter 6, the lowest surface state density is expected when the Fermi energy is in the middle of the band gap, with higher densities when E_F moves towards either band edge. The experimental observations can then be explained if the surface Fermi energy in undoped a-Si:H is shifted towards the conduction band. High boron doping compensates for the shift and moves the Fermi energy back to the middle of the gap, so reducing the induced defect density. The photoemission data find that the Fermi energy of a clean surface is high in the gap, being 1.2 eV above the valence band edge, and that E_F moves down with boron doping. However, other experiments find that the bands bend upwards at the surface of oxidized a-Si:H. It seems likely that the Fermi energy effect

Fig. 9.12. Variation of the surface work function (and Fermi energy) of undoped and boron-doped a-Si:H when exposed to oxygen (Winer and Ley 1989).

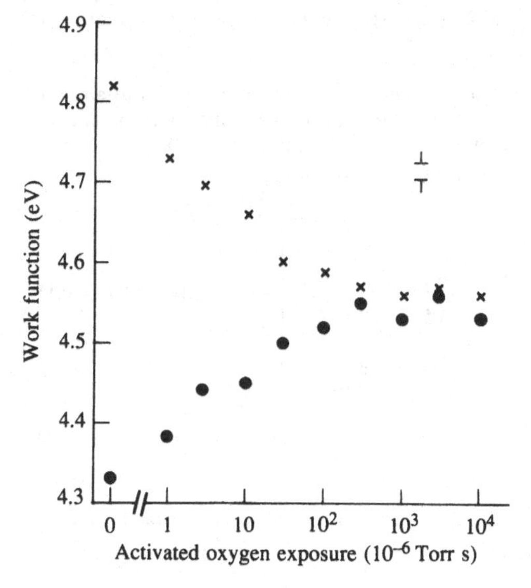

is important, but may not be the only factor in defining the surface states. The thermodynamic models also predict that the neutral defect density depends on the slope of the band tail and the position of the defect states in the gap, so that any change in the disorder or bonding structure near the surface also contributes to the surface state density.

9.2.2 *Gases adsorbed on the surface*

The surface Fermi energy and band bending in samples exposed to the ambient atmosphere are complicated by the effects of adsorbed molecules. The influence of gas exposure on the conductivity of a sample is an excellent illustration of how the surface can completely dominate over what is supposedly a 'bulk' measurement and highlights the caution needed in interpreting conductivity data (Tanielian 1982). The room temperature conductance changes dramatically when the surface is exposed to water vapor or other gas molecules. The effects depend on doping, as is illustrated for different types of samples in Fig. 9.13. There is a rapid initial rise in conductivity after exposure to water vapor of n-type a-Si:H, followed by a slow decay. The changes are reversible by annealing the sample above about 100 °C which drives off the water vapor. Fig. 9.13(*b*) shows an even larger conductance change after exposure to ammonia and dimethyl ether of p-type samples and illustrates that the effects are quite general. Note that the sign of the conductivity change depends on the doping of the sample.

The effects are due to surface band bending induced by charge from the adsorbed molecules. It is known that water acts in this way, with

Fig. 9.13. Change in the conductivity of a-Si:H when exposed to a gas; (*a*) water vapor exposure of n-type material; (*b*) water vapor, ammonia, and dimethyl ether exposure of p-type material (Tanielian 1982).

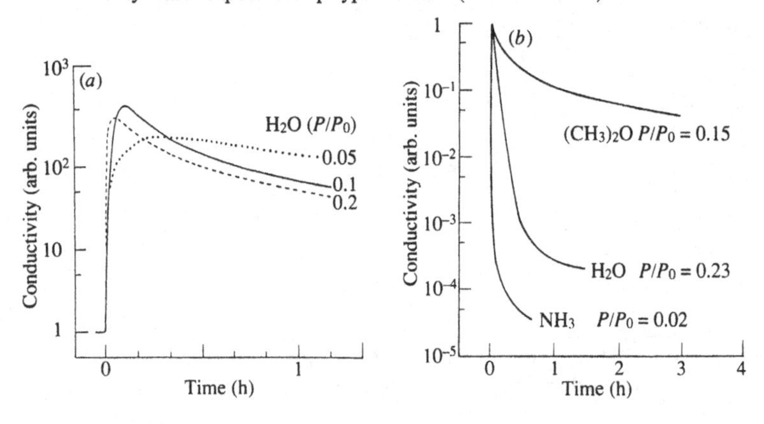

properties of an electron donor. Nor should it be too surprising that the conductivity changes by a large amount. The technology of a-Si:H field effect transistors, described in Chapter 10, is based on the fact that large conductance changes occur when a charge is induced at the surface.

The mechanism of band bending for adsorbed water vapor is illustrated in Fig. 9.14. The water molecule causes a fixed positive charge at the surface which is compensated by an equal negative charge in the a-Si:H film. Material with a low defect density has a wide accumulation layer and the shift of the Fermi energy at the surface is large enough to cause the large changes in the conductance. If the defect density is high, the extra charge resides in traps and the conductance changes much less. The surface effects are therefore largest in the highest quality films. The effects on n-type material are smaller, because the Fermi energy is at the conduction band edge and the defect density is high. Therefore the accumulation layer is narrow and the change in conductivity small. In p-type a-Si:H, the water molecules act as compensating centers and decrease the conductivity.

The data show that the effect of adsorbed molecules is more complicated than is described above, as there are fast and slow processes with opposite conductance changes (see Fig. 9.13(*a*)). One possible explanation is that the band bending induces defect states, through the defect equilibration process described in Chapter 6. This might explain the slow decrease of the conductance after the initial

Fig. 9.14. Illustration of the mechanism of surface band bending by adsorbed water molecules which act as electron donors.

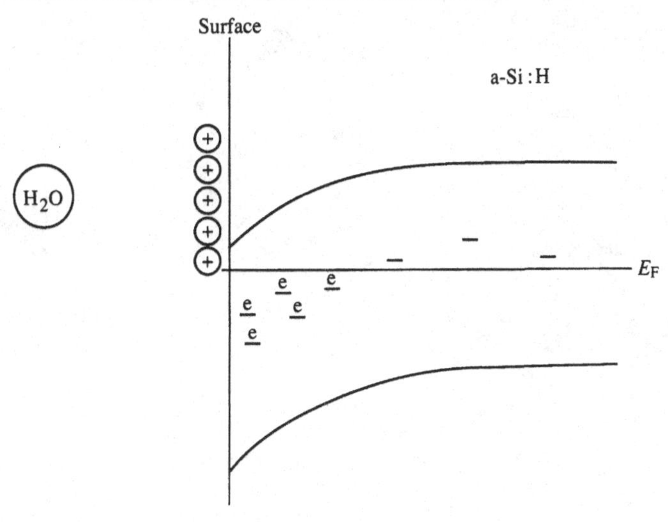

band bending, although the time constant of about an hour at room temperature is faster than the bulk equilibration time of undoped a-Si:H, which takes a week or more to be noticeable. It is not known whether the equilibration effects are more rapid near a surface because of its different structure, or whether a completely different mechanism is in operation.

9.3 Interfaces with dielectrics and semiconductors

The ability to grow smooth and abrupt interfaces between a-Si:H and its various alloys is necessary for most of the technological applications and for the growth of thin multilayer structures. The property follows from the smooth conformal CVD growth and because it is easy to change the reactor gas composition in a time less than that required to deposit one monolayer. It is also essential that atomic interdiffusion at the interface is negligible at the growth temperature. Fig. 9.15 is a transmission electron micrograph of alternating a-Si:H and silicon nitride layers. The thinner layers are 25 Å thick and have a sharp interface. The view is a cross-section extending through about

Fig. 9.15. Transmission electron micrograph showing the cross-section of a multilayer structure containing alternate layers of a-Si:H and a-Si_3N_4:H (Tsai *et al.* 1986b).

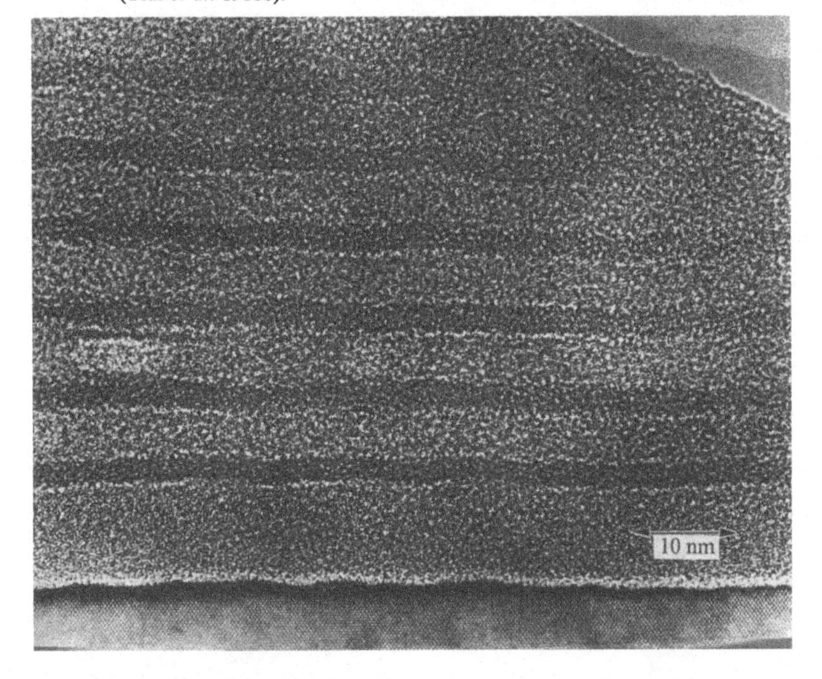

100 Å of material, so that the sharpness of the interface also implies that there is little roughness of the interface.

More precise measurements of the abruptness of the interface are obtained from Raman spectroscopy and X-ray scattering. Some Raman spectra of a-Si:H and a-Ge:H multilayers are shown in Fig. 9.16 (Persans, Ruppert, Abeles and Tiedje 1985). The TO phonon modes of the two materials are at 480 cm^{-1} (silicon) and 280 cm^{-1} (germanium) and the vibrations of Si—Ge bonds are at the average frequency of 380 cm^{-1}. Fig. 9.16 compares data for the individual materials and alloys and for multilayers of different thickness. The relative intensity of the different vibrational frequencies in the spectrum allows an estimate of the fraction of Si—Ge bonds. Few Si—Ge bonds are observed until the multilayer spacing is reduced to a few atomic spacings, showing that there is virtually no intermixing of the layers. The areal density of Si—Ge bonds at the interface is measured to be $(2\text{–}3) \times 10^{15}$ cm^{-2} which corresponds to an intermixing of between one and two monolayers. Other a-Si:H interfaces, such as with silicon nitride are similarly abrupt.

These measurements are not, however, the complete story. Electron

Fig. 9.16. Raman spectra of a-Si:H/a-Ge:H multilayers with repeat distances of 8 Å, 32 Å and 160 Å and the spectra of the corresponding bulk material and the uniform alloy (Persans *et al.* 1985).

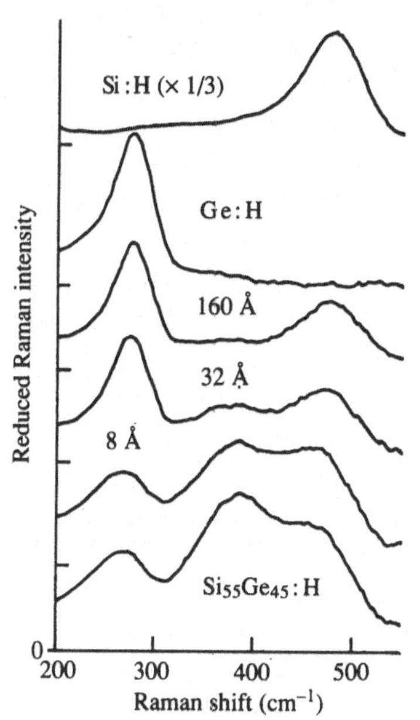

microscopy, Raman and X-ray scattering give information about the heavier elements, silicon, germanium, nitrogen etc., but not about the hydrogen. The hydrogen content is considerably different near an interface and the influence of the interface extends for many atomic layers. Abeles, Wronski, Persans and Tiedje (1985) observe that the total hydrogen concentration in a-Si:H/Si_3N_4 multilayers increases as the layer spacing decreases and there is an excess hydrogen concentration of about 10^{15} cm^{-2} at each interface, extending as much as 20 Å from the interface. The change in composition is a significant problem in interpreting multilayer optical data (see Section 9.4.2), since the increase in hydrogen content alters the band gap and possibly also the electrical properties.

9.3.1 *Band offsets*

The electronic properties of the interface depend in part on how the conduction and valence bands line up when they come into contact. The energy of the band offsets are obtained from X-ray

Fig. 9.17. Upper curves are photoemission data for (*a*) the valence band and (*b*) the conduction band of thin a-Si_3N_4:H layers on a-Si:H. The dotted curves are fits to the spectra using the measurements of the individual bands shown in the lower curves. The band offsets, ΔE_V and ΔE_C, used to fit the data, are indicated (Iqbal *et al.* 1987).

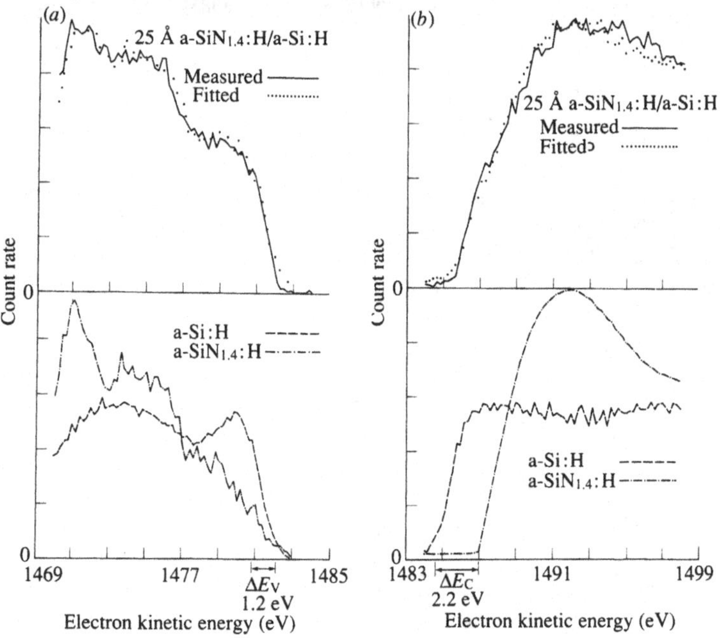

photoemission measurements of samples grown with thin overlayers (Karcher, Ley and Johnson 1984, Iqbal *et al.* 1987). The photoemission spectrum contains the valence bands of both materials, provided the overlayer thickness is less than the electron escape depth of about 50 Å. Some examples of a-Si:H/a-Si_3N_4 layers are shown in Fig. 9.17. Both valence bands are visible in the spectrum and a band offset of 1.2 eV is obtained by a deconvolution using the separately measured valence bands of the two materials. Similar measurements of the conduction bands using inverse photoemission give an offset of 2.2 eV and provide a consistency check, because

$$E_{G1} = E_{G2} + \Delta E_C + \Delta E_V \tag{9.22}$$

E_{G1} and E_{G2} are the band gaps of the two materials (E_{G1} being the larger) and ΔE_C and ΔE_V are the band offsets. For the nitride overlayers, the sum on the right hand side agrees with the silicon nitride band gap of 5.3 eV within about 0.2 eV, which is as accurate as can be expected for this type of measurement. The conduction band offset is nearly twice that of the valence band and the interface is illustrated in Fig. 9.18.

Similar measurements have been made for the interface with a-Ge:H and a-Si:C:H, although there is more disagreement between the various sets of data (Evangelisti 1985). The band offset between a-Si:H and crystalline silicon is particularly interesting. Photoemission

Fig. 9.18. The electronic structure of the interface between a-Si:H and a-Si_3N_4:H, showing the band offsets and the band bending due to charged interface states.

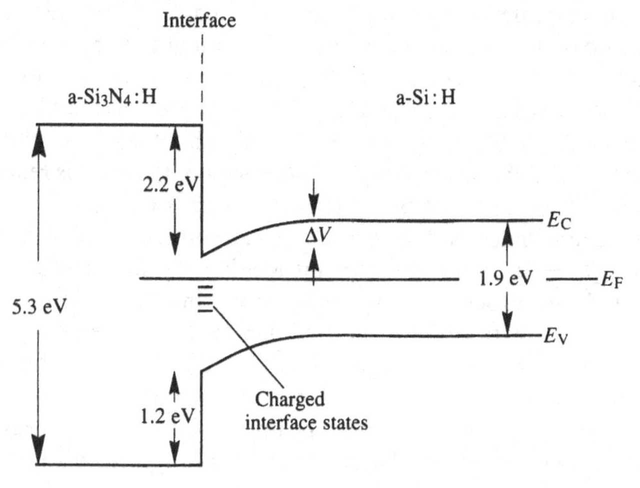

measurements find that the hydrogen in a-Si:H removes states from the top of the valence band and is one reason for the increased band gap compared to the crystal. Accordingly, most of the band offset is at the valence band edge and the conduction bands line up quite well.

9.3.2 *Electronic properties*

Interest in the electronic properties of interfaces centers around a-Si:H/Si_3N_4, because this combination is used in multilayers (Section 9.4) and field effect transistors (Section 10.1.2). The electronic structure of the interface is illustrated in Fig. 9.18. Apart from the band offset which confines carriers to the a-Si:H layer, the distribution of localized interface states and the band bending are the main factors which govern the electronic properties of the interface. The large bulk defect density of the Si_3N_4 also has an effect on the electronic properties near the interface. Band bending near the interface may result from the different work functions of the two materials or from an extrinsic source of interface charge – for example, interface states.

Transient photoconductivity measurements of the depletion width, as described in Section 9.1.3, show that there is an electron accumulation layer at the interface with Si_3N_4 (Street *et al.* 1985b). In contrast, an oxide interface (either a native or deposited oxide) has a depletion layer (Aker, Peng, Cai and Fritzsche 1983). The band bending causes similar changes in the conductance of the films as is described for adsorbed molecules in Section 9.2.2.

The interface band bending depends on the order in which the films are deposited. Fig. 9.19 compares the transient photoconductivity charge collection for structures in which the a-Si:H film is deposited either before or after the nitride layer (referred to as top nitride or bottom nitride). The sign of the charge collection at zero bias indicates that both types of interface have electron accumulation layers, although the top nitride gives a larger charge, indicative of greater band bending according to Eq. (9.19). A reverse bias of only about 0.5 V is sufficient to deplete the bottom nitride sample, whereas more than 5 V is required for the top nitride. The zero bias interface charge is $C_N V_A$, where C_N is the nitride capacitance and V_A is the applied voltage at the flat band condition ($Q_C = 0$). Thus the top nitride has both a larger band bending and an order of magnitude more interface states. This dependence on growth order cannot be due to the difference in work functions between a-Si:H and Si_3N_4 and shows that extrinsic effects are dominating. The origin of the band bending is unknown, but must involve a source of positive charge at the interface equal to the negative charge of the electron accumulation. Other measurements also find an asymmetry, but with the bottom nitride interface having the larger interface state density (Roxlo, Abeles and Tiedje 1984). Apparently the effects depend on the details of growth conditions.

A simple way of studying the interfaces is to use multilayer samples, which contain many interfaces and to apply the techniques normally used for bulk measurements. Fig. 9.20 shows ESR data of the silicon nitride interface for such samples (Tsai, Street, Ponce and Anderson 1986b). The total thickness of each sample, 0.6 μm, and the thicknesses of the nitride and a-Si:H layers are equal; the only difference is in the thicknesses of the individual layers and therefore in the number of interfaces. The dependence of the ESR spin density on this parameter shows that the interfaces dominate most of the spin measurements. It is not possible, however, to separate out the effects of the top and bottom interfaces, both of which are present in equal number. The top nitride interface is presumably the main source of the signal, since it has the larger interface state density.

Both dark and light-induced ESR are shown in Fig. 9.20. The dark spin density first drops as the number of layers increases, reaching a minimum at 20–30 periods, and then increases. The *g*-value of the decreasing spin signal identifies it as the bulk dangling bond defect in the nitride. The reduction is explained by the band bending which transfers charge from the nitride to the interface and changes the electron occupancy of the defects. The increasing spin density has a different *g*-value, and its dependence on the number of layers identifies

Fig. 9.19. Transient photoconductivity measurements of the band bending at the interface of a-Si:H and a-SiN$_4$:H, showing the effects of growth order (Street and Thompson 1984).

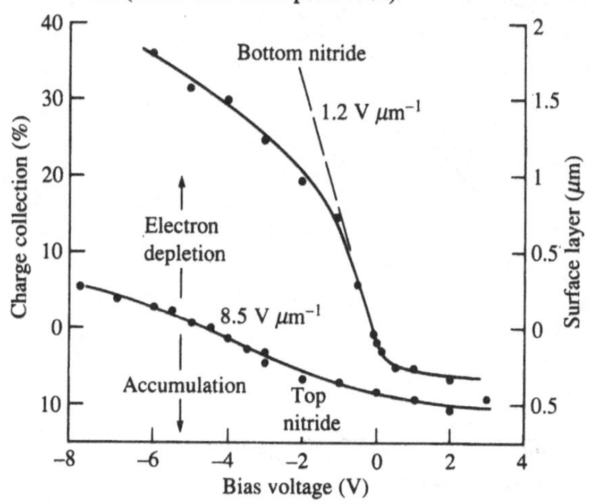

it as interface states. The LESR signal also increases with the number of layers, thus identifying it as an interface effect and has a much larger spin density than the dark signal. Most of the interface states are therefore not paramagnetic until their charge state is changed by the light.

The interface has electronic properties similar to n-type a-Si:H. The conduction band moves towards the Fermi energy due to the band bending, and there is a high density of negatively charged dangling bond defects. The top nitride, which has stronger band bending, also has a higher defect density. The similarity with the doping process suggests that the interface states arise from defect equilibration effects and are caused by the band bending rather than by bond strain at the interface. The positive charge which leads to band bending is equivalent to the donor states and tends to move the surface Fermi energy towards the conduction band edge. As a result, charged dangling bond defects are induced at the interface. Section 6.4.2 shows that an accumulation layer induces dangling bonds at the silicon/nitride interface through the

Fig. 9.20. Dark and light-induced ESR spin density for a-Si:H/a-Si_3N_4:H multilayers with different numbers of repeat periods (Tsai *et al.* 1986b).

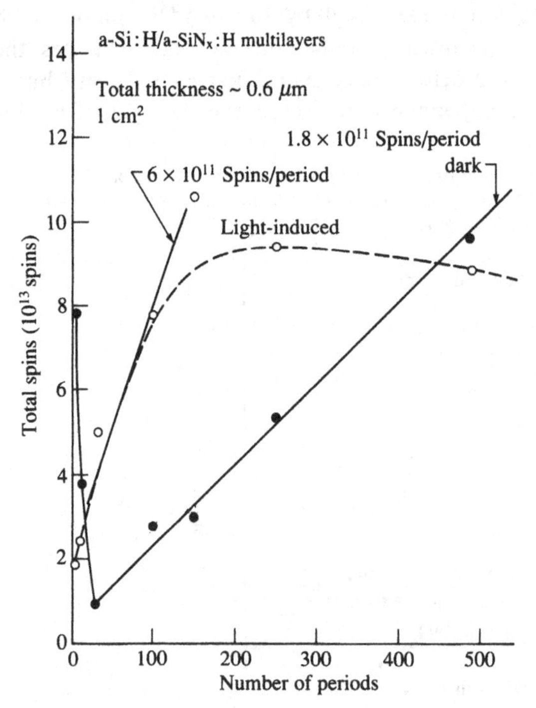

movement of the Fermi energy and the physical mechanism of interface states that are discussed here is essentially the same.

The origin of the positive charge at the interface is unclear, but might be ammonia molecules at the interface. Many different gases are adsorbed on the surface of a-Si:H, contributing fixed charge and causing band bending (see Section 9.2.2). Ammonia adsorption results in a positive fixed charge, which is consistent with the observed band bending.

The electronic properties of carriers near the interface are also influenced by bulk traps in the nitride or other dielectric material. Trapping at interface states and in the bulk dielectric are differentiated by their respective time constants. The distinction between 'fast' interface states and 'slow' bulk states is well known in crystalline silicon. Similarly, the injection of charge from the a-Si:H layer into the nitride has a range of time constants which can extend for hours at room temperature.

The extent of the slow state trapping depends on the quality of the dielectric layer (Street and Tsai 1986). Fig. 9.21 shows the slow charge trapping for nitrides which have a thin compositionally-graded layer at the interface consisting of material with a lower nitrogen concentration and therefore a smaller band gap than Si_3N_4. The concentration of slow

Fig. 9.21. Density of slow charge trapping states at an interface of a-Si:H with a-Si_3N_4:H when there is a thin graded-composition layer at the interface (Street and Tsai 1986).

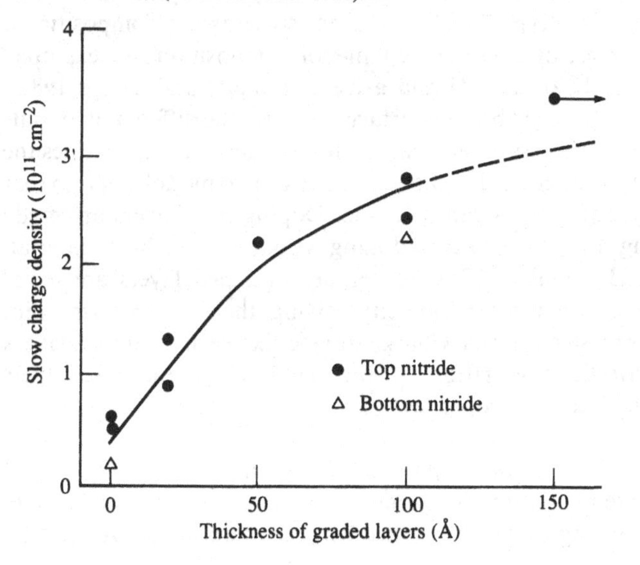

states increases with the thickness of this layer, eventually saturating at 100–200 Å, which represents the farthest distance that charge can tunnel into the dielectric.

Minimization of the density of slow states and interface states is of technological importance for field effect transistors. Presently available nitrides are of sufficient quality that the slow states have no significant effect on electrons at normal temperatures, but some hole trapping occurs. The use of a bottom nitride structure reduces the interface state density to an almost negligible amount. The remaining problem associated with the interface of the transistors is the generation of metastable interface defects in the a-Si:H film due to the accumulation bias.

9.4 Multilayers

The growth of a multilayer structure creates a new material with an imposed periodicity of the layer spacing. Such structures are familiar in crystalline semiconductors; GaAs/GaAlAs multilayers have been particularly widely studied. Both the electronic and vibrational states of the material are influenced by the multilayer structure. The quantum confinement of the electrons and holes in the narrow wells of the multilayer is of special interest because it is not obvious that such effects can occur in amorphous semiconductors. Quantum effects require that the coherence length is larger than the size of the confining well. The short mean free path of the carriers in a-Si:H implies that quantum effects are observed only in very narrow wells.

There are two types of multilayer structures. Compositional multilayers consist of a periodic change of composition, for example, a-Si:H and Si_3N_4, or a-Si:H and a-Ge:H (Abeles and Tiedje 1983). Such structures require abrupt interfaces between the different materials when the layer thickness is very small. The previous section decribes the evidence that the interface is confined to one or two monolayers, so that thin multilayer structures can be made. Doping multilayers are made by alternating n-type and p-type doping within a single host material, usually a-Si:H (Döhler 1984). Sometimes undoped layers are interleaved to give a n–i–p–i periodicity, giving the generic term 'nipi structure' for these materials. Charge transfer between the doped layers causes a periodic modulation of the band edges and gives their characteristic electrical properties.

9.4.1 *Recombination in compositional multilayers*

Low temperature non-radiative recombination in bulk a-Si:H occurs by tunneling to defects over a distance of about 100 Å (Section

8.4.1). The recombination is modified in a multilayer structure whose layer spacing is similar to the carrier tunneling distance and is observed in photoluminescence measurements (Tiedje 1985). Fig. 9.22(*a*) shows that the luminescence intensity of a-Si:H/nitride multilayers decreases as the layer thickness drops below about 500 Å. The interface states and bulk nitride defect states cause non-radiative recombination because the electron–hole pairs are never far from an interface. The model of non-radiative tunneling developed in Section 8.4.1 can be adapted for recombination in thin layers. When the layer thickness is less than the critical transfer radius, R_C, the luminescence efficiency is (see Eq. (8.52)).

$$y_L = \frac{1}{L}\exp(-\pi R_C{}^2 N_I)\int_0^L \exp(-\pi x^2 N_I)\,dx \tag{9.23}$$

where N_I is the density of interface states and L is the layer thickness. There is a modified expression when the layer thickness is greater than R_C and the solid line in Fig. 9.22(*a*) shows that the model accounts well for the data. The fit obtains an interface state density of 1.5×10^{12} cm^{-2}, which is a little higher than the other estimates.

The opposite effect happens in multilayer films of materials in which the luminescence is strongly quenched by bulk defects, for example, a-Ge:H. The luminescence intensity increases in a-Ge:H/a-Si:H

Fig. 9.22. Dependence of the luminescence intensity on layer spacing in (*a*) a-Si:H/a-Si_3N_4:H multilayers and (*b*) a-Si:H/a-Ge:H multilayers. The luminescent material is indicated (Tiedje 1985).

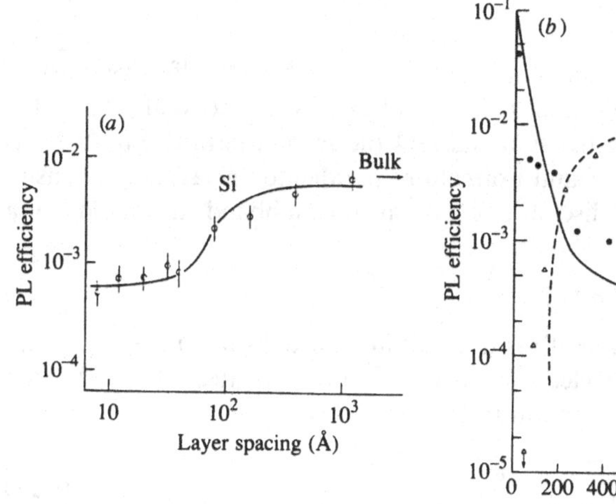

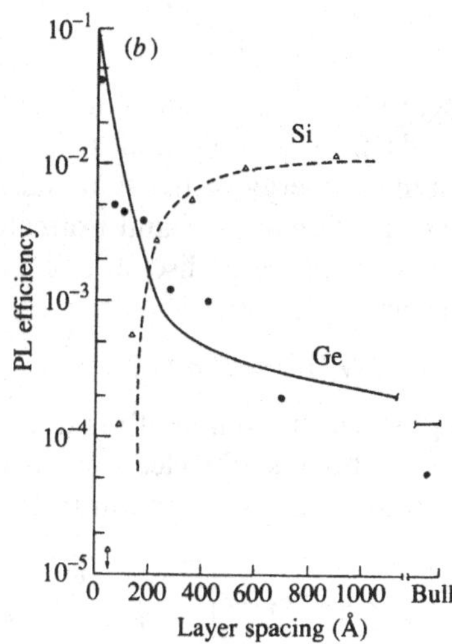

multilayers as the layer width is decreased, as shown in Fig. 9.22(*b*). The increase occurs because the defects are confined in a two-dimensional layer. The luminescence efficiency changes from Eq. (8.52) to

$$y_L = \exp(-\pi R_C{}^2 L N_D) \tag{9.24}$$

which is larger than the bulk value when $L < 4R_C/3$. The efficiency of the a-Ge:H luminescence increases by four orders of magnitude when the multilayer thickness decreases and fits Eq. (9.24) with a critical transfer radius of 200 Å.

9.4.2 *Quantum confinement of electronic states*

Fig. 9.23 shows a schematic diagram of the bands of a compositional superlattice, composed of a-Si:H and nitride. The large difference in the band gaps (1.8 versus 5.3 eV) and comparable band offsets give deep two-dimensional confining potentials for both electrons and holes in the a-Si:H layers. The potential well causes quantization of the energy levels in the direction of growth. The electronic states split into a series of subbands which, in a crystal, have an energy dispersion in momentum space along the other two dimensions. For a crystal with parabolic bands and an infinite well depth, the energy of the subbands above the bottom of the well is (Dingle 1965)

$$E = E_{sn} + \frac{\hbar^2}{2m^*}(k_x{}^2 + k_y{}^2) \tag{9.25}$$

$$E_{sn} = \frac{\hbar^2}{2m^*}\left(\frac{n\pi}{L}\right)^2 \tag{9.26}$$

where n is the quantum state ($n = 1,2,\ldots$), m^* is the effective mass of the electron or hole and L is the thickness of the well. E_{sn} is the minimum energy of the subband and the second term in Eq. (9.25) is the dispersion in the x and y directions parallel to the layer. The density of states changes discontinuously at the subband threshold to a constant value of

$$N_s(E) = 2\pi m^*/\hbar^2 \tag{9.27}$$

In practice the well depth is not infinite and the subband energies are reduced because the electron wavefunction penetrates into the barrier material. The wavefunction in the z direction is,

$$\Psi = c_1 \sin\left[\frac{(2m^* E_{sn})^{\frac{1}{2}}}{\hbar} + \delta\right] \tag{9.28}$$

in the well and in the barrier material

$$\Psi = c_2 \exp\left\{\frac{[2m^*(V_0 - E_{sn})]^{\frac{1}{2}} z}{\hbar}\right\} \tag{9.29}$$

where V_0 is the barrier height and δ is a constant obtained from boundary conditions. A limited number of quantum levels can be contained within a non-infinite well.

Momentum is not a good quantum number in amorphous semiconductors, so that the validity of the above expressions is questionable. Just as the energy bands of a three-dimensional solid are replaced by a density of states distribution, so the subband dispersion relations (Eq. (9.25)) are replaced by a sequence of two-dimensional density of states distributions starting at the subband thresholds. It is generally assumed that the form of the density of states is the same as given by Eq. (9.27) for the crystalline case, but there has not been a detailed theoretical analysis. The three-dimensional bulk density of states in a-Si:H near the band edge is more nearly linear than parabolic, so that the subband density of states is probably also not accurately described by Eq. (9.27). Whatever the exact form, the discontinuity in $N(E)$ at the subband threshold energy remains, roughly as illustrated in Fig. 9.23.

Observation of the subbands is the crucial test of quantization effects and the most common experimental measurement is optical absorption. The subband discontinuities give a joint density of states with a stepped

Fig. 9.23. Illustration of the subband structure and density of states distribution of a quantum well.

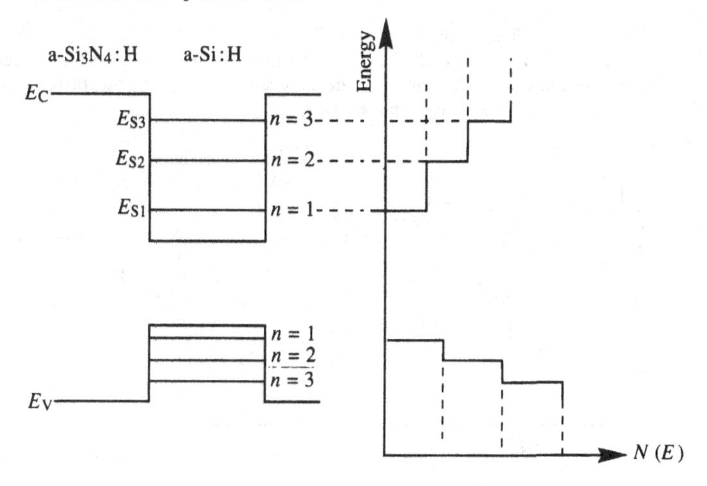

structure as shown in Fig. 9.24(*a*). The wavefunction in the z direction is such that only transitions between electrons and holes at the same values of n are allowed. This rule is the same for amorphous and crystalline states provided that the wavefunctions are described by Eq. (9.28). Exciton effects in the crystal lead to a series of peaks in the optical spectra, as illustrated in Fig. 9.24(*b*). It is unlikely that these would be seen in the amorphous multilayers, as excitons are not observed in the bulk material. However, the step structure should be present, although probably smeared out by the disorder (see Fig. 9.24(*b*)). The quantum confinement raises the energy levels of the electron and hole and shifts the absorption edge to higher energy.

Many attempts have been made to observe the quantization effects in a-Si:H. Fig. 9.25 shows some early measurements of the optical absorption edge in a-Si:H/nitride multilayers (Abeles and Tiedje 1983). As the a-Si:H layer thickness is reduced, the optical absorption moves to higher energies, as predicted by the quantization effects. The shift is about 0.1 eV with a layer thickness of 40 Å, increasing to 0.5 eV for an 8 Å layer. The magnitude of the change is consistent with the predictions for the subband energies of Eq. (9.26), with an effective mass of order unity. However, the results in Fig. 9.25 are not fully convincing because there is no sign of any discontinuities in the spectrum at the subband thresholds. The shift of the absorption may simply be due to the compositional changes which occur near an interface. The increase in the hydrogen concentration near the a-Si:H/nitride interface (see Section 9.3) could be responsible.

More recent measurements have identified the subband thresholds by modulated absorption and by resonant tunneling. The thermo-

Fig. 9.24. Illustration of the multilayer optical absorption: (*a*) the one-particle joint density of states; (*b*) the ideal absorption spectrum including exciton effects (solid line) and the expected broadened absorption of a disordered material (dashed line).

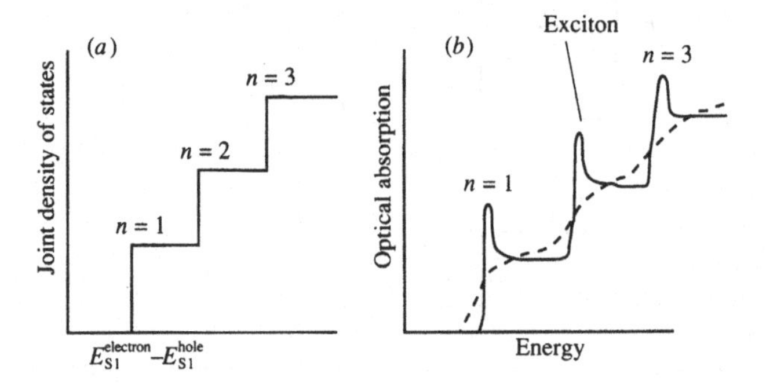

modulation experiment measures the change in optical absorption due to a small oscillatory temperature variation, achieved by pulsed laser heating (Hattori, Mori, Okamoto and Hamakawa 1988). The experiment measures the temperature derivative of the absorption, which is related to the energy derivative through the temperature dependence of the band gap. Thus

$$\frac{\mathrm{d}a}{\mathrm{d}T} = \frac{\mathrm{d}a}{\mathrm{d}E}\frac{\mathrm{d}E}{\mathrm{d}T} \tag{9.30}$$

The derivative spectra should have sharp features at the subband thresholds which are easier to detect than in the normal optical spectrum.

Fig. 9.26 shows data for multilayer structures of different layer spacing measured at a temperature of 100 K. Several features in the spectra are observed whose energy depends on the quantum well thickness and are identified as the subband thresholds. The energies of the subband transitions are shown in Fig. 9.26(*b*) as a function of the layer spacing. Three quantum levels are detected and the solid lines show the theoretical values based on Eq. (9.26). The effective masses used to fit the data are $0.3m_0$ for electrons and m_e for holes.

Quantum confinement is also observed by resonant tunneling, which is a common technique for observing the effect in crystals (Miyazaki,

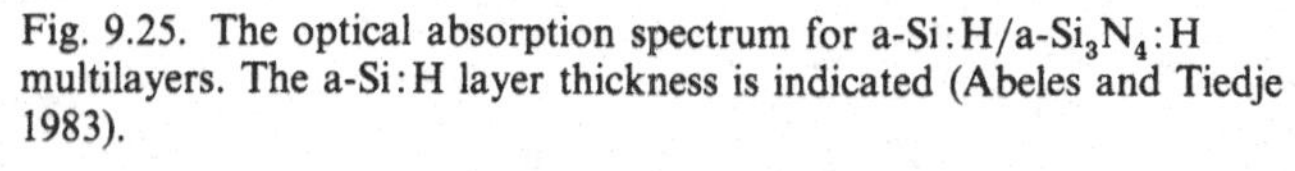

Fig. 9.25. The optical absorption spectrum for $a\text{-Si:H}/a\text{-Si}_3N_4\text{:H}$ multilayers. The a-Si:H layer thickness is indicated (Abeles and Tiedje 1983).

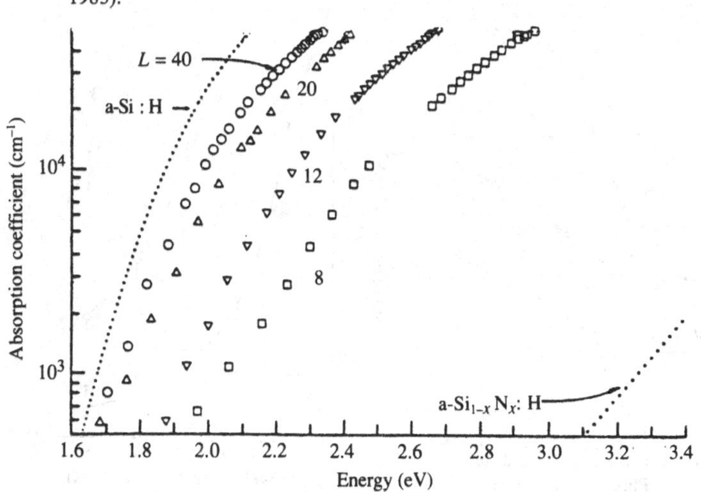

Ihara and Hirose 1987). The structure consists of two nitride barriers, separated by a well made of n-type a-Si:H. At certain values of the applied voltage, the quantum states in the well are resonant with the electron levels outside the well and the current through the nitride barrier is enhanced. The derivative spectrum is expected to show features at the onset of a tunneling threshold. Fig. 9.27 shows this structure in the low temperature current–voltage characteristics. The predicted voltage peaks are shown in the figure and agree well with the data. The features have the correct dependence on the layer thickness and the data agree with the expected quantum levels, assuming an effective mass of $0.6m_0$.

Both the absorption and the resonant tunneling experiments find quantization effects for layer thicknesses of 50 Å or less. It is, however, not immediately obvious why the quantum states should be observed even in these thin layers. The discussion of the transport in Chapter 7 concludes that the inelastic mean free path length is about 10–15 Å at the mobility edge. The rapid loss of phase coherence of the wavefunction should prevent the observation of quantum states even in a 50 Å well, but there are some factors that may explain the observations. The mean free path increases at energies above the

Fig. 9.26. (*a*) Thermomodulated optical absorption data for a-Si:H/Si_3N_4:H multilayers, with the subband thresholds indicated; (*b*) dependence of the subband threshold energies on the well layer thickness. The solid lines are theoretical fits to the data (Hattori *et al.* 1988).

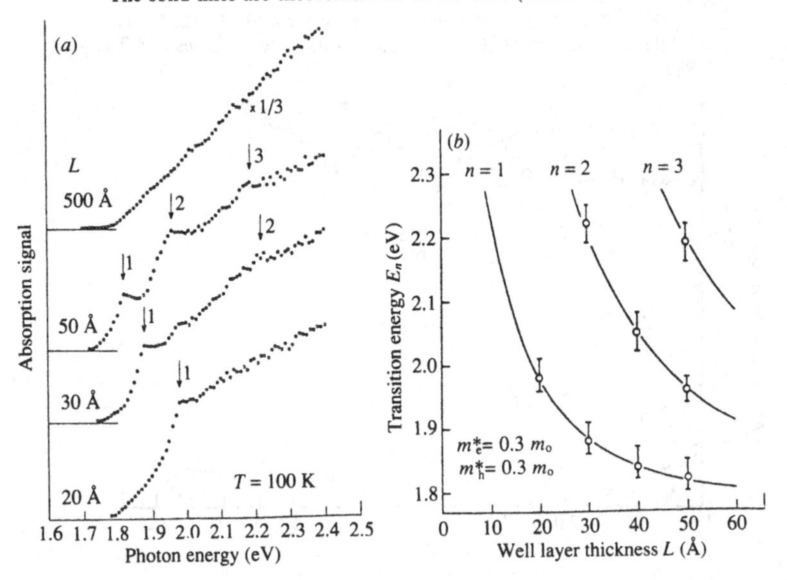

mobility edge and may be substantially larger than 15 Å at the energy of the subbands. The mean free path may also be larger at low temperature, and both measurements were made at 100 K or less. Lastly, the quantum effects give very weak features in the spectra of Figs. 9.26 and 9.27, compared to the equivalent behavior of crystals. The effects may therefore be due to a small subset of carriers that happen to have particularly long mean free paths. One might imagine that the mean free path has a distribution of values with a width approximately equal to the average. A small fraction of the carriers with 2–3 times the average may be sufficient to to account for the quantization effects.

The quantum effects give an opportunity for some detailed information about the electronic properties of a-Si:H. The measurements do not yet agree well for the values of the effective mass and more detailed calculations of the subband energies are needed to take into account the barrier material. It may also be possible to obtain more accurate values of the inelastic scattering length as the magnitude of the quantum effect is better understood.

Fig. 9.27. Current–voltage characteristics of a double barrier structure with well thickness of 40 Å, showing the resonant tunneling features. The predicted voltages of the peaks are indicated (Miyazaki *et al.* 1987).

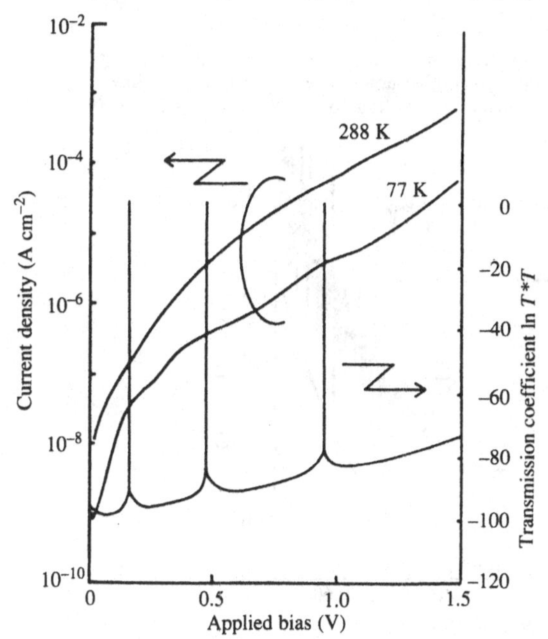

An interesting feature of the resonant tunneling structures is the phenomenon of telegraphic noise (Arce and Ley 1989). Its primary characteristic is that the current through the barrier randomly switches between two or more well-defined values. Fig. 9.28 shows an example of the current–voltage relations which exhibits large noise fluctuations and the insert shows the time dependence of the current at one voltage. In this example the current switches by about 0.5% with a time constant of about 0.1 s. The switching time is thermally activated with an energy which depends on the specific noise feature. The effect is explained by a change in occupancy of a single trap in the barrier, which either changes the barrier height or its potential profile and so causes a different tunneling current in a small area surrounding the trap.

9.4.3 *Quantum confinement of phonons*

There is an analogous quantization effect of the phonon vibrations of multilayer structures (Santos, Hundhausen and Ley 1986). The condition for observation is that the phonon coherence length is larger than the layer thickness. Low frequency acoustic modes fulfill this condition because they are an in-phase motion of a large number of atoms and are not strongly influenced by the disorder – instead reflecting the average bulk elastic properties of the materials.

Fig. 9.28. Current–voltage characteristics for a resonant tunneling multilayer exhibiting telegraphic noise. The insert shows the time dependence of the current at a voltage of 0.5 V (Arce and Ley 1989).

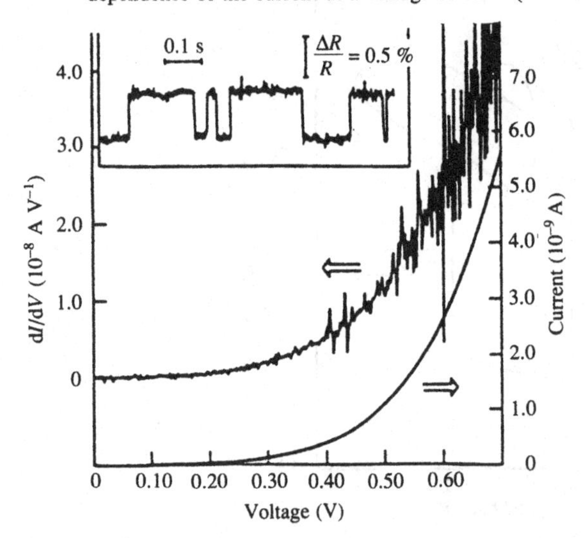

The periodicity of the multilayer provides an additional scattering of the phonons and introduces an artificial Brillouin zone. This leads to a folding of the phonon dispersion curves at wave vectors π/L, where L is the multilayer spacing.

The dispersion of acoustic modes in a uniform material is given by

$$\omega_m(k) = v_s k \tag{9.31}$$

where v_s is the sound velocity and k is the wave vector. A first approximation to the dispersion in a multilayer structure is

$$\omega_m(k) = v_s(k + 2\pi m/L) \tag{9.32}$$

where $m = 0$, ± 1, etc. Fig. 9.29 shows a schematic diagram of the folded dispersion curves for the phonons. The actual dispersion curves are not just a simple folding of the bulk material dispersion, but contain gaps at the zone center and zone boundary because of the different material properties of the a-Si:H and the nitride. The splitting of the dispersion curves arises from phonon modes having different amplitudes in the two materials.

Raman backscattering occurs at twice the wave vector of the incident light, $2k_p = 4\pi n/\lambda_p$, where λ_p is the wavelength of the light and n the

Fig. 9.29. The folded phonon dispersion curves for a multilayer structure. Raman scattering occurs at the intersection with the light wave vector k_p.

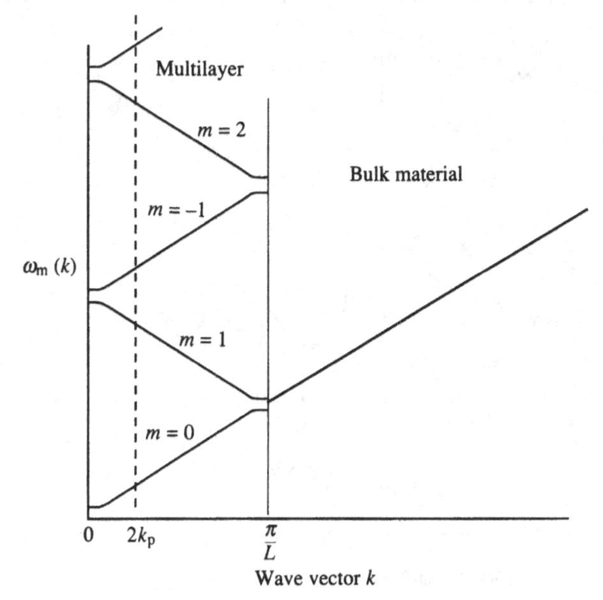

refractive index. The wave vector of the light is small compared to the size of the Brillouin zone, $k_B = \pi/a$ of the normal bulk material – $2k_p$ is approximately 10^6 cm^{-1}, which is about 1% of π/a. The Raman scattering therefore usually measures phonons which are very close to the zone center. The Brillouin zone is greatly reduced in the multilayers because $L \gg a$, so that the k_p is a larger fraction of the folded zone. The ratio of the scattering wave vector and the Brillouin zone is

$$2k_p/k_B = 4nL/\lambda_p \tag{9.33}$$

Fig. 9.29 illustrates the intersection of the light wave vector and phonon dispersion curves and predicts pairs of Raman lines. The Raman shift is 10–100 cm^{-1} for layer widths of order 100 Å.

Fig. 9.30 shows examples of the Raman scattering of a-Si:H/nitride multilayers. Pairs of lines are observed and are attributed to the $m = \pm 1$ lines of the folded zone. The $m = 0$ line is too close to the laser wavelength to be observed and the higher order lines are too broadened out by the disorder. The Raman shift decreases with increasing

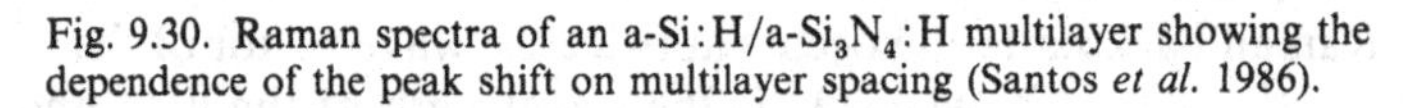

Fig. 9.30. Raman spectra of an a-Si:H/a-Si$_3$N$_4$:H multilayer showing the dependence of the peak shift on multilayer spacing (Santos *et al.* 1986).

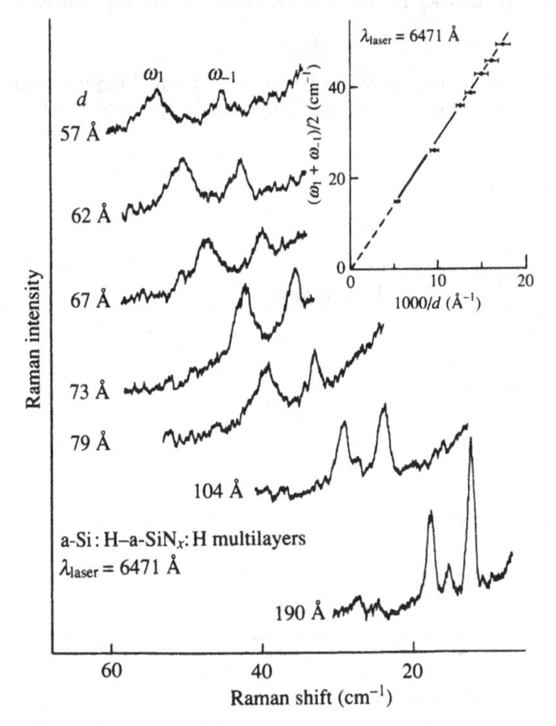

thickness of the layers, since this compresses the Brillouin zone. The average Raman shift of the pairs of peaks in Fig. 9.30 gives the $q = 0$ value and the inset of Fig. 9.30 shows that the average shift scales inversely with the layer thickness. The slope of the data gives the correct longitudinal sound velocity of $(8.5 \pm 0.5) \times 10^5$ cm s^{-1}, confirming that the multilayer is the origin of the Raman peaks.

The magnitude of the gap is measured by choosing the layer spacing so that the wave vector of the incident light exactly matches that of the multilayer (Santos, Ley, Mebert and Koblinger 1987). This condition is obtained with a layer thickness of 370 Å when λ_p is 5145 Å, and the splitting of the gap is found to be 0.5 cm^{-1}. The zone center gap is measured by forward scattering and is a little smaller than the zone boundary gap. The magnitudes of the gaps depend on the differences in the acoustic impedance of the two layers and is calculated by Santos *et al.* (1987).

9.4.4 *Doping multilayers*

Doping multilayers are formed by alternating n-type and p-type doping. This results in a smooth modulation of the bands, as illustrated in Fig. 9.31, because electrons transfer from the n-type to the p-type layer. The maximum amplitude of the modulation is equal to the shift of the Fermi energy between the n-type and the p-type material. There is incomplete transfer when the layer width is less than the depletion width, so that the modulation amplitude is smaller. For this

Fig. 9.31. Illustration of the potential modulation of the bands of a doping superlattice. The dashed line illustrates the screening of the potential by photoexcited charge carriers. Recombination of electrons and holes is by tunneling between layers.

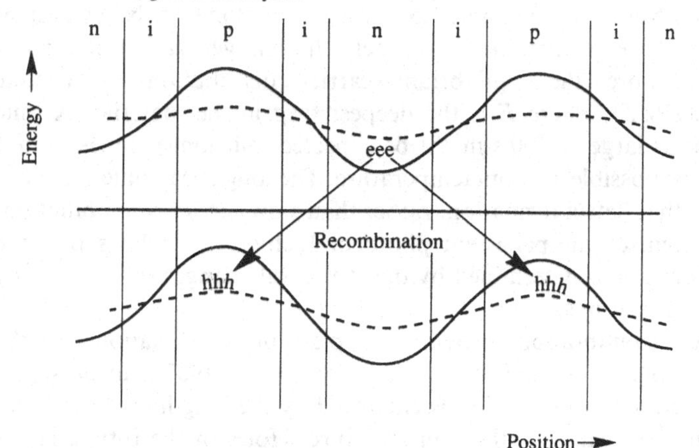

reason, nipi structures usually have layer spacing greater than 100 Å and no quantum confinement effects are seen. There are, however, several other interesting phenomena.

When mobile carriers are excited by illuminating a nipi structure, the internal field causes the electron and holes to separate. The charges screen out the internal field, reducing the modulation of the bands, as illustrated in Fig. 9.31. The response to illumination gives the nipi structure most of its characteristic properties (Döhler 1972). The carriers recombine from the n-type to the p-type layer and the spatial separation increases the recombination lifetime, because the transition is by tunneling. The recombination energy depends on the amplitude of the modulation and varies with the amount of charge transferred between the layers. The slow luminescence decay and a spectrum which changes energy with excitation intensity are both found in crystalline nipi structures. These are difficult to observe in a-Si:H because the luminescence spectrum is already broad, the lifetime is long due to the radiative tunneling recombination and the band tail luminescence is quenched in doped material. A slight shift of the luminescence to higher energy has been reported in nipi structures, the origin of which has, so far, not been explained (Kakalios, Fritzsche and Narasimhan, 1984).

The photoconductivity response of a-Si:H nipi structures has an extremely long recombination lifetime. A brief exposure to illumination causes an increase in the conductivity which persists almost indefinitely at room temperature (Kakalios and Fritzsche 1984). An example of this persistent photoconductivity is shown in Fig. 9.32. The decay time exceeds 10^4 s at room temperature and decreases as the temperature is raised, with an activation energy of about 0.5 eV.

Although the observation of a long recombination time is expected for nipi structures, the time constant is too long to be consistent with the carrier recombination model. Thermal generation of carriers should restore the equilibrium carrier distribution in a time $\omega_0^{-1}\exp(E/kT)$, where E is the deepest trap in the material. A time constant as large as 100 s might be expected, but longer times do not seem to be possible at room temperature. The long decay time therefore suggests that defect formation, rather than slow carrier recombination, is the origin of the persistent photoconductivity. It follows that the relaxation time is determined by the structural changes rather than by electron transitions.

Defect equilibration provides a plausible explanation of the persistent photoconductivity (Kakalios and Street 1987). According to this model, defect generation is enhanced by the long lifetime of holes in the p-type region of the nipi structure. Holes in the p-type layers induce defects at room temperature faster than the creation of defects by electrons in the n-type layers. The extra defects in the p-layers lie above the Fermi energy and consequently are positively charged. The corresponding negative charge remains as band edge electrons in the n-type layer and causes an increase in conductivity which persists as long as the excess defects are present. The annealing of the persistent photoconductivity corresponds to the thermal equilibration of the p layer and occurs at a temperature which is close to that observed in bulk p-type a-Si:H.

The charge-induced defect creation mechanism is too slow to be significant at low temperature and the electronic recombination effects reestablish themselves. Low temperature measurements (0–100 K) have been performed using an IR probe beam to modulate the excess carrier density that is in the band tail states (Hundhausen, Ley and Carius 1984). The carrier lifetime is longer in the nipi structures than for bulk a-Si:H, but the excess carrier lifetimes decrease below a few minutes when the temperature is raised above 50 K. It is concluded that the tunneling recombination mechanism is present at low temperatures, but is obscured by the defect creation mechanism at elevated temperatures.

Even more curious phenomena occur in doping superlattices which are cooled from elevated temperature with an applied bias (Fritzsche, Yang and Takada 1988). Fig. 9.33 shows that the material is non-

Fig. 9.32. (*a*) Time dependence of the conductivity of a nipi multilayer after a brief exposure to light, showing persistent photoconductivity; (*b*) temperature dependence of the relaxation time, τ_R, of the persistent photoconductivity (Kakalios and Fritzsche 1984).

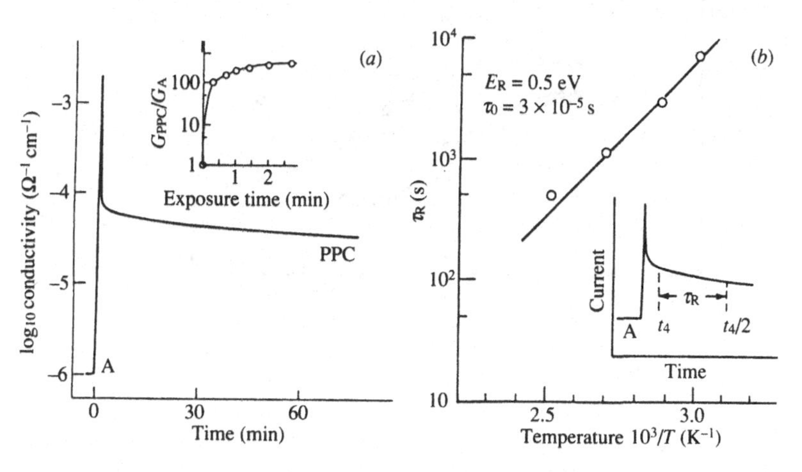

ohmic in a manner which depends on the bias applied during cooling from about 200 °C. Cooling without bias leads to a high conductance state, whereas cooling with bias leads to a conductance which is as much as three orders of magnitude smaller. Furthermore, the low conductance state slowly transforms into the high conductance state after about 1000 s at room temperature. The high conductance state is also removed by annealing to about 100 °C.

The origin of these effects is not at all clear. The bias effects and the room temperature persistent photoconductivity have similar annealing properties. There is also an obvious similarity between the annealing curves and those for the frozen-in excess conductivity of bulk doped a-Si:H (e.g. Fig. 6.3). It is therefore probable that carrier-induced defect creation is the origin of the changes in conductivity and that annealing to the equilibration temperature restores the initial state. However there is as yet no complete explanation for the non-ohmic behavior and why it depends on the applied bias.

Fig. 9.33. Room temperature bias dependence of the conductance of a nipi multilayer after cooling from 200 °C with the indicated bias voltage (Fritzsche *et al.* 1988).

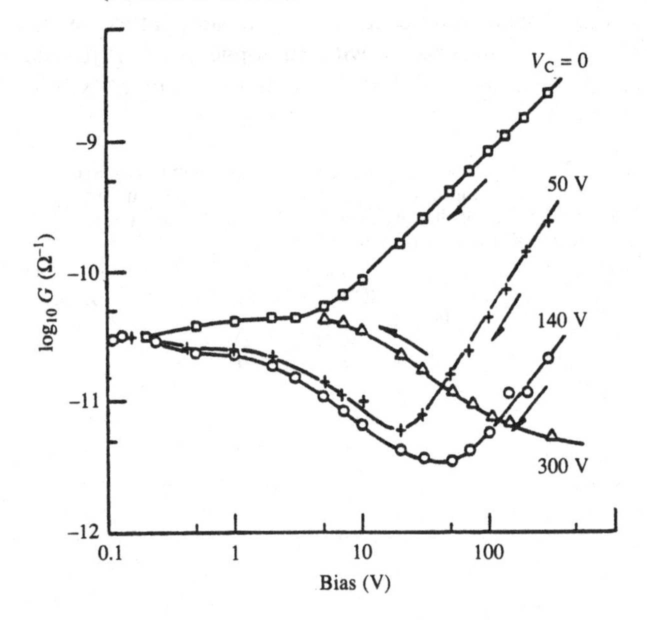

10 Amorphous silicon device technology

Although amorphous silicon has poorer electronic properties than crystalline silicon, it offers the important technical advantage of being deposited inexpensively and uniformly over a large area. The applications are therefore almost entirely in situations in which either a large device or a large array of devices is needed. The technology which has received the most attention is the photovoltaic solar cell – large scale power production obviously depends on the ability to cover very large areas at a low cost. Input and output devices such as displays, photocopiers and optical scanners also take advantage of the large area capability. Each of these applications requires an electronic device whose size matches the interface with human activity – either a display screen or a sheet of paper – with typical dimensions of 25 cm or larger.

The electronic devices are made up of a few different circuit elements, such as transistors, sensors, light emitting diodes etc. Sections 10.1–10.3 describe how the design of these elements is adapted to the specific properties of a-Si:H. A few of the actual and potential applications are then discussed.

10.1 Light sensors

Light sensors made from a-Si:H are either p–i–n or Schottky barrier structures. Unlike crystalline silicon, a p–n junction is ineffective without the undoped layer, because of the high defect density in doped a-Si:H. Illumination creates photoexcited carriers which move to the junction by diffusion or drift in the built-in potential of the depletion layer and are collected by the junction. A photovoltaic sensor (solar cell) operates without an externally applied voltage and collection of the carriers results from the internal field of the junction. When the sensor is operated with a reverse bias, the charge collection generally increases and the main role of the doped layers is to suppress the dark current. A Schottky device replaces the p-type layer with a metal which provides the built-in potential.

The relative importance of drift and diffusion is different in a-Si:H compared to the crystalline semiconductors. The ratio of the depletion width, W, and the diffusion length, L, is from Eqs. (9.6) and (8.56)–(8.58)

$$\frac{W}{L} = \left(\frac{2\varepsilon\varepsilon_0 V}{eN_D}\right)^{\frac{1}{2}} (D\tau)^{-\frac{1}{2}} \simeq \left(\frac{\text{const.}\, V}{\mu_0}\right)^{\frac{1}{2}} \tag{10.1}$$

where the constant depends on the capture cross-sections and is roughly 20 in a-Si:H. Eq. (10.1) highlights the result that the low free carrier mobility in a-Si:H suppresses the effects of diffusion. Virtually all the charge collection occurs within the depletion layer, unlike the equivalent crystalline device, in which there is efficient collection by diffusion from the field-free region.

10.1.1 *p–i–n sensors*

The structure of a p–i–n device is shown in Fig. 10.1. The depletion layer width of low defect density undoped a-Si:H at zero bias is about 1 μm, but is less than 100 Å in heavily doped material (see Fig. 9.9). The p and n layers provide the built-in potential of the junction but contribute virtually nothing to the collection of carriers. Therefore the doped layers need be no more than the width of the depletion layer to establish the junction – any additional thickness unnecessarily reduces the charge collection by absorbing incident light. An efficient sensor usually requires that the undoped layer be as thick as possible to absorb the maximum flux of photons, but it cannot be thicker than

Fig. 10.1. Schematic diagram of a p–i–n sensor at zero bias showing the internal field of the undoped layer and the built-in potential V_B.

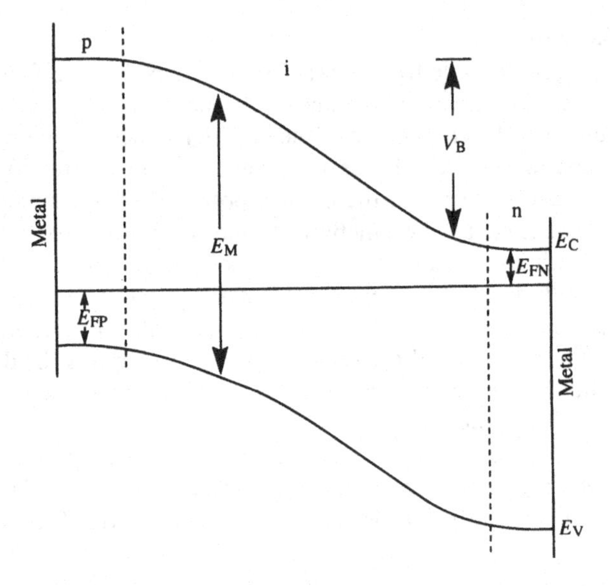

the depletion width, without losing collection efficiency. The depletion width is about 1 μm in the photovoltaic mode.

The depletion layer increases rapidly in reverse bias and widths of up to 50 μm are possible at high bias (Perez Mendez *et al.* 1989). The depletion layer extends from the p-type contact rather than the n layer, because the Fermi energy of undoped a-Si:H is slightly above midgap. Fig. 10.2 demonstrates the depletion of a 38 μm device by comparing the charge collection of electron–hole pairs resulting from excitation which is uniformly absorbed across the sample or absorbed close to the n-type contact. The collection efficiency of uniformly distributed carriers increases linearly with voltage, saturating at 600 V, but there is virtually no collection of holes excited near the n-type contact until about 500 V. The explanation is that the holes are only collected when the depletion field reaches the n-type contact. The depletion layer extends from the p-layer and reaches the opposite contact at a bias of 500 V. Application of Eq. (9.6) shows that full depletion at this bias corresponds to an ionized defect density in the undoped layer of $7\times10^{14}\,\text{cm}^{-3}$. The depletion charge is less than the dangling bond density because the state is sufficiently deep in the gap that not all the defects are ionized.

Fig. 10.2. Voltage dependence of charge collection of a thick p–i–n sensor when illuminated through the n-layer with strongly absorbed (665 nm) or weakly absorbed (880 nm) light. The onset of hole collection at 450 V occurs when the depletion layer reaches the n-type contact (Perez Mendez *et al.* 1989).

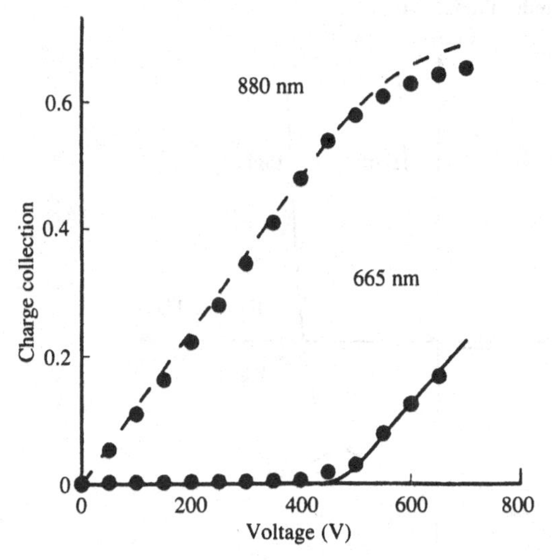

10.1.2 *Response to illumination*

A collection width of 1 μm would give a poor device for visible light sensing, if made from crystalline silicon. There is complete absorption of light in such a thin device only above about 3 eV because of the indirect band gap of the crystal. The amorphous material, however, has a higher absorption coefficient because of the relaxation of the momentum selection rules, and a 1 μm thick device absorbs light across most of the visible spectrum.

Fig. 10.3 illustrates the current–voltage characteristics of a p–i–n sensor in the dark and in light. The photocurrent rapidly saturates in reverse bias when there is full collection of the incident absorbed photon flux. The photovoltaic properties in slight forward bias are characterized by the short circuit current J_{sc}, the open circuit voltage, V_{oc}, and the fill factor, F. The maximum power delivered by the device is the product of the three terms

$$P_{max} = J_{sc} V_{oc} F \tag{10.2}$$

and defines the fill factor. A more detailed discussion of these parameters is given in Section 10.3.1.

The spectral dependence of the charge collection, $Q(\lambda)$, of a photovoltaic p–i–n device is shown in Fig. 10.4. The collection efficiency peaks at about 80–90 % at wavelengths between 5000–6000 Å, demonstrating that there is little loss due to recombination. The

Fig. 10.3. Schematic of the current–voltage characteristics of a p–i–n sensor in the dark and under illumination.

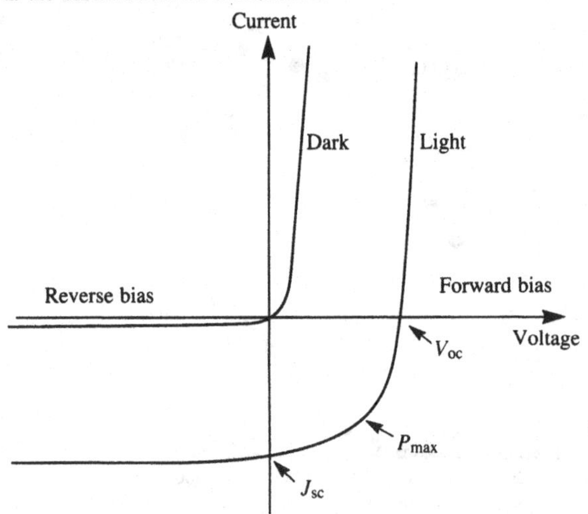

reduction on the long wavelength side is due to the decreasing absorption coefficient of a-Si:H, so that a smaller fraction of photons is absorbed in the film. Low band gap alloys such as a-Si:Ge:H, extend the spectrum to lower energy, but currently have the problem of a large defect density, which reduces the thickness over which carriers can be collected.

The decreasing collection efficiency at high energy has several origins. One cause is the absorption of light in the p or n layer, whichever is directly exposed to the incident light. The resulting photoexcited carriers give virtually no contribution to the charge collection because of the low minority carrier lifetime in the doped material. The absorption coefficient of a-Si:H reaches $3 \times 10^5\ \mathrm{cm}^{-1}$ at a wavelength of 4000 Å and a doped layer of thickness 100 Å absorbs about 30% of the light.

The diffusion of carriers in the undoped layer against the internal field also reduces the collection efficiency. A thermalized carrier can, on average, diffuse over a potential of kT. Backward diffusion can therefore occur over a distance, x_D, of approximately

$$x_D = kTd/V_B \tag{10.3}$$

where d is the thickness of the i layer, V_B is the built-in potential at the operating voltage, and it is assumed for simplicity that the internal field

Fig. 10.4. Typical spectral dependence of the charge collection efficiency of a 0.5 μm thick p–i–n solar cell. The dashed line shows the reduction of the blue response with a poor p-type contact.

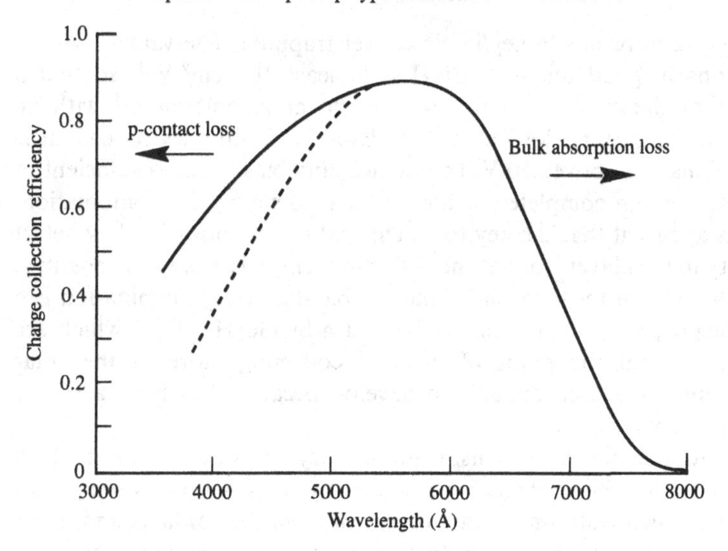

is uniform. With values of $d = 0.5$ μm and $V_{bi} = 0.25$ V, the estimated diffusion distance is 500 Å. In practice the distance is smaller than this because the field is not uniform and is largest near the contacts. Devices operating under reverse bias are less susceptible to the loss of charge collection because the internal field is larger. The effect of back diffusion is most significant at high photon energy when the absorption length is less than 1000 Å and the charge collection is reduced at wavelengths below 5000 Å. The back diffusion is further enhanced by hot carrier effects. At high photon energies, electron–hole pairs are created with a large excess energy and rapidly diffuse 50–100 Å during thermalization to the band edges (Section 8.1.3). This distance is independent of the applied field because the drift distance within a time of 10^{-12} s is negligible.

These light sensors have a reasonably fast response time. A device of thickness 1 μm operated at 5 V reverse bias has an electron transit time of 2×10^{-9} s and a hole transit time of 2×10^{-7} s. The device can always be configured so that at least half of the signal is due to electron collection, by illuminating through the p-type layer. Faster devices have been made using amorphous silicon with a high defect density. The response time is then determined by the deep trapping time which can be in the range 10–100 ps. The deep trapping, of course, reduces the sensitivity of such devices because of recombination.

A figure of merit of the sensors is the ratio of the deep trapping or recombination time, τ, to the transit time, τ_T,

$$R_m = \tau/\tau_T = \mu\tau V_A/d^2 \tag{10.4}$$

A large ratio results in negligible carrier trapping. The value of $\mu\tau$ for electrons in good quality a-Si:H is at least 10^{-7} cm^2 V^{-1}, so that a device of thickness 1 μm has no significant recombination with an internal voltage as low as 0.1 V. Even a 50 μm device has little recombination above 250 V. In practice, any bias which is sufficient to deplete the film completely is also enough to suppress recombination.

It is apparent that the key to an efficient p–i–n sensor is a low defect density in the i layer, so that the collection length is as large as possible. For this reason the creation of metastable defects by illumination is a significant problem for solar cells; and a-Si:Ge:H alloys, which are desirable from the point of view of collecting more of the solar spectrum, have been difficult to develop because they have a higher defect density.

The reverse bias p–i–n sensor is a primary photoconductor in which injection from the contacts is prevented by the junction. A secondary photoconductor allows charge to flow in from the contacts and offers the possibility of photoconductive gain. Gain occurs when the recombination lifetime exceeds that of the transit time and is given by R_m in Eq. (10.4). The gain increases the response time of the sensor by a factor R_m under ideal conditions. Photoconductive sensors with gain can be made from a-Si:H using doped ohmic contacts in forward bias. However, such devices have not proved to be useful in circuits. One reason is that the response time increases much faster than the gain, for reasons which are not clearly understood, but may be due to contact effects. Ohmic contacts also cause a large dark current, which reduces the dynamic range of the sensor and is usually undesirable.

Crystalline silicon devices take advantage of avalanche multiplication to enhance the gain without the problems of the secondary photoconductor. Carriers are accelerated under the field until their energy is high enough to cause impact ionization. The requirement is roughly that

$$eV_A L_i/d > E_G \tag{10.5}$$

where V_A is the applied voltage, L_i is the carrier mean free path and E_G is the band gap. Avalanche effects are observed at fields of about 10^5 V cm^{-1} in crystalline silicon. A rough estimate suggests that avalanche gain is unlikely in a-Si:H because the mean free path of only about 10 Å requires a field of 2×10^7 V cm^{-1}, which is above the breakdown strength. However avalanche gain is observed in amorphous selenium at a field of 10^6 V cm^{-1} and it remains to be seen whether it is possible in a-Si:H (Tsuji, Takasaki, Hirai and Taketoshi, 1989).

10.1.3 *Electrical characteristics*

A sensor with a large dynamic range requires a dark current which is low and reproducible. The current in a typical 1 μm thick p–i–n device is in the range 10^{-11}–10^{-9} A cm^{-2} but is time-dependent. Examples of the slow changes in current after the application of a bias are illustrated in Fig. 10.5 for different bias voltages. The current decreases over a period of a few minutes at low bias, but increases at high bias. The steady state current increases approximately exponentially with increasing voltage. The data in Fig. 10.5 are only one example of a diverse set of observations which depend on the details of the sample structure and on the quality of the junctions to the doped layer.

There are bulk and contact components to the dark current. The bulk current originates from charge generation through gap states. Electrons are excited from the valence band to empty gap states and from filled traps to the conduction band. These excitations generate electron–hole pairs which are separated and collected by the internal field. This type of charge generation also determines the position of the trap quasi-Fermi energy, E_{qF}, under conditions of deep depletion (see Section 9.1.2). The thermal generation current is (Eq. (9.15))

$$J_{th} = eN(E_{qF})kT\omega_0 \exp[-(E_C - E_{qF})/kT]Ad \tag{10.6}$$

where A is the sample area and d the thickness. E_{qF} is approximately at midgap, so that $E_C - E_{qF} \approx E_m/2 = 0.9$–$1.0$ eV. The exact position of E_{qF} depends on the ratio of the electron and hole capture cross-sections.

There is a larger thermal generation current when the voltage is first applied, because the equilibrium E_F is above E_{qF},

$$J_{th}(0) = J_{th} \exp(\Delta E_F/kT) \tag{10.7}$$

where $\Delta E_F = E_F - E_{qF}$ and a constant density of states is assumed. The

Fig. 10.5. Examples of the time dependence of the reverse bias current of a p–i–n sensor at different applied voltages. The different regions, A–C, are discussed in the text.

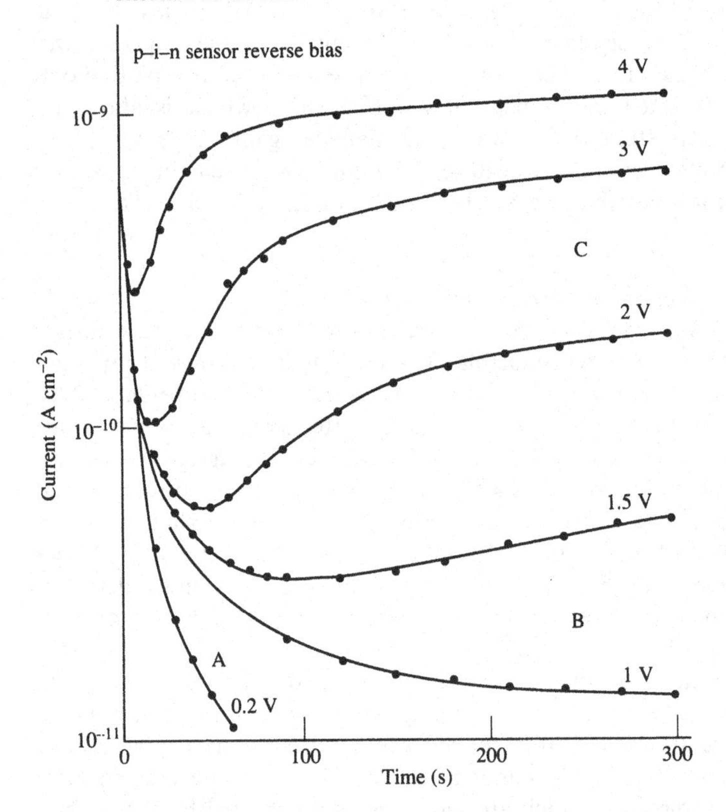

shift of E_F corresponds to the depletion of the undoped layer and explains the time-dependent decrease in the leakage current. The depletion charge, Q_D, is

$$Q_D = eN(E_F)\Delta E_F \tag{10.8}$$

The time taken for the depletion to occur is

$$\tau_D = \omega_0^{-1} \exp[(E_C - E_{qF})/kT] \tag{10.9}$$

Thus,

$$J_{th} = Q_D kTAd/\Delta E_F \tau_D \tag{10.10}$$

The depletion charge in low defect density a-Si:H is 7×10^{-4} C cm^{-3}, ΔE_F is about 0.1 eV and the leakage current has a decay time constant of about 100 s. Thus the thermal generation current is estimated to be 2.5×10^{-11} A cm^{-2} for a 1 μm thick detector. This current corresponds

Fig. 10.6. Examples of the forward bias current–voltage characteristics of p–i–n sensors, compared to a Schottky barrier device.

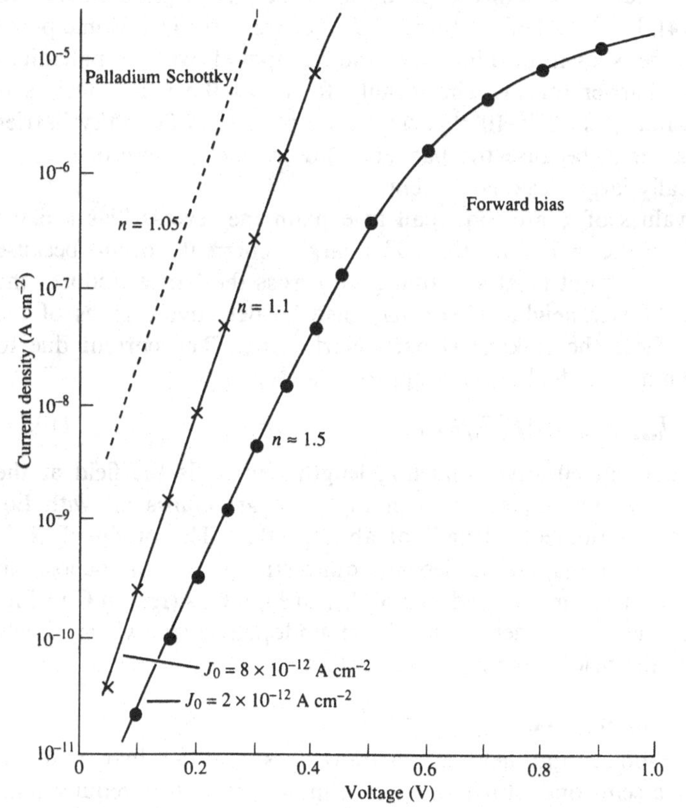

to a deep depletion Fermi energy position of 0.95 eV, consistent with a mobility gap of 1.85 eV.

The low voltage data in Fig. 10.5 are explained by the thermal generation current. As formulated above, J_{th} has no voltage dependence. However, the generation current is reduced when the bias voltage is too small to deplete the sensor, because not all the electron–hole pairs are collected. This is indicated by region A in Fig. 10.5. Region B corresponds to full depletion for which the current is given by Eq. (10.10). The thermal generation current represents the lowest dark current possible in a fully depleted sensor and can only be reduced by lowering the defect density or raising the band gap. Thick sensors have a correspondingly higher thermal generation current because J_{th} is proportional to the thickness. The larger voltage-dependent leakage current in region C is caused by leakage across the contacts.

The contact leakage current for an ideal Schottky barrier is the saturation current J_0, which depends on the barrier height according to Eq. (9.14). Examples of the forward J–V characteristics of some p–i–n sensors are shown in Fig. 10.6 and compared with a palladium Schottky barrier sensor. The ideality factor of the p–i–n devices is 1.1–1.5 and J_0 is 10^{-12}–10^{-11} A cm^{-2}. The palladium Schottky barrier has a larger J_0 because the barrier is lower, giving these devices an intrinsically larger leakage current.

The values of J_0 are too small to explain the reverse bias leakage current in the p–i–n devices. The larger currrents occur because the junction is not ideal and tunneling across the barrier reduces the effective barrier height. There may also be defective regions of the barrier where the leakage is particularly large. The current due to tunneling across the barrier is approximately,

$$J_{leak} = J_0 \exp(E_t R_t / kT) \qquad (10.11)$$

where R_t is an effective tunneling length and E_t is the field at the contact. The steady state data in Fig. 10.5 are consistent with Eq. (10.11) for a tunneling length of about 100 Å. The origin of R_t is unclear and its magnitude depends quite strongly on the deposition conditions. The time dependence of J_{leak} at high bias (region C in Fig. 10.5) is due to the depletion of the undoped layer, which slowly increases the field at the injecting contact.

10.2 Thin film transistors (TFTs)

Bipolar and field effect transistors are the two standard crystalline semiconductor structures. Bipolar transistors require minority carrier diffusion through a doped base layer, which is not easily achieved in a-Si:H because of the very small diffusion length in doped material. The field effect transistor, being an intrinsically planar device, is well suited to a-Si:H and was demonstrated shortly after the first p–i–n solar cells were made (Snell, Mackenzie, Spear, LeComber and Hughes, 1981). The specific material properties of a-Si:H again govern the design of the device.

The usual TFT structure is shown in Fig. 10.7 and comprises the a-Si:H channel, a gate dielectric, and source, drain, and gate contacts. N-channel accumulation mode operation using an undoped a-Si:H channel is the only structure widely used. Depletion mode devices are prevented by the high defect density of doped material, which makes it difficult to deplete the channel. The much lower mobility of holes compared to electrons gives p channel devices a lower current by about a factor 100, which is undesirable.

The main material requirement of the dielectric is a low growth temperature, consistent with the a-Si:H film and the use of glass substrates. Silicon nitride is an effective dielectric and can be deposited in the same reactor, which adds to the convenience. Care is needed to grow the nitride close to the stoichiometric composition, when the dielectric strength is high and there is no significant charge injection from the channel of the gate contact. Charge trapping was a problem in early devices, but has been nearly eliminated by improved deposition techniques. There are some reports of TFTs with deposited silicon dioxide, and other dielectrics, such as TaF_6, have been investigated.

Chapter 9 discusses the interface states and band bending at the interface of a-Si:H and nitride. The interface states degrade the performance of the TFT and band bending causes a shift in the electrical characteristics. The problems are particularly severe when the

Fig. 10.7. Side and top views of the structure of a TFT.

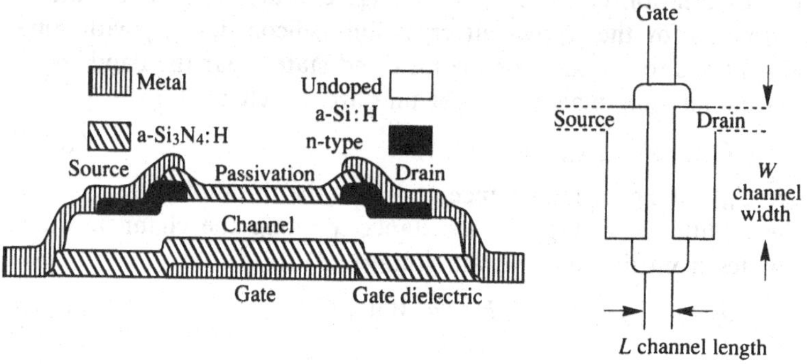

nitride is grown on top of a-Si:H. For this reason virtually all TFT designs use the inverted structure shown in Fig. 10.7, with the gate deposited first.

Ohmic source and drain contacts provide the carriers for the channel and are doped n-type. The contact layer is usually deposited on top of the undoped a-Si:H channel material, mainly for reasons of processing convenience. The electron accumulation layer is only about 100 Å wide so that the a-Si:H layer can be thin and is sometimes as little as 500 Å. The dimensions of the devices along the plane are governed by the capabilities of photolithography patterning. Channel lengths in large area circuits are commonly 5–20 μm, but are decreasing as the technology improves. The channel width is determined by the current requirements and can be from 10 μm upwards.

The metastable defect creation mechanism affects the TFTs, since defects are created when the device is held in accumulation for an extended time. Fortunately the defect creation rate is low near room temperature, so that the resulting threshold shift is less than 0.1 V and is not much of a problem. The threshold voltage shifts very rapidly at elevated temperatures, by several volts in a few minutes at about 100 °C, so that the TFTs cannot be used in high temperature applications.

10.2.1 *TFT electrical characteristics*

Fig. 10.8 shows the transfer and output characteristics of a typical TFT, which are plots of source-drain current, I_D, as a function of gate bias, V_G, and source-drain voltage, V_D. The maximum on-current is about 10^{-4} A, while the current at zero gate bias is about 10^{-11} A, giving a wide dynamic range. The threshold voltage is typically about 1 V. There is virtually no p channel conduction at a negative gate bias, because of the low hole mobility and the blocking n-type source and drain contacts.

The description of the current–voltage characteristics is essentially the same as for the equivalent crystalline silicon devices, with some modifications to account for the localized states near the band edges. The accumulation charge, Q_A, per unit area of channel is

$$Q_A(x) = C_G[V_G - V_T - V(x)] \tag{10.12}$$

where C_G is the gate capacitance and V_T is the threshold voltage. $V(x)$ is the additional voltage at a distance x along the channel, which originates from the source-drain voltage. J_D is

$$J_D(x) = WQ_A(x)\mu_{FE}\,E(x) = W\mu_{FE}\,C_G[V_G - V_T - V(x)]\,\mathrm{d}V(x)/\mathrm{d}x \tag{10.13}$$

where μ_{FE} is the effective carrier mobility, $E(x)$ is the source-drain field, and W is the channel width. Integration gives

$$J_D = \mu_{FE} C_G[(V_G - V_T)V_D - V_D^2] W/L \tag{10.14}$$

where L is the channel length. Low values of V_D correspond to the linear region of the output characteristics and in this limit the channel conductance is

$$g_D = J_D/V_D = \mu_{FE} C_G(V_G - V_T) W/L \tag{10.15}$$

Saturation of J_D occurs when $dJ_D/dV_D = 0$ and from this condition the saturation current is

$$J_{Dsat} = \mu_{FE} C_G(V_G - V_T)^2 W/2L \tag{10.16}$$

Fig 10.9 shows the output current plotted in both forms, Eqs (10.14) and (10.16). The mobility is about 1 cm^2 V^{-1} s^{-1} which is comparable to the drift mobility of bulk a-Si:H. It is evident that the Fermi energy in the channel remains within the band tail states and that the band tails are not much perturbed by the close proximity of the dielectric layer. If a nitride thickness of 3000 Å with a dielectric constant of 7 is assumed, then the gate capacitance is 2×10^{-8} F cm^{-2}. With parameters $W/L = 10$ and $\mu_{FE} = 1$ cm^2 V^{-1} s^{-1}, the channel conductance is 10^{-6} Ω^{-1} at a gate voltage of 5 V. This value is about 1000 times smaller than crystalline silicon, because of the lower mobility.

The noise associated with the TFT current at frequency f is,

$$\langle J \rangle^2 = 2kTg_D + A/f \approx 5 \times 10^{-8} + 5 \times 10^{-13}/f \quad \text{A}^2\,\text{Hz}^{-1} \tag{10.17}$$

Fig. 10.8. (*a*) Transfer and (*b*) output characteristics of a typical TFT (Snell *et al.* 1981).

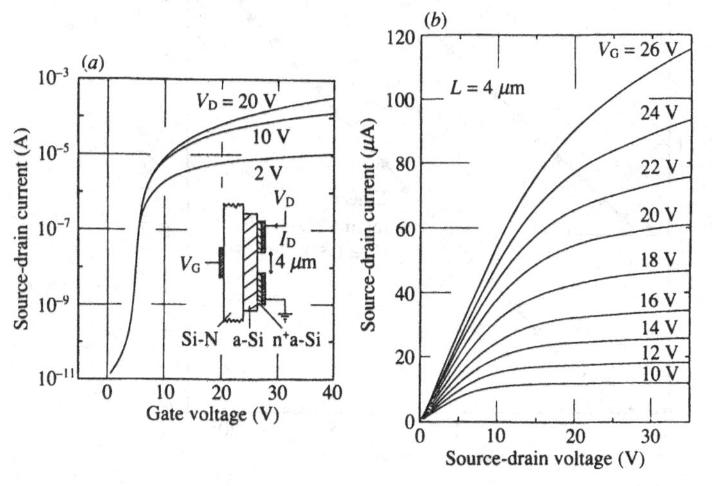

where $\langle J \rangle^2$ is the mean square noise current and the values are measured at room temperature for a typical TFT. The first term is the Nyquist resistance of the TFT channel and the second term is $1/f$ noise, which dominates at frequencies below 10^5 Hz. The $1/f$ noise in a crystalline silicon field effect transistor arises from the thermal excitation of electrons from interface states to the conduction band. Presumably the same mechanism applies in a-Si:H.

A much higher mobility and g_D would be possible if the Fermi energy in the channel crossed the mobility edge, and there was accumulation of free carriers. According to the density of states developed in Chapter 3, there are about 10^{20} cm^{-3} band tail localized states, so that an accumulation charge of 10^{14} cm^{-2} in a channel extending 100 Å is required to generate free carriers. For the parameters given above, a gate voltage of 500 V would be needed, which is well above the breakdown strength of the nitride. There has been one report of free carrier motion in the channel of a TFT, but the result has not been confirmed by others (Leroux and Chenevas-Paule 1985).

The only other approach to increasing the channel conductance

Fig. 10.9. Transfer characteristics of a TFT in the linear and saturation regimes, showing the fits to Eqs. (10.15) and (10.16). The field effect mobility is approximately 1 cm^2 V^{-1} s^{-1}.

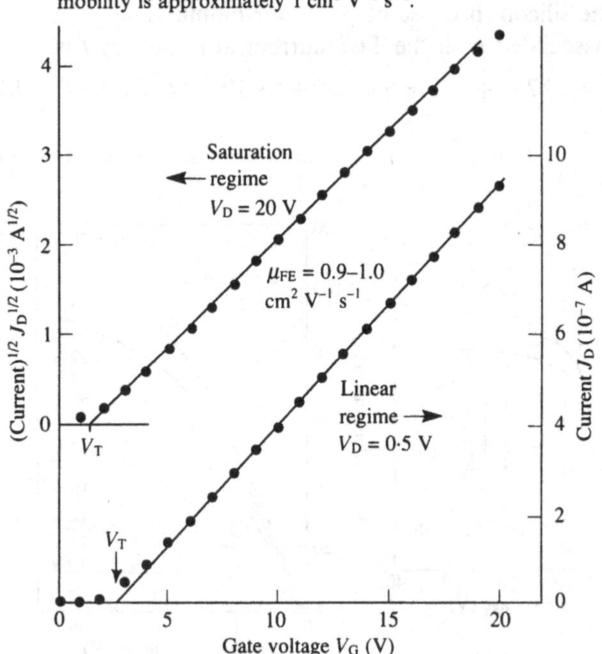

significantly is by increasing the geometrical factor, W/L in Eq. (10.15). There is an obvious disadvantage in making W very large, because there is often a constraint on the total area WL, of the device. Present photolithography places a lower limit on L of about 5 μm in large area circuits, but this dimension will no doubt decrease as the technology develops further.

10.2.2 *Other TFT structures*

One strategy for making L smaller without the need for more precise processing is to reconfigure the device as a vertical structure, and two examples are illustrated in Fig. 10.10. The channel length is now defined by the thickness of the film rather than by the accuracy of the photolithographic patterning. A film thickness of about 1000 Å potentially increases the transconductance by a factor of 100. Improved values of g_D have been demonstrated in vertical TFT structures,

Fig. 10.10. Two alternative designs for vertical TFT structures (Yaniv, Hansell, Vijan and Cannella 1984, Hack, Shaw and Shur 1988).

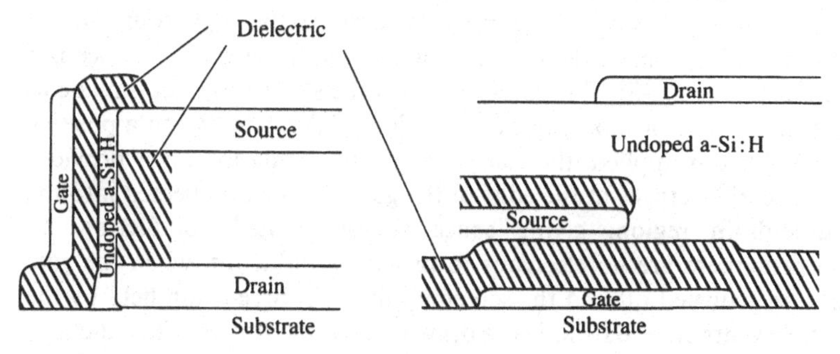

Fig. 10.11. (*a*) Structure of a high voltage TFT capable of switching up to 400 V; and (*b*) transfer and output characteristics of the high voltage TFT (Tuan 1986).

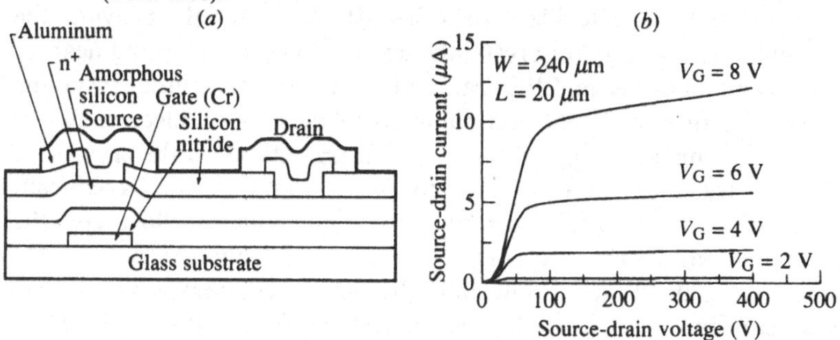

although further development is needed to make this a viable technology. The structure shown in Fig. 10.10(*a*) is made by depositing a metal–dielectric–metal structure to form the source and drain (Yaniv *et al.* 1984). The a-Si:H channel, nitride dielectric and the metal gate are deposited after etching, taking advantage of the conformal growth morphology of these films. The problem with this device is that the on–off current ratio is low. In the alternative vertical structure shown in Fig. 10.10(*b*), the geometry is such that accumulation at the gate induces charge flow between source and drain (Hack *et al.* 1988). A high off-current is potentially the main problem, but is suppressed by surrounding the source electrode with a p-type blocking layer.

The material properties of a-Si:H allow the design of TFTs which switch high voltages (Tuan 1986). The device structure, which is shown in Fig. 10.11(*a*), is the same as the conventional TFT except that the drain contact is moved away from the gate. The output characteristics in Fig. 10.11(*b*) show that the device can switch up to 400 V with a gate bias of less than 10 V. The device operates because the undoped a-Si:H film is highly resistive and can easily support the high source-drain voltage with a channel length of about 25 μm. In effect the device is a planar undoped a-Si:H resistor, in which the source contact is made to be either ohmic or blocking by the action of the gate. When a positive gate voltage is applied, the charge in the accumulation layer provides a source of electrons at the edge of the gate. These are injected into the source-drain region, giving space charge limited conduction. In contrast, there are no electrons to provide conduction when the gate voltage is turned off and the screening of the source-drain field by the gate prevents charge from being drawn out of the source. Thus the gate region acts as a blocking contact and the current is low.

10.3 Other devices

10.3.1 *Light emitting diodes* (*LEDs*)

Gallium arsenide and other III–V materials provide the dominant technology for crystalline semiconductor visible and near IR LEDs and diode lasers. LEDs cannot be made from intrinsic crystalline silicon because of the indirect band gap. Amorphous silicon does not have the same selection rules for optical transitions as the crystal, so that it is feasible to make LEDs from a-Si:H and its alloys (Kruangam *et al.* 1986, 1987). The larger band gap and range of alloys give the possibility of LEDs spanning the visible spectrum. At present the efficiency of the devices is low, but it has improved greatly over the last few years and may eventually become suitable for display applications.

Most of the interest has been directed at making visible LEDs for which a-Si:H itself is not a candidate. Alloying with carbon increases the gap, and moves the luminescence into the visible region. The emission band is broad, and the room temperature emission is white at a carbon concentration of about 50 at%.

Fig. 10.12 shows the design of the best LED device made to date. It is a p–i–n structure, operated in forward bias. The i layer has a thickness of about 500 Å and a band gap of 2.5 eV to give yellow emission. The doped contact layers are deposited using an electron cyclotron resonance plasma, with the gases diluted with hydrogen. This technique results in microcrystalline contacts which give better current injection than with a completely amorphous structure. The band gap of the doped layers is smaller than of the amorphous i layer. Fig. 10.13 shows the brightness and peak emission energy as a function of the optical gap of the p-type layer. The fact that the colour depends on the p layer indicates that the operation mechanism is complex and is not simply the recombination of electron–hole pairs in the undoped layer. The maximum brightness of 10 cd m^{-2} is about an order of magnitude less than in a typical TV screen. Thus the intensity is low, but not out of reach of present commercial devices.

The light output of a LED is given by the injection current and the radiative efficiency and is governed by two primary loss mechanisms. A large fraction of the recombination at room temperature is by non-radiative transitions at defects (see Section 8.3.5). The thermal quenching of the photoluminescence is lower in the alloys than in

Fig. 10.12. Structure of a LED fabricated from a-Si:C:H (Kruangam *et al.* 1986).

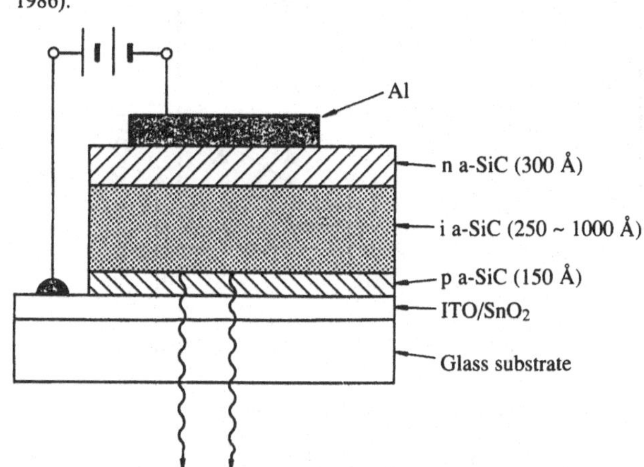

a-Si:H, so that their room temperature intensity is usually higher than that of a-Si:H. However, the reduced thermal quenching arises from the broader band tail density of states, which also causes a lower carrier mobility. This represents a problem for the LED structure because of the need to inject a large current density.

The other main loss mechanism in these LEDs is from carriers which do not recombine in the i layer, but instead are transported completely through the film. Ideally, the p-type contact should comprise a blocking layer preventing electrons from being collected, but easily injecting holes, and vice versa for the n-type contact. Perhaps when the band discontinuities between the different alloys are better understood, some new and more efficient structure can be designed.

10.3.2 *Memory switching devices*

A-Si:H memory switches do not have counterparts in standard crystalline semiconductor technology, unlike the sensors, TFTs and LEDs. However, amorphous semiconductor switching devices have a

Fig. 10.13. Brightness and peak emission energy of a LED at a current of 100 mA cm^{-2}, as a function of the optical gap of the p-type layer (Kruangam *et al.* 1987).

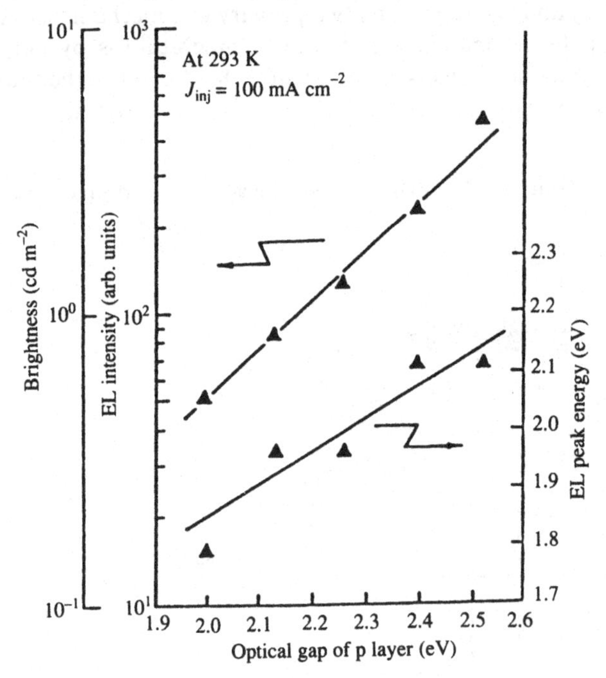

long history; devices made from chalcogenide glasses were discovered many years ago (Ovshinsky 1968). The memory effect is also found in a-Si:H devices (Owen, LeComber, Spear and Hajto 1983, LeComber *et al.* 1985). The best structure is a p^+–n–i layer, although the oppositely doped n^+–p–i device also works. The same structure gives a two-terminal switch in crystalline silicon, but with the important difference that the amorphous device has a memory and retains its state even when the voltage is removed, but the crystalline one is volatile.

The current–voltage characteristics of the device are shown in Fig. 10.14. The switching voltage is near 4 V, above which there is an abrupt change to a low resistance state. The ratio of the on- and off-resistances is more than 10^3 and varies with the geometry of the device, as discussed shortly. The device switches back to the high resistance states at a voltage of about -1 V.

Amorphous semiconductor memory devices involve a structural change in the material and usually have two characteristic properties. One of these is the forming process; the first time the device is switched, the I–V characteristic is different from the subsequent behavior. The a-Si:H device has the odd property that the forming time has a huge range. The device holds a larger voltage without switching when the voltage is first applied than in subsequent operation. However, switching occurs if the voltage is applied for long enough. Fig. 10.15 shows that about 100 s is required for an applied voltage of 10 V, but that the forming time drops to 10^{-8} s at a voltage of 20 V. Subsequent switching occurs within 10^{-8} s, which is fast enough to make the device interesting for computer memories.

Fig. 10.14. Current–voltage characteristics of the a-Si:H memory switch, showing the reversible transition from a high resistance to a low resistance state (Owen *et al.* 1983).

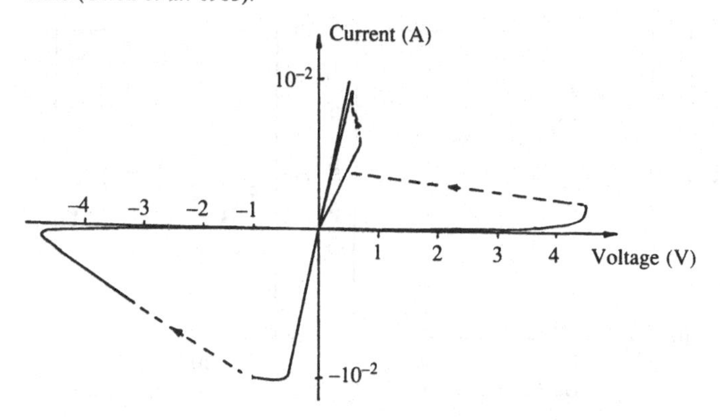

The second characteristic of a memory device is filamentary conduction. Fig. 10.16 shows the on- and off-resistances as functions of the area of the device. The off-resistance scales inversely with area, as would be expected for a uniformly conducting material. The on-

Fig. 10.15. The delay time for memory switching versus applied voltage. V_{CR} is the critical switching voltage (Owen *et al.* 1983).

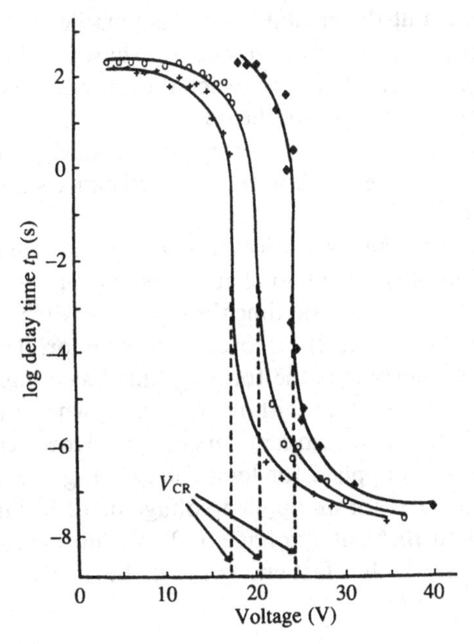

Fig. 10.16. Dependence of the memory switch off-resistance and on-resistance on device area, showing evidence of filamentary conduction in the on-state (LeComber *et al.* 1985).

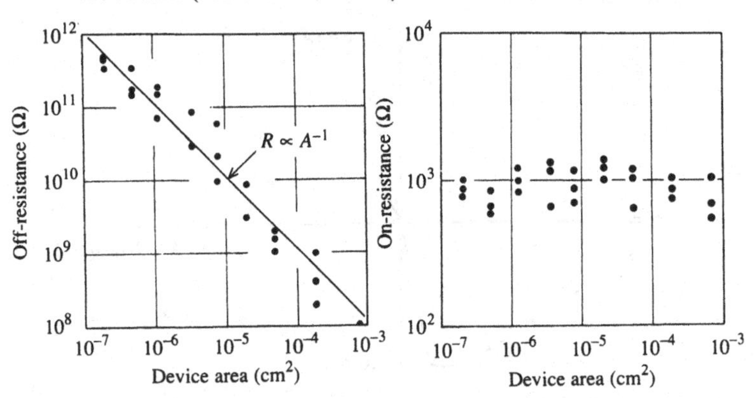

resistance, however, is independent of area, indicating conduction in a single small filament. The size of the filament is evidently no more than a few microns in diameter, this being the minimum area shown in Fig. 10.16. The existence of the filament was confirmed by overcoating a device with liquid crystal, which is sensitive to the heat developed by the current flow. These measurements found the filament size to be only 0.5 μm, from which a current density in the on-state of about 10^6 A cm^{-2} is estimated.

There is presently no complete explanation of how the device works. There is evidently a reversible structural change in the filament, which may have as its origin either the high field, high current flow, or local heating. Crystallization of the filament has been ruled out for several reasons – for example this would not give a reversible effect.

10.4 Applications of amorphous silicon devices

Many applications of a-Si:H have been proposed or are under development, and it is impossible to describe them all here. Almost all the applications depend on the ability to deposit a-Si:H in a large area. The next few years will tell which of the suggestions prove to be viable and marketable. The following is not intended to be a complete list, but rather an illustration of some different possibilities.

10.4.1 *Solar cells*

Photovoltaic solar power conversion was the first major application proposed for a-Si:H and to date is the largest in production. The first devices were reported by Carlson and Wronski in 1976 and had an efficiency of only 2–3%. Some of the early devices were Schottky barrier cells, but were quickly discarded in favour of p–i–n cells. Since the first report, there has been a remarkable increase in the efficiency of the cells, increasing by roughly 1% conversion efficiency per year, to a present value of 14%, as is shown in Fig. 10.17. The increase has resulted from a variety of innovations in the design, materials, and structure of the cells. The electronic properties of the solar cell are described next and then these innovations are outlined more or less in the order in which they occurred.

The solar cell efficiency, η, is from Eq. (10.2),

$$\eta = J_{sc} V_{oc} F / W_{inc} \qquad (10.18)$$

where W_{inc} is the incident solar power (about 100 mW cm^{-2}). A typical current–voltage characteristic is shown in Fig. 10.18.

The short circuit current is the product of the photon flux of the incident solar spectrum, $S(\lambda)$, and the wavelength-dependent collection efficiency, $Q(\lambda)$

$$J_{sc} = \int S(\lambda)\, Q(\lambda)\, d\lambda \qquad (10.19)$$

The function $Q(\lambda)$ is illustrated in Fig. 10.4. The atmospheric solar spectrum peaks at 5000 Å, drops rapidly below 4000 Å but extends far into the infra-red. There are collection losses in the solar cells at both ends of the visible spectrum. Most of the loss is at long wavelength where the collection efficiency is unavoidably reduced by the optical absorption edge and by the limited thickness of the cell. On the short wavelength side of the spectrum there are collection losses due to

Fig. 10.17. Improvement of the solar cell conversion efficiency from 1975 to 1989. Data are shown for small area single and stacked cells and for large area modules (Sabisky and Stone 1988).

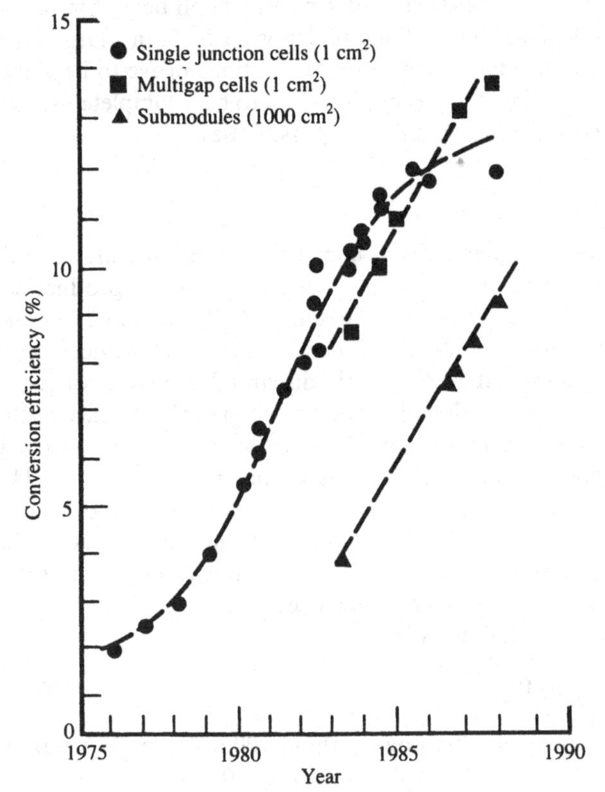

absorption in the doped layers and to the diffusion of carriers against the internal field (see Section 10.1.1).

The open circuit voltage is related to the built-in potential, V_B, and to the electrical quality of the junction. V_B varies with the operating voltage, V, of the cell and is given by (see Fig. 10.1),

$$eV_B(V) = E_M - E_{FN} - E_{FP} - eV \tag{10.20}$$

where E_M is the mobility gap energy of a-Si:H, E_{FN} and E_{FP} are the positions of the Fermi energies of the n- and p-type contacts from the mobility edges (see Fig. 10.1). Values of $E_M = 1.85$ eV, $E_{FN} = 0.25$ eV, $E_{FP} = 0.40$ eV give an estimated built-in potential of 1.2 V under short circuit conditions. The open circuit voltage cannot exceed this value and as a rule is limited to about 80% of the maximum V_B, which would be 0.95 V. In high efficiency a-Si:H cells V_{oc} is usually in the range 0.85–0.95 V.

V_{oc} is also related to the current–voltage characteristics of the p–i–n device. The illumination generates electron–hole pairs which result in a reverse current equal to J_{sc}. A simple model assumes that as the operating voltage increases, the forward current offsets the photocurrent and that these balance at V_{oc}. Thus from the diode characteristics of Eqs. (9.12) and (9.14),

$$J_{sc} = J_0 \exp(eV_{oc}/nkT) = A^*T^2 \exp(eV_{oc}/nkT - eV_B/nkT) \tag{10.21}$$

Fig. 10.18. Representative current–voltage data for sunlight illumination of a solar cell with a conversion efficiency of 11%.

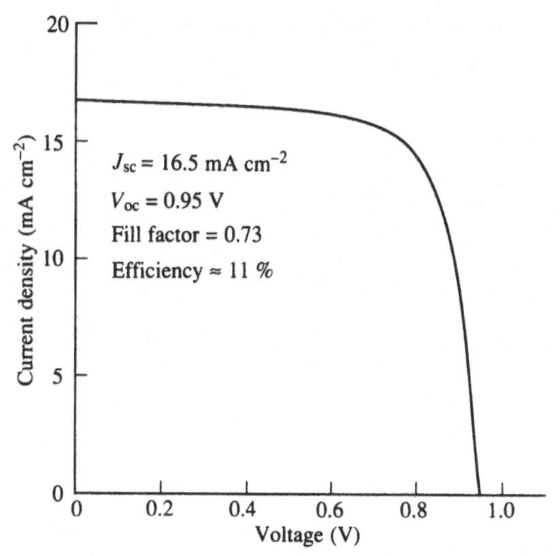

Hence,

$$V_B - V_{oc} = (nkT/e)\ln(A^*T^2/J_{sc}) \simeq 9nkT/e \qquad (10.22)$$

J_0 is the reverse bias saturation current density and n is the ideality factor under illumination. Typical values of J_0 of 10^{-12} A cm^{-2} and $n = 1.5$ give values of $V_B - V_{oc}$ of 0.2–0.25 eV under sunlight conditions, which is consistent with the observations.

The fill factor in a crystalline silicon solar cell is largely determined by the contact resistance, but in a-Si:H the charge collection efficiency is more important. The built-in potential decreases as the operating voltage increases (Eq. (10.20)) and there is a loss of charge collection. The lower current is reflected in a reduced fill factor. The main parameter governing the charge collection is the collection length,

$$d_c = \mu_D \tau V_B/d \qquad (10.23)$$

where d is the thickness of the cell; d_c is the average distance which a carrier moves before being captured into a deep trap. Charge collection is complete when $d_c \gg d$ and is smaller otherwise. The collection efficiency is derived from the Hecht expression (Eq. (8.61)), which approximates to

$$Q/Q_0 = 1 - d^2/2\mu_D \tau V_B = 1 - d/2d_c \qquad (10.24)$$

when the collection efficiency is large ($d_c > d$). Charge collection is reduced as the operating voltage is raised and there is a corresponding decrease in the current.

Fig. 10.19 shows the relation between the collection length measured under short circuit conditions and the fill factor for many different solar cells (Faughnan and Crandall 1984). The data are described empirically by

$$F = 0.35 + 0.15\ln(d_c/d) \qquad (10.25)$$

and confirm that a high collection efficiency is necessary for a large fill factor.

Eqs. (10.23)–(10.25) for the charge collection do not differentiate between the electrons and holes which are collected in the solar cell. Does an efficient cell need a large value of $\mu_D \tau$ for electrons, holes, or both? Faughnan and Crandall (1984) argue that the correct value in Eq. (10.24) is the sum of the electron and hole $\mu_D \tau$, because the electrons and holes are collected separately. If this is true, then it is sufficient that one or other carrier has a large $\mu_D \tau$ to give an efficient cell. This argument, however, ignores the fact that the electron and holes are created in pairs and by charge neutrality must also be collected in pairs otherwise there would be a continual increase in the space charge within the device. There is an imbalance between the collection of electrons and holes when illumination is first applied. A space charge develops which causes a non-uniform internal field, such that the collection of electrons and holes exactly balance. Numerical solutions of the transport equations give the details of the internal fields and find that the charge collection is limited by the carrier with the lowest value of $\mu_D \tau$, not the highest (Hack and Shur 1985). Eqs. (10.23) and (10.24) assume a uniform internal field and so are only approximations to the actual charge collection. The fabrication of efficient solar cells has stringent material requirements because even light doping of the undoped layer greatly suppresses $\mu_D \tau$ for the minority carrier, due to the defect creation mechanism.

The basic a-Si:H solar cell is a p–i–n device of thickness about 0.5 μm, with p and n layer contacts of thickness about 50–100 Å. A common design is shown in Fig. 10.20(*a*), in which the light is incident through a glass substrate coated with a tin oxide transparent conducting film. The top metal is designed to be highly reflecting to increase the absorption of light in the cell. An alternative design uses a stainless steel substrate, with light incident through the transparent conducting layer on top of the a-Si:H film.

Fig. 10.19. The dependence of the solar cell fill factor on the carrier collection length (Faughan and Crandall 1984).

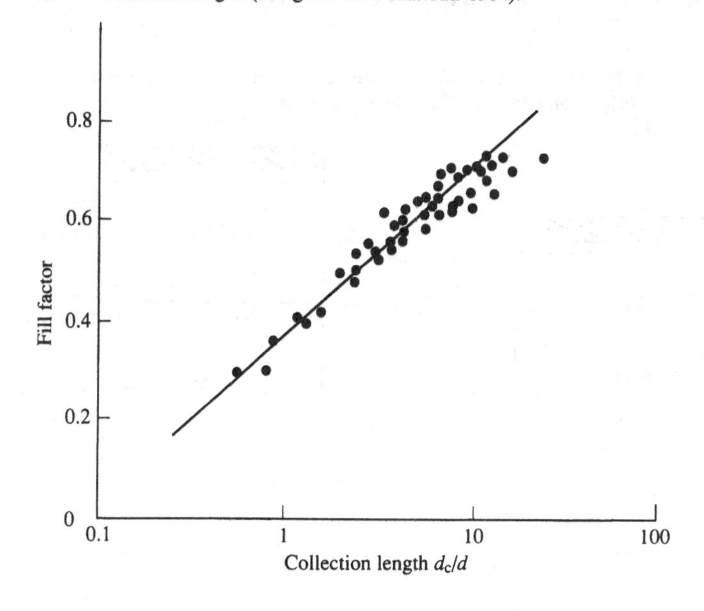

The fabrication of an efficient solar cell involves much more than the optimization of the a-Si:H material. Some of the innovations that have helped to increase the efficiency are as follows.

(1) *Textured substrates*

One of the first significant improvements was to increase the light absorbed in the cell by the use of a textured substrate. This causes the incident light to be scattered along the plane of the film, increasing the effective path length for absorption. As a result the collection efficiency is enhanced at long wavelengths and there is a 20–30% improvement in the total solar cell efficiency. Under the right deposition conditions the tin oxide transparent conductor layer naturally assumes a rough texture with the morphology needed to scatter the light, but maintaining a high conductivity.

(2) *Doped contact layers*

The next improvement to the efficiency came from a change in the material used for the contacts. The three sources of the loss of efficiency associated with the contacts are the absorption of light, the collection of carriers diffusing against the internal field, and resistive loss. The first two losses are improved by the use of a-Si:C:H p-type layers, particularly in those devices deposited on glass where the light is incident on the p layer. There is less absorption loss because the band gap of a-Si:C:H is larger than that of a-Si:H. The conductivity is a little

Fig. 10.20. Typical structure of (*a*) single-junction and (*b*) stacked solar cells, with illumination through the glass substrate.

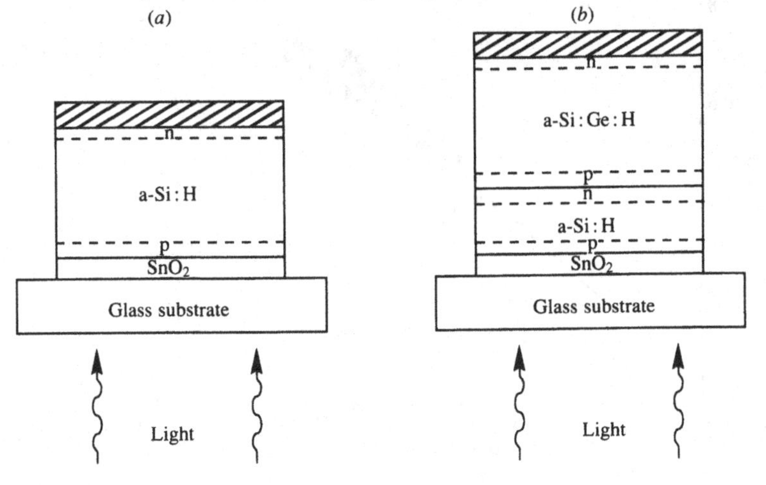

lower than that of a-Si:H, but at alloy levels of 20–30 %, the doping is sufficiently good and does not introduce any further resistive losses. The larger band gap of the SiC alloy also provides a heterojunction barrier to the back-diffusion of carriers, so that the short wavelength response is further improved. Microcrystalline silicon or SiC p layers give a further improvement in the performance of the contact.

The n-type layer is improved by making it microcrystalline. Such films can be made in the same plasma deposition reactor as for the amorphous films, but with the silane gas heavily diluted with hydrogen (see Fig. 2.5). The microcrystalline silicon has a higher conductivity than a-Si:H, which tends to improve the fill factor by reducing resistive loss. Although the band gap of crystalline silicon is smaller than a-Si:H, it has less absorption throughout the visible spectrum because the gap is indirect. Hence this type of contact also reduces the absorption loss, which is important in devices in which the light is incident through the n layer.

(3) *Stacked cells*

Present efforts to improve the cell efficiency further are aimed at multiple stacked cells. The maximum theoretical efficiency of single-junction cells is estimated to be about 14–15 %, but that of multiple cells is over 20 %. Present single-junction cells are at about 80–90 % of the theoretical efficiency, but the corresponding factor for multiple cells is lower. This type of cell therefore offers a greater opportunity for improvement.

Stacked cells have the advantage of allowing high energy photons to be collected with a high V_{oc}, while also collecting the lower energy photons. A single cell is a compromise between the high V_{oc} of a material with a large band gap and the high charge collection over a wide portion of the solar spectrum from a material with low band gap. Multiple cells split up the solar spectrum and collect each part with the largest possible V_{oc}. The optimum configuration for a double cell based on a-Si:H is to have the second junction with a smaller band gap and for a triple cell to have cells with both larger and smaller band gaps. The primary materials need is therefore for a lower band gap material and a-Si:Ge:H is the obvious candidate. A frustrating property of the general class of these hydrogenated materials is that the alloys of a-Si:H all have a much larger defect density than a-Si:H, which adversely affects their solar cell performance. There have been significant improvements in the properties of a-Si:Ge:H alloys, so that at the time of writing, stacked cells are just beginning to outperform the single-junction cells.

The structure of the tandem cell is shown in Fig. 10.20(*b*). A two-terminal device is convenient to manufacture, but complicates the design because the current must flow through both cells. Electrons which are created in the bottom cell flow to the middle junction and recombine with holes that originate in the top cell. The recombination takes place at the p–n junction dividing the two cells and takes advantage of the fact that heavy doping causes the junction to collapse completely due to the short depletion lengths. The individual currents in the two cells exactly balance and there is loss of efficiency if they are not properly designed for equal charge collection. Great care is therefore needed to match the thicknesses of the devices properly to the solar spectrum. Similarly, in a triple cell the three individual currents are identical.

(4) *Stability*

The metastable creation of defects by prolonged exposure to sunlight remains one of the most important material problems of solar cells. The effect was first discovered by Staebler and Wronski in 1977 and its physical origin is now quite well understood (see Chapter 6). However with the understanding has come the realization that the problem has no simple solution and instead seems to be a fundamental property of a-Si:H. The solar cell technology has been obliged to accept the effect as a limitation and to design cells around the problem. Fortunately this has been quite successful, so that there are high efficiency cells which degrade no more than 10% over their lifetime.

The Staebler–Wronski effect is the result of the light-induced creation of defects which can trap electrons or holes. The defects reduce $\mu_D \tau$, which suppresses charge collection and reduces the cell efficiency. Fig. 10.19 shows that the fill factor increases approximately logarithmically with the short circuit collection length, and that a high fill factor requires $d_c > 10d$. Undoped a-Si:H has a defect density of about 3×10^{15} cm^{-3} before exposure to light, and the $\mu_D \tau$ value for holes is greater than 10^{-7} cm^2 V^{-1}. The criterion of $d_c > 10d$ can be met with a film thickness of 1 μm. However, the defect density increases to about 10^{17} cm^{-3} after prolonged illumination and the same criterion applies to a film thickness of only 2000 Å.

Thin solar cells are therefore much more stable than thick ones. On the other hand the initial efficiency of the thicker cells is larger because more of the solar spectrum is absorbed. The optimum design is a compromise between these two effects and the typical thickness used is about 0.5 μm. Tandem or triple cells are more efficient at absorbing over a wide spectral range because each cell is designed to cover a different portion of the spectrum. Consequently these cells can be made thinner. Thus the stacked cells have the advantage of being inherently more stable than single-junction cells.

An empirical relation describing the time dependence of the fill factor under illumination is derived from Eq. (10.23) (Smith *et al.* 1985). Inserting the time dependence of the defect creation rate from Eq. (6.95) and the relation between $\mu_D \tau$ and the defect density given by Eq. (8.58) results in a predicted time dependence of the fill factor,

$$F(t) = 0.7\text{–}0.08 \log t \tag{10.26}$$

where t is measured in hours of sunlight exposure. The logarithmic time dependence agrees well with the observations, as does the predicted slope.

10.4.2 *Active matrix arrays*

Active matrix arrays are used for imaging and display. Examples are liquid crystal displays (Snell *et al.* 1981), optical scanners (Kaneko, Kajiwara, Okumura and Ohkubo 1985), printer heads and radiation imaging arrays (Perez Mendez *et al.* 1989). These devices contain many individual elements (pixels) which must be addressed or read out. The general structure of the array is illustrated in Fig. 10.21 and represents either a liquid crystal display or a two-dimensional sensor array, depending on the design of the pixel. The problem of addressing or reading out data with large numbers of pixels is solved by having a grid structure of interconnecting lines. Each pixel is at the

Fig. 10.21. Diagram showing the structure of a two-dimensional matrix-addressed array. The pixel designs for a liquid crystal and for an image sensor are shown.

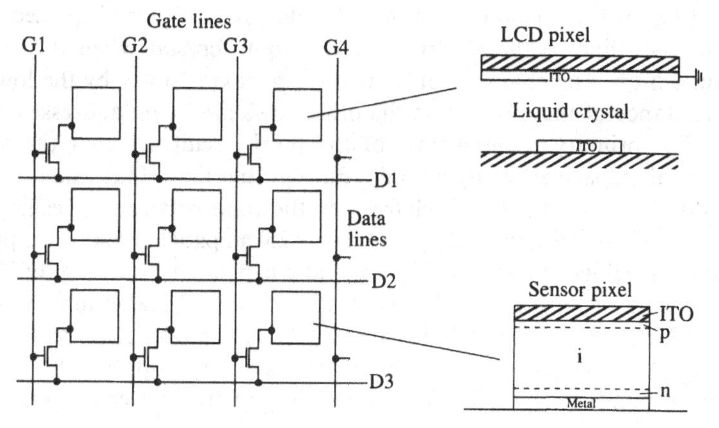

intersection of a gate and data line, which are connected to a pass transistor at the pixel. Application of a voltage to a gate line activates a column of pass transistors, and allows the associated pixels to be activated. The complete array is accessed by sequentially addressing all the gate lines.

(1) *Liquid crystal displays*
Most types of liquid crystal contain rod-shaped molecules which cause a rotation of the polarization of transmitted light when they are aligned by an applied electric field. The thin film of liquid crystal is placed between crossed polarizers so that the light transmittance is modulated by the field. The display is either viewed in reflection or by illumination from the back, which gives a brighter image. Color displays are made with individual color filters on the pixels. The response of the liquid crystal to a voltage is highly non-linear and the device usually switches on over a 2–3 V interval at an applied voltage of 5–10 V. These voltages depend on the type of liquid crystal and on the thickness of the layer (usually about 10 μm), but are typical of normal usage. The design of the pixel is shown in Fig. 10.21. The liquid crystal is an insulator and lies between a ground plane and a metal pad connected to the pass transistor. A display comparable with a television contains about 500 × 500 pixels for a total number of 2.5×10^5.

It is possible to drive the array without any active devices at each pixel, because the liquid crystal turns on above a threshold voltage, so that the device may be activated only when there is a voltage applied to both the column and row. However, each pixel is addressed for only a fraction of the total time and the contrast is poor for large arrays. An active matrix array improves the contrast by allowing each pixel to be held on while it is not being directly addressed. Charge is passed from the data line to the electrode of the liquid crystal when the gate is turned on. The charge is prevented from leaking away by the low off-resistance of the TFT while the other rows are being addressed.

The primary requirements of the pixel circuit are that the liquid crystal capacitor charges fully during the time that each row is addressed and holds the charge for the time between refreshing the display. Consider the example of a 25 × 25 cm panel of 500 × 500 pixels. Each pixel has an area of 500 μm and a capacitance of roughly 10 pF. The panel is refreshed at the usual video rate of 30 Hz, so that each row is addressed for 60 μs (30 ms/500). Each TFT must deliver a charge of 10^{-10} C in 60 μs, in order to charge the pixel up to 10 V, which corresponds to a current of 1.5 μA. The TFT transfer characteristics in

Fig. 10.8 confirm that this is well within the capability of the a-Si:H devices. To prevent charge leaking from the panel before the display is refreshed, the *RC* time constant of the TFT and pixel capacitor must be larger than 30 ms, which translates to a requirement that the off-current of the TFT be less than 10^{-10} A. Again this is comfortably within the TFT operating characteristics. Thus active matrix displays with at least 1000 lines are possible with present a-Si:H TFT technology. The low charge requirements of the liquid crystal allows the use of the a-Si:H devices and the low off-current, which is inherent to the a-Si:H TFT, is an essential feature.

(2) *Scanner arrays*

Linear arrays of a-Si:H light sensors are used in optical scanners, for example in FAX machines. A page-wide array eliminates the optics necessary to reduce the image size when using a crystalline silicon CCD imaging chip. The resolution of the scanner is typically 200–400 spots per inch (spi), so that 2000–3000 pixels are required in a standard 8.5 inch page-wide array. The matrix-addressing arrangement is essentially identical to that of the liquid crystal panels, except that it is configured as a one-dimensional array. The circuit comprises a pass transistor and a p–i–n or Schottky sensor as shown in Fig. 10.21. Charge is transferred to the bottom electrode of the sensor during illumination and accumulates while the TFT is turned off. Turning on the TFT allows the charge to flow out to the data line and be read by the external electronics. The matrix-addressing scheme turns on a row of pass transistors while simultaneously reading out the data lines – a matrix of 50 gate and data lines is sufficient to read out a 300 spi linear array. The sensor integrates the signal over the whole time that the sensor is not being addressed and is read out for only a small fraction of the time, giving efficient collection of charge.

The pixel circuit requirements for the scanner are similar to those of the liquid crystal display. A 300 spi array scanning an image at 3 inch s^{-1} must read each line in a time of 1 ms. For a matrix of 50 gate and data lines, each pixel is addressed for 20 μs. The capacitor of the sensor is about 1 pF (note that 300 spi corresponds to a pixel of about 80 μm which is smaller than in the liquid crystal example). The required TFT current of 1 μA is easily obtained. The low off-current of the TFT also ensures that the charge will not leak away between being read out.

In both the liquid crystal and scanner a-Si:H arrays, the time limitation of the circuit is not the intrinsic switching speed of the TFT, but is the *RC* time constant of the pixel circuit, comprising the capacitance of the sensor and the resistance of the TFT. Complete charging takes about three times the RC time constant. There is also a voltage change when the gate is turned on and off, which is caused by the gate-to-drain capacitance of the TFT. This must be much smaller than the load capacitance to minimize the voltage drop.

(3) *High voltage TFT arrays*

Arrays of high voltage TFTs have found applications in electrographic printing. In this technology, charge is transferred to a dielectric coated paper by direct breakdown of air. This requires a voltage of about 400 V and, before the availability of a-Si:H devices, the print heads consisted of individually contacted fine wires. The a-Si:H TFTs allow the possibility of integrating the print head onto a monolithic substrate and addressing it with a low voltage active matrix array. The circuit and response is shown in Fig. 10.22. The high voltage TFTs are part of an inverter circuit, which pulls down the voltage to the print head when a low gate voltage is applied. The figure shows that a voltage of over 400 V can be switched by a 5 V gate. As with the other array applications the device requires a very small current which is easily met by the a-Si:H TFTs.

(4) *Array fabrication*

The fabrication of these one- and two-dimensional arrays follows the device processing technology developed for crystalline silicon integrated circuits. Since the feature size of the elements is of

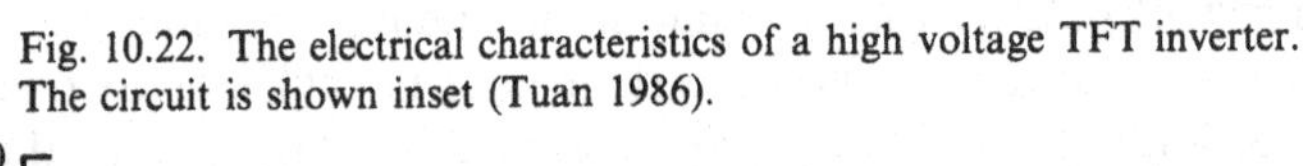

Fig. 10.22. The electrical characteristics of a high voltage TFT inverter. The circuit is shown inset (Tuan 1986).

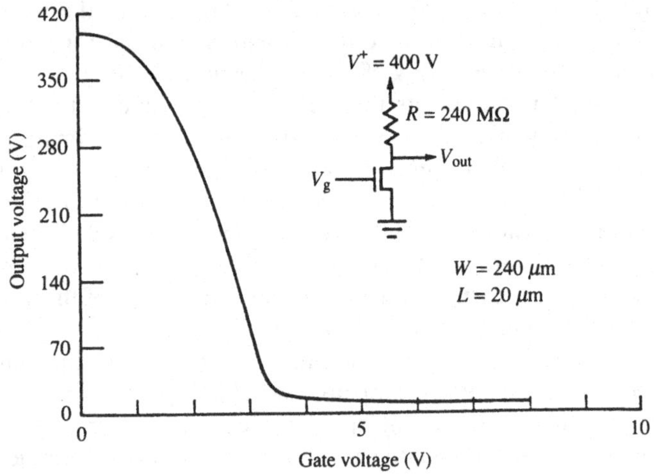

order 10 μm, photolithography is the only viable processing technique. The device is made by sequential thin film deposition and patterning of a-Si:H, metals, dielectrics, and passivation layers. Fortunately a-Si:H has similar etching properties to crystalline silicon, so that the a-Si:H technology can take advantage of the accumulated experience with crystalline silicon. The size of the circuits is limited by the largest area that can be patterned, which at present is about 30 cm square. The minimum feature sizes used over this area are about 10 μm, but no doubt will continue to decrease as the yield improves.

Fig. 10.23 shows a cross-sectional view of a typical circuit for a scanner, and shows the p–i–n sensor and the pass transistor. This particular circuit takes nine mask levels. The metals used are chromium and aluminum, the former for contacts to the TFTs and sensors and the latter for the interconnecting lines. The transparent conducting contact to the sensor is made with ITO and polyimide is used for passivation and isolation of the devices.

The metastability phenomena influence the performance of the active matrix arrays. Defect creation in the channel causes a threshold voltage shift when a TFT is held on for an extended time and results in a slow drift of the on-current. Fortunately the rate of defect creation is low at room temperature and represents a minor problem. There is a larger effect on the characteristics of the high voltage TFTs. Resistors fabricated from n^+ a-Si:H change their resistance slowly because of defect equilibration and can affect the gain of amplifier circuits.

The a-Si:H arrays are rapidly increasing in size and in the total number of TFTs. From the first report of an individual TFT in 1976, the arrays have now reached up to 10^6 transistors, which represents a doubling in number each year. Up to 1990 the increase in TFT number has primarily been driven by the increase in the total number of pixels.

Fig. 10.23. Cross-sectional view of a sensor array (Ito *et al.* 1987).

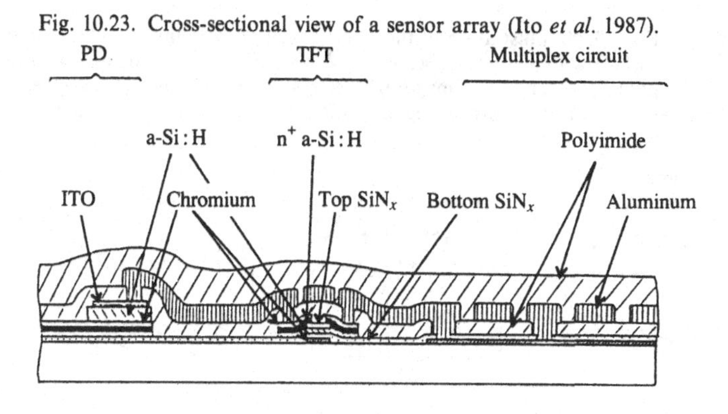

However, it seems likely that the pixel number will not increase rapidly in the future and any increase in the number of TFTs will arise from an increasing complexity of the pixel circuitry.

10.4.3 *Photoreceptors*

The first major commercial application of amorphous semiconductors was as the photoreceptor in xerographic copiers and subsequently in laser printers. The early photoreceptors were selenium films, but several other materials were subsequently developed, including As_2Se_3 and various organic films. Amorphous silicon is a good material for a xerographic photoreceptor (e.g. Shimizu (1985), Pai (1988)) and is used in some commercial copying machines.

The principles of xerography are illustrated in Fig. 10.24. The surface of the photoreceptor is charged by a corona discharge. The image of the page to be copied is then projected onto the surface. The illumination causes photoconductivity in the photoreceptor and discharges the surface, leaving an electrostatic image on the drum. This latent image is developed by first attaching ink particles (toner) to the surface by the electrostatic force, and then transferring the ink to paper where it is fused by heat or pressure to make it permanent.

The time-dependent discharge of the surface voltage of a photoreceptor is shown in Fig. 10.25. The voltage decays slowly in the absence of light and is referred to as the dark discharge. The decay is much faster during illumination and depends on the carrier generation efficiency. Incomplete discharge leaves a residual voltage on the surface. The requirements of the photoreceptor material are rather similar to those of the light sensor described in Section 10.1.1. A low dark current is needed to hold the surface charge and a high photoconductivity efficiency is needed to give a rapid discharge. The

Fig. 10.24. Illustration of the principles of xerography; (*a*) charging the surface with a corona discharge; (*b*) selective discharge by exposing the photoreceptor to the image; (*c*) electrostatic attachment of ink-laden toner particles to the charged regions. The toner is subsequently transferred to paper.

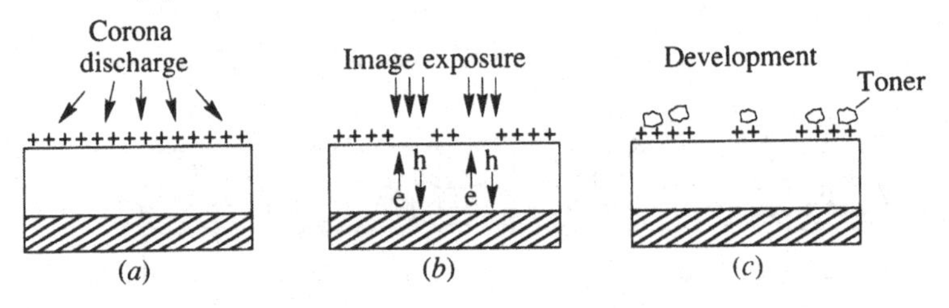

dark conductivity must be below $10^{-12}\ \Omega^{-1}\ cm^{-1}$, because the entire development takes about 1 s. A-Si:H is a better photoconductor than selenium, because there is no room temperature geminate recombination and light is absorbed and generates carriers across virtually the entire visible spectrum.

The residual voltage at the end of the discharge is caused by carrier trapping and is minimized by having a large value of $\mu_D \tau$ for the photoconducting carrier. The residual voltage, V_R, is approximately the voltage at which the distance that the carrier moves before trapping, $\mu_D \tau V/L$, is equal to the thickness, L, of the photoreceptor, so that,

$$V_R \approx L^2/\mu_D \tau \tag{10.25}$$

A photoreceptor of thickness 20 μm, with $\mu_D \tau = 10^{-6}\ cm^2\ s^{-1}$, has a residual voltage of only 4 V which is small enough to be unimportant.

The development process usually requires that the surface is charged to about 400 V by the corona discharge. In order to prevent the dark discharge, three sources of current must be avoided – the thermal generation of bulk carriers, injection from the back contact, and dielectric breakdown. A dielectric blocking layer is used to suppress injection and a slight boron doping of the photoconductor layer

Fig. 10.25. Charging and discharging characteristics of a xerographic photoreceptor.

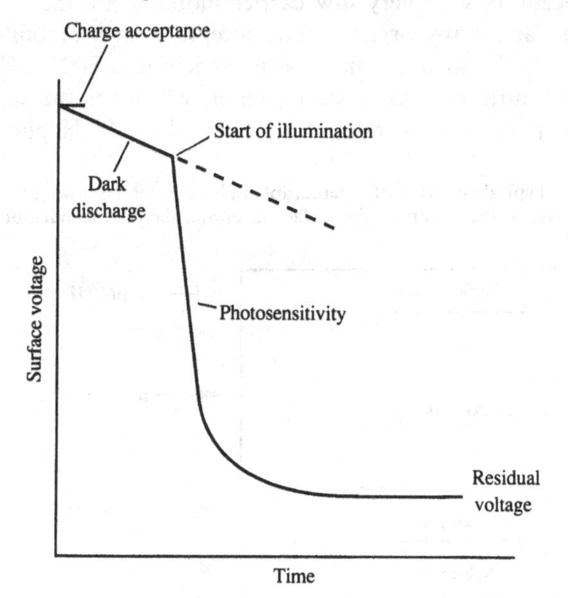

reduces the bulk thermal generation of carriers, by aligning the Fermi energy accurately with the middle of the band gap (Eq. (10.6)). The typical photoreceptor structure is shown in Fig. 10.26. The need to limit the internal electric field to prevent breakdown typically constrains the thickness of the photoreceptor to at least 10–20 μm.

The exact mechanism by which charge is held at the surface of a photoreceptor is poorly understood, but there is always a maximum voltage to which the surface can be charged and this is termed the charge acceptance. In a-Si:H this is maximized by light boron doping. There is also a requirement of low lateral conduction of the charge across the surface of the photoreceptor. The creation of the latent image establishes large surface electric fields, so that any significant surface conduction obliterates the image. The ability to fabricate a-Si:H TFTs implies that there can be a large surface conduction under conditions of surface charge accumulation. These effects were found in early a-Si:H photoreceptors and, furthermore, the surface was shown to be particularly susceptible to scratching. However, an overcoat of silicon nitride or carbide passivates the surface and removes the surface conduction.

The requirement on the response time of the photoreceptor is minimal, even at high copy rates. The image is developed entirely in parallel, so that the response time need only be that of the development time for the entire page, which is 0.5 s for a copy speed of 120 pages per minute. Photoreceptors with very low carrier mobility are therefore adequate and, in fact, many organic photoreceptors have mobilities below 10^{-5} cm^2 V^{-1} s^{-1}, much less than both carriers in a-Si:H. (This situation can be contrasted with a laser printer, which for the same printing speed requires a data rate of 1 GHz.) The a-Si:H photo-

Fig. 10.26. Typical structure of a xerographic photoreceptor showing the blocking layer at the substrate, the surface layer and the photoconductor (Pai 1988).

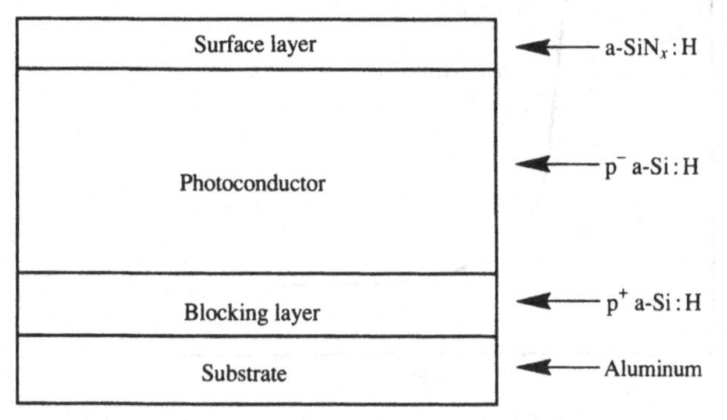

receptor can therefore be designed to transport either electrons or holes depending on the desired sign of the surface charge.

10.4.4 *Vidicon*

The principle of the vidicon is similar to that of the photoreceptor. A photoconductive material is used to discharge a surface charge upon illumination. In this case the image is obtained by the current flow as the surface is raster-scanned by an electron beam. The vidicon material needs a high photoconductivity and a low dark conductivity, like the photoreceptor, and hence is constructed with a blocking layer at the substrate. One difference from the photoreceptor is that the image is read out serially, so that a photoconductor with a fast response is essential. The typical requirement of a TV image is about 2.5×10^5 pixels per frame, giving a data rate of about 1 MHz. This sets an upper limit on the carrier transit time of 1 μs. Electrons in a-Si:H easily satisfy this requirement in a film of thickness 5 μm, at a voltage of 50 V, which are typical operating conditions.

Recent work on amorphous selenium vidicons has shown that avalanche gain occurs at fields of more than 10^6 V cm^{-1}. Avalanche gain is a useful feature in any type of sensor, but has not been demonstrated in a-Si:H. The general condition for avalanching is that the energy gained during a mean free path should be enough to cause ionization of carriers. The short mean free path of carriers in all amorphous semiconductors makes the observation of avalanching quite surprising. It is possible that the inelastic scattering events only give up a small energy and so the carrier energy continues to increase even with frequent scattering events.

10.4.5 *High energy radiation imaging*

There are many applications for large area imaging of high energy radiation and particles. The most widely used are in the medical field for X-ray diagnostics or radiation therapy. Present X-ray diagnostics either use film, which is an effective medium but does not lend itself to real time display or to electronic analysis and storage, or fluoroscopy in which a visible image is created on a fluorescent screen and imaged by a TV camera. Imaging of radiation is also used in biological measurements, for example in gel electrophoresis, in which fragments of biological molecules (e.g. DNA) are diffused in a gel and are detected by adding a radioactive tracer and imaging the emitted radiation. High energy physics experiments are another radiation imaging application in which the tracks of particles emanating from a nuclear interaction are identified.

Sensing of high energy radiation is performed either directly in the a-Si:H film or using an intermediate converter material, as illustrated in Fig. 10.27. Direct detection requires a thick film, typically 10–100 μm, because most ionizing radiation interacts weakly with matter. The converter material is usually a phosphor which absorbs the radiation and emits visible light. Thin a-Si:H sensors are excellent detectors of such visible light.

The requirements of the detectors depend on the specifics of the radiation. X-rays of energy up to about 100 keV interact with matter by ejecting electrons out of core levels to states high up in the conduction bands. The emitted electrons cause ionization by creating electron–hole pairs as they lose energy. The electrons and holes are collected in the same way as for visible light sensors. High energy γ-rays cause the production of high energy electrons by Compton scattering and these electrons give secondary electron–hole pairs. The absorption depth of X-rays and γ-rays tends to decrease with increasing energy, but with structure in the absorption spectrum due to the absorption thresholds for the core levels. The absorption also increases with the atomic mass of the detector element. Silicon is a light element and so is not a particularly good absorber. A-Ge:H films would be much better if they could be made with a low defect density.

High energy charged particles directly cause ionization and create electron–hole pairs. The ionization rate is given by,

$$\frac{\mathrm{d}E}{\mathrm{d}x} = \frac{Zz^3M}{E}(\text{relativistic terms}) \tag{10.27}$$

where Z is the atomic number and z, M, and E are the particle charge, mass, and energy. The relativistic terms increase with energy such that

Fig. 10.27. Illustration of the imaging of high energy radiation using (*a*) a thick p–i–n detector or (*b*) a phosphor to convert the radiation into visible light.

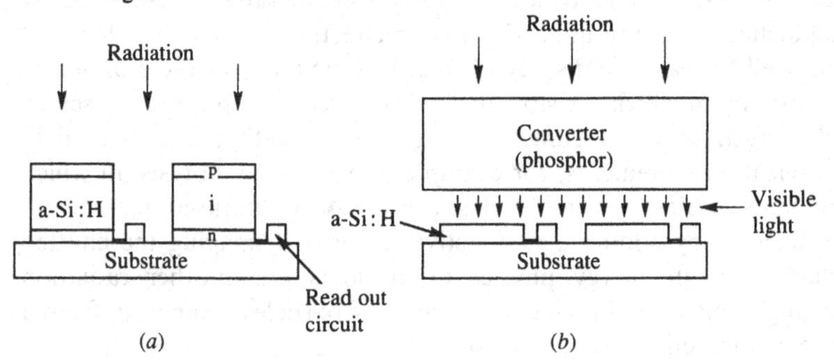

there is a minimum ionization rate which occurs at an energy of about 1 MeV for electrons. These particles are referred to as 'minimum ionizing particles', detection of which ensures that all other possible particles can also be detected.

The number of electron–hole pairs created by the passage of the particle through a film of thickness $\mathrm{d}x$ is

$$\mathrm{d}N = \frac{\mathrm{d}E}{\mathrm{d}x}\frac{1}{W_\mathrm{i}}\mathrm{d}x \tag{10.28}$$

W_i is the average energy loss associated with the creation of an electron–hole pair, and is usually about three times the band gap energy; W_i is about 6 eV in a a-Si:H. The value of $\mathrm{d}E/\mathrm{d}x$ varies from about 200 keV μm^{-1} for an α-particle with energy 4 MeV, to 600 eV μm^{-1} for a minimum ionizing particle. The α-particle has a high ionization rate because of its large mass and charge.

Direct detection of a single ionizing particle by a-Si:H therefore requires a thick film and a sensitive detector. p–i–n devices with undoped layer thickness of up to 50 μm can be fully depleted in reverse bias. Furthermore the leakage currents are low enough that the noise level just allows the detection of minimum ionizing particles. Particles with higher ionization rates are easily detected, so that the use of a-Si:H imaging arrays appears to be feasible. The manufacture of large arrays with this large thickness is, however, a challenging problem.

The response of the detector to α-particles is considerably smaller than is predicted from the ionization rate, because of recombination due to high carrier densities. Although the ionizing particle produces only about 3×10^4 electron–hole pairs per micron, these are within a column of material along the track of the particle, having a radius of about 1000 Å. The carrier concentration is therefore 3×10^{18} cm^{-3}, which is sufficiently high to cause recombination during charge collection. The effect is reduced when the particle is incident at an angle and is not colinear with the collection field, and also is reduced for less strongly ionizing particles.

In applications involving a fluorescent converter screen (Fig. 10.27(*b*)), the a-Si:H array detects visible light. This arrangement has the advantage of needing much thinner detectors and the structure is essentially identical to that used in optical scanners. The fluorescent screen can be chosen to optimize the conversion of the radiation, although there is an inevitable loss of efficiency in the two-stage detection process. The maximum collection of light is achieved by placing the sensor array directly in contact with the fluorescent screen, which is possible with the a-Si:H devices because of their large size. In contrast, imaging of the phosphor onto a small sensor array greatly reduces the detection ability. The optical transfer efficiency is

$$T = M^2/8f^2 \tag{10.29}$$

where M is the magnification and f is the f-number of the collection optics. Imaging of a 30 cm phosphor onto a 1 cm sensor with $f = 1$ optics, results in a transfer efficiency of only 10^{-4}. Thus, the large size of the a-Si:H arrays confers a great advantage in imaging from an extended source.

A critical requirement for the applications to ionizing radiation is the resistance to radiation damage of the a-Si:H films. A-Si:H TFTs have virtually no degradation even after 5 Mrad of γ-ray exposure. Similarly, p–i–n sensors are more radiation resistant than the equivalent crystalline silicon devices. It is not immediately obvious why this should be so. The already disordered network of a-Si:H may be less susceptible to the additional disordering effect of the radiation damage. However, damage which causes bond breaking generates defects in either type of material. The thin a-Si:H sensors are probably less influenced by defects than the crystalline devices. The passivation of defects by hydrogen may also play an important role in suppressing defect formation. The annealing properties of a-Si:H, in which damage

Fig. 10.28. Structure of the position-sensing device made from a-Si:H (Takeda and Sano 1988).

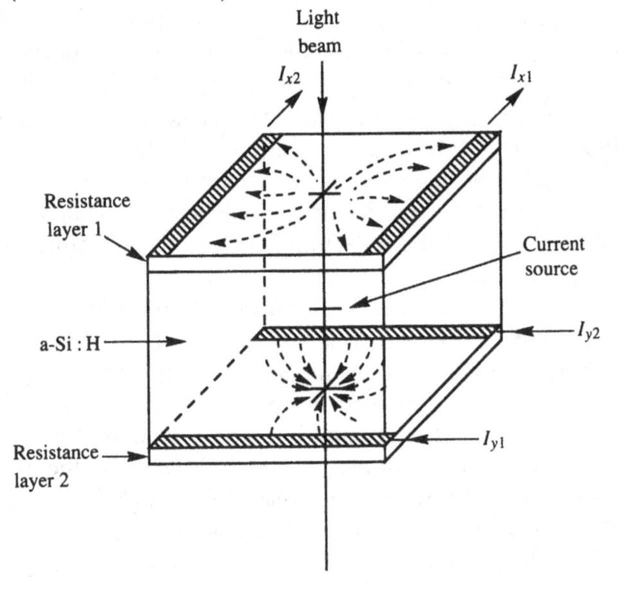

is removed at about 150 °C, restoring the original properties of the films, means that any radiation damage produced in the devices can be easily reversed.

10.4.6 *Position sensors*

A position-sensing device made from a-Si:H is described by Takeda and Sano (1988). The structure consists of a continuous film of undoped a-Si:H between two transparent conducting electrodes as illustrated in Fig. 10.28. The electrodes are contacted along opposite edges, with the top and bottom terminals at right angles, giving four terminals. When the device is illuminated at a single spot by a light pen, a photocurrent flows across the a-Si:H film at that point. The top and bottom contacts form resistive dividers, so that the ratio of the currents out of the two terminals at the top contact gives the position of the spot across the electrode. A similar location is deduced from the ratio of the current into the sensor at the other two terminals, so that the coordinate of the spot is obtained. This device is relatively simple to fabricate because it requires no photolithography and has only four terminals. It has a positioning accuracy of 300 μm over a 10 cm surface. The position sensor is intended for an input device in a telephone display, so that a simple image can be written, transmitted down a telephone connection, and displayed at the other receiver.

References

Abeles, B. and Tiedje, T. (1983) *Phys. Rev. Lett.* **51**, 2003.

Abeles, B., Wronski, C. R., Persans, P. D., and Tiedje, T. (1985) *Amorphous Metals and Semiconductors*, eds. P. Haasen and R. I. Jaffee (Pergamon, Oxford) 289.

Abkowitz, M. (1985). *Physics of Disordered Materials*, eds. D. Adler, H. Fritzsche, and S. R. Ovshinsky (Plenum, New York) 483.

Abrahams, E., Anderson, P. W., Licciardello, D. C., and Ramakrishman, T. V. (1979) *Phys. Rev. Lett.* **42**, 673.

Abrahams, E., Anderson, P. W., and Ramakrishman, T. V. (1980) *Philos. Mag.* **B42**, 827.

Adler, D. (1978) *Phys. Rev. Lett.* **25**, 1755.

Adler, D. (1981) *J. Phys. (Paris)* **42**, C4, 3.

Adler, D. and Joffa, E. J. (1976) *Phys. Rev. Lett.* **36**, 1197.

Aker, B., Peng, S.-Q., Cai, S.-Y., and Fritzsche, H. (1983). *J. Non-Cryst. Solids* **59 & 60**, 509.

Anderson, P. W. (1958) *Phys. Rev.* **109**, 1492.

Anderson, P. W. (1975) *Phys. Rev. Lett.* **34**, 953.

Apsley, N., Davis, E. A., Troup, A. P., and Yoffe, A. D. (1977) *Proc. 7th Int. Conf. on Amorphous and Liquid Semiconductors*, ed. W. E. Spear (CICL, Edinburgh) 447.

Arce, R. and Ley, L. (1989) *Proc. MRS Symp.* **149**, 675.

Atkins, P. W. (1978) *Physical Chemistry* (Oxford University Press, Oxford).

Austin, I. G., Jackson, W. A., Searle, T. M., Bhat, P. K., and Gibson, R. A. (1985) *Philos. Mag.* **B52**, 271.

Austin, I. G. and Mott, N. F. (1969) *Adv. Phys.* **18**, 41.

Bar-Yam, Y., Adler, D., and Joannopoulos, J. D. (1986) *Phys. Rev. Lett.* **57**, 467.

Bar-Yam, Y. and Joannopoulos, J. D. (1986) *Phys. Rev. Lett.* **56**, 2203.

Baum, J., Gleason, K. K., Pines, A., Garroway, A. N., and Reimer, J. A. (1986) *Phys. Rev. Lett.* **56**, 1377.

Beyer, W. and Fischer, R. (1977) *Appl. Phys. Lett.* **31**, 850.

Beyer, W. and Overhof, H. (1979) *Solid State Commun.* **31**, 1.

Beyer, W. and Overhof, H. (1984) *Semiconductors and Semimetals* (Academic Press, Orlando), 21C, Chapter 8.

Beyer, W. and Wagner, H. (1982) *J. Appl. Phys.* **53**, 8745.

Biegelsen, D. K. (1980) *Solar Cells* **2**, 421.

Biegelsen, D. K. (1981) in *Nuclear and Electron Resonance Spectroscopies Applied to Material Science*, eds Kaufmann and Shenoy (North Holland, Amsterdam).

Biegelsen, D. K., Street, R. A., and Jackson, W. B. (1983) *Physica B*, **117B & 118B**, 899.

Biegelsen, D. K., Street, R. A., Tsai, C. C., and Knights, J. C. (1979) *Phys. Rev.* **B20**, 4839.

Biegelsen, D. K. and Stutzmann, M. (1986) *Phys. Rev.* **B33**, 3006.

Biswas, R. and Hamann, D. R. (1987) *Phys. Rev.* **B36**, 6434.

Biswas, R., Wang, C. Z., Chan, C. T., Ho, K. M., and Soukoulis, C. M. (1990) to be published.
Boyce, J. B., Ready, S. E., and Tsai, C. C. (1987) *J. Non-Cryst. Solids* **97 & 98**, 345.
Boyce, J. B. and Stutzmann, M. (1985) *Phys. Rev. Lett.* **54**, 562.
Brodsky, M. H. (1980) *Solid State Commun.* **36**, 55.
Brodsky, M. H. and Title, R. S. (1969) *Phys. Rev. Lett.* **23**, 581.
Bustarret, E., Vaillant, F., and Hepp, B. (1988) *Proc. MRS Symp.* **118**, 123.
Caplan, P. J., Poindexter, E. H., Deal, B. E., and Razouk, R. R. (1979) *J. Appl. Phys.* **50**, 5847.
Carasco, F. and Spear, W. E. (1983) *Philos. Mag.* **B47**, 495.
Carlos, W. E. and Taylor, P. C. (1980) *Phys. Rev. Lett.* **45**, 358.
Carlos, W. E. and Taylor, P. C. (1982) *Phys. Rev.* **B25**, 1435.
Carlson, D. E. (1986) *Appl. Phys.* **A41**, 305.
Carlson, D. E. and Magee, C. W. (1978) *Appl. Phys. Lett.* **33**, 81.
Carlson, D. E. and Wronski, C. R. (1976) *Appl. Phys. Lett.* **28**, 671.
Ching, W. Y., Lam, D. J., and Lin, C. C. (1980) *Phys. Rev.* **B21**, 2378.
Chittick, R. C., Alexander, J. H., and Sterling, H. F. (1969) *J. Electrochemical Soc.*, **116**, 77.
Chittick, R. C. and Sterling, H. F. (1985) *Tetrahedrally Bonded Amorphous Semiconductors*, eds. D. Adler and B. B. Schwartz (Plenum, New York), p. 1.
Cloude, C., Spear, W. E., LeComber, P. G., and Hourd, A. C. (1986) *Philos. Mag. Lett.* **54**, L113.
Cody, G. D. (1984). *Semiconductors and Semimetals* (Academic Press, Orlando) Vol. 21B, Chapter 2.
Cody, G. D., Tiedje, T., Abeles, B., Brooks, B., and Goldstein, Y. (1981) *Phys. Rev. Lett.* **47**, 1480.
Cohen, J. D. (1989) Proc. ICALS Conf to be published.
Cohen, J. D., Harbison, J. P., and Wecht, K. W. (1982) *Phys. Rev. Lett.* **48**, 109.
Cohen, M. H., Economou, E. N., and Soukoulis, C. M. (1983) *J. Non-Cryst. Solids* **59 & 60**, 15.
Collins, R. W. and Cavese, J. M. (1989) *MRS Symp. Proc.* **118**, 19.
Collins, R. W., Paesler, M. A., and Paul, W. (1980) *Solid State Commun.* **34**, 833.
Connell, G. A. N. and Temkin, R. J. (1974) *Phys. Rev.* **B12**, 5323.
Conradi, M. S. and Norberg, R. E. (1981) *Phys. Rev.* **B24**, 2285.
Crandall, R. S. (1981) *Phys. Rev.* **B24**, 7457.
Cutler, M. and Mott, N. F. (1969) *Phys. Rev.* **181**, 1336.
Davis, E. A. and Mott, N. F. (1970) *Philos. Mag.* **22**, 903.
Delahoy, A. E. and Tonon, T. (1987) *AIP Conf. Proc.* 157, 263.
Dersch, H., Schweitzer, L., and Stuke, J. (1983) *Phys. Rev.* **B28**, 4678.
Dersch, H., Stuke, J. and Beichler, J. (1980) *Appl. Phys. Lett.* **38**, 456.
Dingle, R. (1965) *Proc. Int. Conf. On the Physics of Semiconductors*, ed. F. G. Fumi (Tipografia Marves, Rome) p. 965.
Döhler, G. H. (1972) *Phys. Stat. Sol.* **B52**, 79.
Döhler, G. H. (1979) *Phys. Rev.* **B19**, 2083.
Döhler, G. H. (1984) *Proc. Int. Conf. on Physics and Semiconductors*, ed. J. D. Chadi and W. A. Harrison (Springer–Verlag, Berlin), p. 491.
Dow, J. D. and Redfield, D. (1970) *Phys. Rev.* **B1**, 3358.
Dunstan, D. J. (1981) *J. Phys. C.* **14**, 1363.
Dunstan, D. J. and Boulitrop, F. (1981) *J. de Physique*, **42**, Suppl. 10, C4–331.

Dunstan, D. J. and Boulitrop, F. (1984) *Phys. Rev.* **B30**, 5945.
Emin, D. (1977) *Proc. 7th Int. Conf. on Amorphous and Liquid Semiconductors*, ed. W. E. Spear (CICL, Edinburgh) 249.
Emin, D., Seager, C. H., and Quinn, R. K. (1972) *Phys. Rev. Lett.* **28**, 813.
Evangelisti, F. (1985) *J. Non-Cryst. Solids* **77 & 78**, 969.
Fauchet, P. M., Hulin, D., Migus, A., Antonetti, A., Kolodzey, J., and Wagner, S. (1986) *Phys. Rev. Lett.* **57**, 2438.
Faughnan, B. W. and Crandall, R. S. (1984) *Appl. Phys. Lett.* **44**, 537.
Fedders, P. A. and Carlsson, A. E. (1988) *Phys. Rev.* **B37**, 8506; (1989) *Phys. Rev.* **B39**, 1134.
Fenz, P., Müller, H., Overhof, H., and Thomas, P. (1985) *J. Phys. C.* **18**, 3191.
Friedman, L. (1971) *J. Non-Cryst. Solids* **6**, 329.
Fritzsche, H. (1977) *Proc. 7th Int. Conf. on Amorphous and Liquid Semiconductors*, ed. W. E. Spear (CICL, Edinburgh), p. 3.
Fritzsche, H. (1984a) *Phys. Rev.* **B29**, 6672.
Fritzsche, H. (1984b) *Semiconductors and Semimetals*, ed. J. I. Pankove, (Academic Press, Orlando), Vol. 21C, Chapter 9.
Fritzsche, H. (1987) *AIP Conf. Proc.* **157**, 366, and other papers in the same volume.
Fritzsche, H., Yang, S.-H., and Takada, J. (1988) *Proc. MRS Symp.* **118**, 275.
Gallagher, A. (1986) *Mat. Res. Soc. Symp. Proc.* **70**, 3.
Gallagher, A. (1988) *J. Appl. Phys.* **63**, 2406.
Gallagher, A. and Scott, J. (1987) *J. Solar Cells*, **21**, 147.
Goodman, N. and Fritzsche, H. (1980) *Philos. Mag.* **B42**, 149.
Graebner, J. E., Golding, B., Allen, L. C., Knights, J. C., and Biegelsen, D. K. (1984) *Phys. Rev.* **B29**, 3744.
Grant, A. J. and Davis, E. A. (1974) *Solid State Commun.* **15**, 427.
Greeb, K. H., Fuhs, W., Mell, H., and Welsch, H. M. (1982) *Solar Energy Mat.* **7**, 253.
Greenbaum, S. G., Carlos, W. E., and Taylor, P. C. (1984) *J. Appl. Phys.* **56**, 1874.
Hack, M., Shaw, J. G., and Shur, M. (1988) *MRS Symp. Proc.* **118**, 207.
Hack, M. and Shur, M. (1985) *Appl. Phys.* **58**, 997.
Hack, M. and Street, R. A. (1988) *Appl. Phys. Lett.* **53**, 1083.
Hack, M., Street, R. A., and Shur, M. (1987) *J. Non-Cryst. Solids* **97 & 98**, 803.
Hamanaka, H., Kuriyama, K., Yahagi, M., Satoh, M., Iwamura, K., Kim, C., Kim, Y., Shiraishi, F., Tsuji, K., and Minomura, S. (1984) *Appl. Phys. Lett.* **45**, 786.
Han, D. and Fritzsche, H. (1983) *J. Non-Cryst. Solids* **59 & 60**, 397.
Hattori, K., Mori, T., Okamoto, H., and Hamakawa, Y. (1988) *Phys. Rev. Lett.* **60**, 825.
Hayes, T. M., Allen, J. W., Beeby, J. L., and Oh, S. J. (1985) *Solid State Commun.* **56**, 953.
He, H. and Thorpe, M. F. (1985) *Phys. Rev. Lett.* **54**, 2107.
Hecht, K. (1932) *Z. Phys.* **77**, 235.
Henry, C. H. and Lang, D. V. (1977) *Phys. Rev.* **B15**, 989.
Herring, C. (1960) *J. Appl. Phys.* **31**, 1939.
Hong, K. M., Noolandi, J., and Street, R. A. (1981) *Phys. Rev.* **B23**, 2967.
Howard, J. and Street, R. A. (1991) to be published.
Hundhausen, M., Ley, L., and Carius, R. (1984) *Phys. Rev. Lett.* **53**, 1598.
Ioffe, A. F. and Regel, A. R. (1960) *Prog. Semicond.* **4**, 237.
Iqbal, A., Jackson, W. B., Tsai, C. C., Allen, J. W., and Bates, C. W., Jr. (1987) *J. Appl. Phys.* **61**, 2947.

Ito, H., Suzuki, T., Nobue, M., Nishihara, Y., Sakai, Y., Ozawa, T., and Tomiyama, S. (1987) *MRS Symp. Proc.* **95**, 437.
Jackson, W. B. (1982) *Solid State Commun.* **44**, 477.
Jackson, W. B. (1989) Proc. ICALS Conf. to be published.
Jackson, W. B. and Amer, N. M. (1982) *Phys. Rev.* **B25**, 5559.
Jackson, W. B., Biegelsen, D. K., Nemanich, R. J., and Knights, J. C. (1983) *Appl. Phys. Lett.* **42**, 105.
Jackson, W. B. and Kakalios, J. (1988) *Phys. Rev.* **B37**, 1020.
Jackson, W. B., Kelso, S. M., Tsai, C. C., Allen, J. W., and Oh, S.-J. (1985) *Phys. Rev.* **B31**, 5187.
Jackson, W. B., Marshall, J. M., and Moyer, D. M. (1989) to be published.
Jackson, W. B. and Moyer, M. D. (1988) *Proc. MRS Symp.* **118**, 231.
Jackson, W. B. and Nemanich, R. J. (1983) *J. Non-Cryst. Solids* **59 & 60**, 353.
Jackson, W. B., Nemanich, R. J., Thompson, M. J., and Wacker, B. (1986) *Phys. Rev.* **B33**, 6936.
Jang, J., Kim, S. C., Park, S. C., Kim, J. B., Chu, H. Y., and Lee, C. (1988) *MRS Symp. Proc.* **118**, 189.
Johnson, N. M. (1983) *Appl. Phys. Lett.* **42**, 981.
Johnson, N. M. and Biegelsen, D. K. (1985) *Phys. Rev.* **B31**, 4066.
Johnson, N. M., Herring, C., and Chadi, D. J. (1986) *Phys. Rev. Lett.* **56**, 769.
Johnson, N. M., Ponce, F. A., Street, R. A., and Nemanich, R. J. (1987) *Phys. Rev.* **B35**, 4166.
Johnson, N. M., Ready, S. E., Boyce, J. B., Doland, C. D., Wolff, S. H., and Walker, J. (1989) *Appl. Phys. Lett.* **53**, 1626.
Kakalios, J. and Fritzsche, H. (1984) *Phys. Rev. Lett.* **53**, 1602.
Kakalios, J., Fritzsche, H., and Narasimhan, K. L. (1984) *AIP Conf. Proc.* **120**, 425.
Kakalios, J. and Street, R. A. (1987) In: *Disordered Semiconductors*, eds. M. A. Kastner, G. A. Thomas, and R. Ovshinsky (Plenum, New York) p. 529.
Kakalios, J., Street, R. A., and Jackson, W. B. (1987) *Phys. Rev. Lett.* **59**, 1037.
Kamitakahara, W. A., Shanks, H. R., McClelland, J. F., Buchenau, U., Gompf, F., and Pintchovious, L. (1984) *Phys. Rev. Lett.* **52**, 644.
Kaneko, S., Kajiwara, Y., Okumura, F., and Ohkubo, T. (1985) *MRS Symp. Proc.* **49**, 423.
Karcher, R., Ley, L., and Johnson, R. L. (1984) *Phys. Rev.* **B30**, 1896.
Kasap, S. O., Polischuk, B., Aiyah, V., and Yannacopoulos, S. (1989) *J. Non-Cryst. Solids* **114**, 49.
Kastner, M., Adler, D., and Fritzsche, H. (1976) *Phys. Rev. Lett.* **37**, 1504.
Kawabata, A. (1981) *Solid State Commun.* **38**, 823.
Kimerling, L. (1978) *Solid State Electronics* **21**, 1391.
Kivelsen, S. and Gelatt, C. D., Jr. (1979) *Phys. Rev. B* **19**, 5160.
Knights, J. C. (1979) *Jap. J. Appl. Phys.* **18–1**, 101.
Knights, J. C., Hayes, T. M., and Mikkelson, J. C. (1977) *Phys. Rev. Lett.* **39**, 712.
Knights, J. C. and Lucovsky, G. (1980) *CRC Critical Reviews in Solid State and Materials Sciences* **21**, 211.
Knights, J. C., and Lujan, R. A. (1979) *Appl. Phys. Lett.* **35**, 244.
Knights, J. C., Lujan, R. A., Rosenblum, M. P., Street, R. A., Biegelsen, D. K., and Reimer, J. (1981)A., *Appl. Phys. Lett.* **38**, 331.
Kocka, J. (1987) *J. Non-Cryst. Solids* **90**, 91.
Kocka, J., Vanacek, M., and Schauer, F. (1987) *J. Non-Cryst. Solids* **97 & 98** 715.
Kohlrausch, R. (1847) *Ann. Phys. (Leipzig)* **12**, 393.

Kruangam, D., Deguchi, M., Hattori, Y., Toyama, T., Okamoto, H. and Hamakawa, Y. (1987) *Proc. MRS Symp.* **95**, 609.
Kruangam, D., Endo, T., Deguchi, M., Guang-Pu, W., Okamoto, H., and Hamakawa, Y. (1986) *Optoelectronics* **1**, 67.
Kruhler, W., Pfleiderer, H., Plattner, R., and Stetter, W. (1984) *AIP Conf. Proc.*, **120**, 311.
Landsberg, P. T. (1970) *Phys. Stat. Sol.* **41**, 457.
Lang, D. V. (1974) *J. Appl. Phys.* **45**, 3023.
Lang, D. V., Cohen, J. D., and Harbison, J. P. (1982a) *Phys. Rev.* **B25**, 5285.
Lang, D. V., Cohen, J. D., and Harbison, J. P. (1982b) *Phys. Rev. Lett.* **48**, 421.
Lang, D. V., Cohen, J. D., Harbison, J. P., Chen, M. C. and Sergent, A. M. (1984) *J. Non-Cryst. Solids* **66**, 217.
Lannin, J. S. (1984) *Semiconductors and Semimetals* (Academic Press, Orlando) vol. 21B, Chapter 6.
Lannin, J. S. (1987) *J. Non-Cryst. Solids* **97 & 98**, 39.
Leadbetter, A. J., Rashid, A. A. M., Richardson, R. M., Wright, A. F., and Knights, J. C. (1980) *Solid State Commun.* **33**, 973.
LeComber, P. G., Jones, D. I., and Spear, W. E. (1977) *Philos. Mag.* **35**, 1173.
LeComber, P. G., Madan, A., and Spear, W. E. (1972) *J. Non-Cryst. Solids* **11**, 219.
LeComber, P. G., Owen, A. E., Spear, W. E., Hajto, J., Snell, A. J., Choi, W. K., Rose, M. J., and Reynolds, S. (1985) *J. Non-Cryst. Solids* **77 & 78**, 1373.
LeComber, P. G. and Spear, W. E. (1970) *Phys. Rev. Lett.* **25**, 509.
Lee, C., Ohlsen, W. D., and Taylor, P. C. (1987) *Phys. Rev.* **B36**, 2965.
Leidich, D., Linhart, E., Niemann, E., Grueniger, H. W., Fischer, R., and Zeyfang, R. R. (1983) *J. Non-Cryst. Solids* **59 & 60**, 613.
Leroux, T. and Chenevas-Paule, A. (1985) *J. Non-Cryst. Solids* **77 & 78**, 443.
Lewis, A. J., Connell, G. A. N., Paul, W., Pawlik, J., and Temkin, R. (1974) *AIP Conf. Proc.* **20**, 27.
Ley, L. (1984) in *The Physics of Hydrogenated Amorphous Silicon II*, eds. J. D. Joannopoulos and G. Lucovsky (Springer–Verlag, Berlin) Chapter 3.
Ley, L., Reichart, J., and Johnson, R. L. (1982) *Phys. Rev. Lett.* **49**, 1664.
Liedke, S., Lips, K., Bort, M., Jahn, K., and Fuhs, W. (1989) *J. Non-Cryst. Solids* **114**, 522.
Lucovsky, G., Nemanich, R. J., and Knights, J. C. (1979) *Phys. Rev. B* **19**, 2064.
Mackenzie, K. D., LeComber, P. G., and Spear, W. E. (1982) *Philos. Mag.* **46**, 377.
Madan, A., LeComber, P. G., and Spear, W. E. (1976) *J. Non-Cryst. Solids* **20**, 239.
Mahan, A. H., Nelson, B. P., Crandall, R. S., and Williamson, D. L. (1989) *IEEE Trans. Electron Devices* **36**, 2859.
Marshall, J. M. (1977) *Philos. Mag.* **36**, 959.
Marshall, J. M., Berkin, J., and Main, C. (1987) *Philos. Mag.* **B56**, 641.
Marshall, J. M., Street, R. A., and Thompson, M. J. (1984) *Phys. Rev.* **B29**, 2331.
Marshall, J. M., Street, R. A., and Thompson, M. J. (1986) *Philos. Mag.* **54**, 51.
Matsuo, S., Nasu, H., Akamatsu, C., Hayashi, R., Imura, T., and Osaka, Y. (1988). *Jap. J. Appl. Phys.* **27**, L132.
McMahon, T. J. and Crandall, R. S. (1989) *Phys. Rev.* **B39**, 1766.
McMahon, T. J. and Tsu, R. (1987) *Appl. Phys. Lett.* **51**, 412.
Messier, R. and Ross, R. C. (1982) *J. Appl. Phys.* **53**, 6220.
Meyer, W. von and Neldel, H. (1937) *Z. Tech. Phys.* **12**, 588.
Michiel, H., Adriaenssens, G. J., and Davis, E. A. (1986) *Phys. Rev.* **B34**, 2486.
Miyazaki, S., Ihara, Y., and Hirose, M. (1987) *Phys. Rev. Lett.* **59**, 125.

Monroe, D. (1985) *Phys. Rev. Lett.* **54**, 146.
Mott, N. F. (1967) *Adv. Phys.* **16**, 49.
Mott, N. F. (1968) *J. Non-Cryst. Solids* **1**, 1.
Mott, N. F. (1969) *Philos. Mag.* **19**, 835.
Mott, N. F. (1988) *Philos. Mag.* **B58**, 369.
Mott, N. F. and Davis, E. A. (1979) *Electronic Processes in Non-crystalline Materials* (Oxford University Press, Oxford).
Mott, N. F. and Kaveh, M. (1985) *Adv. Phys.* **34**, 329.
Müller, G., Kalbitzer, S., and Mannsperger, H. (1986) *Appl. Phys.* **A39**, 243.
Müller, G., Kalbitzer, S., Spear, W. E. and LeComber, P. G. (1977) *Proc. 7th Int. Conf on Amorphous and Liquid Semiconductors*, ed. W. E. Spear (CICL, Edinburgh) p. 443.
Müller, H. and Thomas, P. J. (1984) *J. Phys. C.* **17**, 5337.
Muramatsu, Y. and Yabumoto, N. (1986), *Appl. Phys. Lett.* **49**, 1230.
Nebel, C. E., Bauer, G. H., Gorn, M., and Lechner, P. (1989) Proc. European Photovoltaic Conf. to be published.
Nemanich, R. J. (1984) *Semiconductors and Semimetals*, ed. J. Pankove (Academic Press, Orlando) Vol 21C, Chapter 11.
Nemanich, R. J., Tsai, C. C., Thompson, M. J., and Sigmon, T. W. (1981) *J. Vac. Sci. Technol.* **19**, 685.
Nichols, C. S. and Fong, C. Y. (1987) *Phys. Rev.* **B35**, 9360.
Northrup, J. E. (1989) *Phys. Rev.* **B40**, 5875.
Okushi, H., Tokumaru, Y., Yamasaki, S., Oheda, H., and Tanaka, K. (1982) *Phys. Rev.* **B25**, 4313.
Onsager, L. (1938) *Phys. Rev.* **54**, 554.
Orenstein, J. and Kastner, M. (1981) *Phys. Rev. Lett.* **46**, 1421.
Overhof, H. and Beyer, W. (1981) *Philos. Mag.* **B43**, 433.
Ovshinsky, S. R. (1968) *Phys. Rev. Lett.* **21**, 1450.
Owen, A. E., LeComber, P. G., Spear, W. E., and Hajto, J. (1983) *J. Non-Cryst. Solids* **59 & 60**, 1273.
Pai, D. M. (1988) *Proc. 4th Int. Conf. of Non Impact Technologies, SPSE, New Orleans*, 20.
Palmer, R. G., Stein, D. L., Abrahams, E., and Anderson, P. W. (1984) *Phys. Rev. Lett.* **53**, 958.
Pankove, J. I., Carlson, D. E., Berkeyheiser, J. E., and Wance, R. O. (1983) *Phys. Rev. Lett.* **51**, 2224.
Pantelides, S. T. (1986) *Phys. Rev. Lett.* **57**, 2979.
Pantelides, S. T. (1987) *Phys. Rev. Lett.* **58**, 1344.
Park, H. R., Liu, J. Z., and Wagner, S. (1990) *Appl. Phys. Lett.* in press.
Perez Mendez, V., Cho, G., Fujieda, I., Kaplan, S. N., Qureshi, S., and Street, R. A. (1989) *MRS Symp. Proc.* **149**, 621.
Persans, P. D. (1988) in *Advances in Amorphous Semiconductors*, ed. H. Fritzsche (World Scientific, Singapore) p.1045.
Persans, P. D. (1989) *Phys. Rev.* **B39**, 1797.
Persans, P. D. and Ruppert, A. F. (1987) *MRS Symp. Proc.* **57**, 329.
Persans, P. D., Ruppert, A. F., Abeles, B., and Tiedje, T. (1985) *Phys. Rev.* **B32**, 5558.
Pfister, G. and Scher, H. (1977) *Phys. Rev.* **B15**, 2062.
Phillips, J. C. (1979) *J. Non-Cryst. Solids* **34**, 153.
Phillips, W. A. (1985) *Amorphous Solids and the Liquid State* (Plenum Press, New York) p. 467.

Pierce, D. T. and Spicer, W. E. (1972) *Phys. Rev.* **B5**, 3017.
Polk, D. E. (1971) *J. Non-Cryst. Solids* **5**, 365.
Pollak, M. (1977) *Philos, Mag.* **36**, 1157.
Ponpon, J. P. and Bourdon, B. (1982) *Solid State Electron.* **25**, 875.
Redfield, D. (1987) *J. Non-Cryst. Solids* **97 & 98**, 783.
Reimer, J. A. and Duncan, T. M. (1983) *Phys. Rev.* **B27**, 4895.
Reimer, J. A., Vaughan, R. W., and Knights, J. C. (1980) *Phys. Rev. Lett.* **44**, 193.
Reinelt, M., Kalbitzer, S., and Moller, G. (1983) *J. Non-Cryst. Solids* **59 & 60**, 169.
Reiss, H. (1956) *J. Chem. Phys.* **25**, 400.
Rose, A. (1955) *Phys. Rev.* **97**, 1538.
Ross, R. C. and Vossen, J. L. (1984) *Appl. Phys. Lett.* **45**, 239.
Roxlo, C. B., Abeles, B., and Tiedje, T. (1984) *Phys. Rev. Lett.* **52**, 1994.
Sabisky, E. S. and Stone, J. L. (1988) *Proc. 20th IEEE Photovoltaic Specialists Conference,* (IEEE, New York), 39.
Santos, P., Hundhausen, M., and Ley, L. (1986) *Phys. Rev.* **B33**, 1516.
Santos, P. V., Ley, L., Mebert, J., and Koblinger, O. (1987) *Phys. Rev.* **B36**, 4858.
Scher, H. and Montroll, E. W. (1975) *Phys. Rev.* **B12**, 2455.
Scholch, H. P., Kalbitzer, S., Fink, D., and Behar, M. (1988) *Mat. Sci. and Eng.* **B1**, 135.
Schulke, W. (1981) *Philos. Mag.* **B43**, 451.
Searle, T. M. and Jackson, W. A. (1989) *Philos. Mag.* **B60**, 237.
Shanks, H., Fang, C. J., Ley, L., Cardona, M., Demond, F. J., and Kalbitzer, S. (1980) *Phys. Stat. Sol.* (*b*) **100**, 43.
Shepard, K., Smith, Z. E., Aljishi, S., and Wagner, S. (1988) *MRS Symp. Proc.* **118**, 147.
Shimizu, I. (1985) *MRS Symp. Proc.* **49**, 395.
Shklovskii, B. I., Fritzsche, H., and Baranovskii, S. D. (1989) *Phys. Rev. Lett.* **62**, 2989.
Shlesinger, M. F. and Montroll, E. W. (1984) *Proc. Nat. Acad. Sci. USA* **81**, 1280.
Silver, M., Giles, N. C., Snow, E., Shaw, M.P., Cannella, V., and Adler, D. (1982) *Appl. Phys. Lett.* **41**, 935.
Smith, Z. E., Aljishi, S., Slobodin, D., Chu, V., Wagner, S., Lenahan, P. M., Arya, R. R., and Bennett, M. S. (1986) *Phys. Rev. Lett.* **57**, 2450.
Smith, Z. E. and Wagner, S. (1987) *Phys. Rev. Lett.* **59**, 688.
Smith, Z. E., Wagner, S., and Faughnan, B. W. (1985) *Appl. Phys. Lett.* **46**, 1078.
Snell, A. J., Mackenzie, K. D., Spear, W. E., LeComber, P. G., and Hughes, A. J. (1981) *Appl. Phys.* **24**, 357.
Spear, W. E. (1968) *J. Non-Cryst. Solids* **1**, 197.
Spear, W. E. (1974) *Proc. Int. Conf. on Amorphous and Liquid Semiconductors*, eds J. Stuke and W. Brenig (Taylor and Francis, London) p. 1.
Spear, W. E., Hourd, A. C., and Kinmond, S. (1985) *J. Non-Cryst. Solids* **77 & 78**, 607.
Spear, W. E. and LeComber, P. G. (1975) *Solid State Commun.* **17**, 1193.
Spear, W. E. and LeComber, P. G. (1977) *Adv. Phys.* **26**, 811.
Spear, W. E., Loveland, R. J., and Al-Sharbaty, A. (1974) *J. Non-Cryst. Solids* **15**, 410.
Staebler, D. L. and Wronski, C. R. (1977) *Appl. Phys. Lett.* **31**, 292.
Stathis, J. H. and Pantelides, S. T. (1988) *Phys. Rev.* **B37**, 6579.
Stoneham, A. M. (1977) *Philos. Mag.* **33**, 983.
Street, R. A. (1976) *Adv. Phys.* **25**, 397.
Street, R. A. (1978) *Philos. Mag.* **B37**, 35.
Street, R. A. (1981a) *Adv. Phys.* **30**, 593.

Street, R. A. (1981b) *Phys. Rev.* **B23**, 861.
Street, R. A. (1982) *Phys. Rev. Lett.* **49**, 1187.
Street, R. A. (1983) *Phys. Rev.* **B27**, 4924.
Street, R. A. (1984) *Philos. Mag.* **B49**, L15.
Street, R. A. (1985) *J. Non-Cryst. Solids* **77 & 78**, 1.
Street, R. A. (1987a) *Proc. MRS Symp.* **95**, 13.
Street, R. A. (1987b) *Proc. SPIE Symp.* **763**, 10.
Street, R. A. (1989) *Philos. Mag.* **B60**, 213.
Street, R. A. and Biegelsen, D. K. (1982) *Solid State Commun.* **44**, 501.
Street, R. A. and Biegelsen, D. K. (1984) *The Physics of Hydrogenated Amorphous Silicon II*, eds. J. D. Joannopoulos, and G. Lucovsky (Springer-Verlag, Berlin) Chapter 5.
Street, R. A., Biegelsen, D. K., Jackson, W. B., Johnson, N. M., and Stutzmann, M. (1985) *Philos, Mag.* **52**, 235.
Street, R. A., Biegelsen, D. K., and Knights, J. C. (1981) *Phys. Rev.* **B24**, 969.
Street, R. A., Biegelsen, D. K., and Weisfield, R. L. (1984) *Phys. Rev.* **B30**, 5861.
Street, R. A., Hack, M., and Jackson, W. B. (1988a) *Phys. Rev.* **B37**, 4209.
Street, R. A., Johnson, N. M., Walker, J., and Winer, K. (1989) *Philos. Mag. Lett.* **60**, 177.
Street, R. A., Kakalios, J., and Hack, M. (1988b) *Phys. Rev.* **B38**, 5603.
Street, R. A., Kakalios, J., Tsai, C. C., and Hayes, T. M. (1987a) *Phys. Rev.* **B35**, 1316.
Street, R. A., Knights, J. C., and Biegelsen, D. K. (1978) *Phys. Rev.* **B18**, 1880.
Street, R. A. and Mott, N. F. (1975) *Phys. Rev. Lett.* **35**, 1293.
Street, R. A. and Thompson, M. J. (1984) *Appl. Phys. Lett.* **45**, 769.
Street, R. A., Thompson, M. J., and Johnson, N. M. (1985b) *Philos. Mag.* **B51**, 1.
Street, R. A. and Tsai, C. C. (1986) *Appl. Phys. Lett.* **48**, 1672.
Street, R. A., Tsai, C. C., Kakalios, J., and Jackson, W. B. (1987b) *Philos. Mag.* **B56**, 305.
Street, R. A., Tsai, C. C., Stutzmann, M., and Kakalios, J. (1987c) *Philos. Mag.* **B57**, 289.
Street, R. A. and Winer, K. (1989) *Phys. Rev.* **B40**, 6236.
Street, R. A., Zesch, J., and Thompson, M. J. (1983) *Appl. Phys. Lett.* **43**, 672.
Stuke, J. (1977) *Proc. 7th Int. Conf. on Amorphous and Liquid Semiconductors*, ed. W. E. Spear (CICL, Edinburgh) 406.
Stuke, J. (1987) *J. Non-Cryst. Solids* **97 & 98**, 1.
Stutzmann, M. (1987a) *Philos. Mag.* **B56**, 63.
Stutzmann, M. (1987b) *Phys. Rev.* **B35**, 735.
Stutzmann, M. (1989) *Philos. Mag.* **B60**, 531.
Stutzmann, M. and Biegelsen, D. K. (1983) *Phys. Rev.* **B28**, 6256.
Stutzmann, M. and Biegelsen, D. K. (1986) *Phys. Rev.* **B34**, 3093.
Stutzmann, M. and Biegelsen, D. K. (1988) *Phys. Rev. Lett.* **60**, 1682.
Stutzmann, M., Biegelsen, D. K., and Street, R. A. (1987) *Phys. Rev.* **B35**, 5666.
Stutzmann, M., Jackson, W. B., and Tsai, C. C. (1985) *Phys. Rev.* **B32**, 23.
Stutzmann, M., Jackson, W. B., and Tsai, C. C. (1986) *Phys. Rev.* **B34**, 63.
Stutzmann, M. and Street, R. A. (1985) *Phys. Rev. Lett.* **54**, 1836.
Swartz, G. A. (1984) *Appl. Phys. Lett.* **44**, 697.
Takada, J. and Fritzsche, H. (1987) *Phys. Rev.* **B36**, 1710.
Takeda, T. and Sano, S. (1988) *Proc. MRS Symp.* **118**, 399.
Tanaka, K. (1976) *AIP Conf. Proc.* **31**, 148.
Tanielian, M. (1982) *Philos. Mag.* **B45**, 435.

Tauc, J. (1982) *Festkorperprobleme* **22**, 85.
Tauc, J., Grigorovici, R. and Vancu, A. (1966) *Phys. Stat. Sol.* **15**, 627.
Thomas, G. A. (1985) *Philos. Mag.* **B52**, 479.
Thomas, D. G., Hopfield, J. J., and Augustiniak, W. M. (1965) *Phys. Rev.* **140**, 202.
Thompson, M. J., Johnson, N. M., Nemanich, R. J., and Tsai, C. C. (1981) *Appl. Phys. Lett.* **39**, 274.
Thomsen, C., Stoddart, H., Zhou, T., Tauc, J., and Vardeny, Z. (1986) *Phys. Rev.* **B33**, 4396.
Thorpe, M. F. and Weaire, D. (1971) *Phys. Rev. Lett.* **26**, 1581.
Thouless, D. J. (1977) *Phys. Rev. Lett.* **18**, 1167.
Tiedje, T. (1984) *Semiconductors and Semimetals* (Academic Press, Orlando) Vol. 21C, Chapter 6.
Tiedje, T. (1985) *Proc. MRS Symp.* **49**, 121.
Tiedje, T., Cebulka, J. M., Morel, D. L., and Abeles, B. (1981) *Phys. Rev. Lett.* **46**, 1425.
Tiedje, T. and Rose, A. (1980) *Solid State Commun.* **37**, 49.
Tsai, C. C. (1979) *Phys. Rev.* **B19**, 2041.
Tsai, C. C., Knights, J. C., Chang, G., and Wacker, B. (1986a) *J. Appl. Phys.* **59**, 2998.
Tsai, C. C. and Nemanich, R. J. (1980) *J. Non-Cryst Solids*, **35 & 36**, 1203.
Tsai, C. C., Street, R. A., Ponce, F. A., and Anderson, G. B. (1986b) *Proc. MRS Symp.* **70**, 351.
Tsang, C. and Street, R. A. (1979) *Phys. Rev.* **B19**, 3027.
Tsuji, K., Takasaki, Y., Hirai, T., and Taketoshi, K. (1989) *J. Non-Cryst. Solids* **114**, 94.
Tuan, H. C. (1986) *MRS Symp. Proc.* **70**, 651.
Urbach, F. (1953) *Phys. Rev.* **92**, 1324.
van Berkel, C. and Powell, M. J. (1987) *Appl. Phys. Lett.* **51**, 1094.
Vardeny, Z. and Tauc, J. (1981) *Phys. Rev. Lett.* **46**, 1223.
Veprek, S. (1989) *J. Appl. Phys.* in press.
Veprek, S., Heintze, M., Sarott, F.-A., Jurcik-Rajman, M., and Willmott, P. (1988) *Proc MRS Symp.* **118**, 3.
von Roedern, B., Ley, L., and Cardona, M. (1977) *Phys. Rev. Lett.* **39**, 1576.
Weaire, D. (1971) *Phys. Rev. Lett.* **26**, 1541.
Weber, W. (1977) *Phys. Rev.* **B15**, 4789.
Williams, F. (1968) *Phys. Stat. Sol.* **25**, 493.
Winer, K. and Ley, L. (1987) *Phys. Rev.* **B36**, 6072.
Winer, K. and Ley, L. (1989) In: *Amorphous Silicon and Related Materials*, ed. H. Fritzsche (World Scientific, Singapore) p. 365.
Winer, K. and Street, R. A. (1989) *Phys. Rev. Lett.* **63**, 880.
Wooten, F., Winer, K., and Weaire, D. (1985) *Phys. Rev. Lett.* **54**, 1392.
Wronski, C. R., Abeles, B., Cody, G. D., and Tiedje, T. (1980) *Appl. Phys. Lett.* **37**, 96.
Wronski, C. R. and Carlson, D. E. (1977) *Solid State Commun.* **23**, 421.
Wronski, C. R. and Daniel, R. E. (1981) *Phys. Rev.* **B23**, 794.
Wronski, C. R., Lee, S., Hicks, M., and Kumar, S. (1989) *Phys. Rev. Lett.* **63**, 1420.
Xu, X., Okumura, A., Morimoto, A., Kumeda, M., and Shimizu, T. (1988) *Phys. Rev.* **B38**, 8371.
Yaniv, Z., Hansell, G., Vijan, M., and Cannella, V. (1984) *Proc. MRS Symp.* **33**, 293.
Zachariasen, W. H. (1932) *J. Am. Chem. Soc.* **54**, 3841.

Index